DOMESTICATED

Also by Richard C. Francis

EPIGENETICS:
HOW ENVIRONMENT SHAPES OUR GENES

WHY MEN WON'T ASK FOR DIRECTIONS:
THE SEDUCTIONS OF SOCIOBIOLOGY

DOMESTICATED

Evolution in a Man-Made World

RICHARD C. FRANCIS

W. W. NORTON & COMPANY

INDEPENDENT PUBLISHERS SINCE 1923

NEW YORK   LONDON

For information about permission to reproduce selections from this book,
write to Permissions, W. W. Norton & Company, Inc.,
500 Fifth Avenue, New York, NY 10110

For information about special discounts for bulk purchases, please contact
W. W. Norton Special Sales at specialsales@wwnorton.com or 800-233-4830

Manufacturing by Quad Graphics Fairfield
Book design by Lovedog Studio
Production manager: Julia Druskin

Library of Congress Cataloging-in-Publication Data

Francis, Richard C., 1953–
Domesticated : evolution in a man-made world / Richard C. Francis.
    pages cm
Includes bibliographical references and index.
ISBN 978-0-393-06460-5 (hardcover)
1. Domestic animals—History. 2. Animals and civilization. I. Title.

SF41.F65 2015
636—dc23

                        2014046934

ISBN 978-0-393-35303-7 pbk.

W. W. Norton & Company, Inc.
500 Fifth Avenue, New York, N.Y. 10110
www.wwnorton.com

W. W. Norton & Company Ltd.
Castle House, 75/76 Wells Street, London W1T 3QT

1 2 3 4 5 6 7 8 9 0

For Andrew

Contents

DOMESTICATED

PREFACE

WITHOUT OUR DOMESTICATED ANIMALS AND PLANTS, human civilization as we know it would not exist. We would still be living at a subsistence level as hunter-gatherers. It was the unprecedented surplus calories resulting from domestication that ushered in the so-called Neolithic revolution, which created the conditions for not only an agricultural economy but also urban life and, ultimately, the suite of innovations we think of as modern culture. The cradle of civilization is, not coincidentally, also the place where barley, wheat, sheep, goats, pigs, cattle, and cats commenced a fatefully intimate association with humans.

At the onset of the Neolithic revolution, an estimated 10 million humans inhabited the earth; now there are over 7 billion of us. The human population explosion has been bad for most other living things, but not for those lucky enough to warrant domestication. They have thrived quite as much as we have. Since the Neolithic, extinction rates have been 100–1,000 times those of the previous 60 million years. Among those lost were the wild ancestors of domesticated species, such as the tarpan (horse) and auroch (cattle), but no domesticated species has ever become extinct. The wild ancestors of camels, cats, sheep, and goats are teetering on the brink of oblivion; their domesticated descendants, though, are among the most common large mammals on earth. In an evolutionary sense, it pays to be domesticated.

The success of domesticates comes at a price, however, in the form of increasing evolutionary submission. We humans have largely wrested from nature control of their evolutionary fate, which, it turns out, renders domesticated creatures uniquely informative for those seeking to understand the evolutionary process. Indeed, domesticated creatures provide some of the most dramatic examples of evolution in the world today. Even a creationist recognizes, at some level, that the transition from wolf to dog is an evolutionary process. That is why the "artificial" selection for particular traits in domesticated breeds—from dogs to pigeons—figured so prominently in Darwin's argument for an analogous process that he called "natural" selection.

The fact that wolves, which competed with and even preyed on Paleolithic humans, were the first animals domesticated is testimony to the power of humans as both a conscious and unconscious evolutionary force. And throughout most of the domestication process for dogs and other mammals, the role of humans as an unconscious evolutionary force was paramount. For that reason, the distinction between natural selection and artificial selection is actually quite blurry. As we will see, the domestication process was often initiated by the domesticated animals themselves, when they sought, for various reasons, human proximity. This process of self-taming occurred primarily through ordinary natural selection. The sort of conscious selection we call artificial selection came only much later in the domestication process. There is a large gray area of transition between natural selection and artificial selection, in which humans became an increasingly important but only partly conscious component of the selection regime.

The combination of natural and artificial selection has proved very powerful indeed. The size range of domestic dogs—from Chihuahuas to Great Danes—far exceeds not only that of wild wolves but that of the entire canine family (wolves, coyotes, jackals, foxes, etc.), both living and extinct, which originated in the Oligocene epoch nearly 40 million years

ago. In a mere 15,000–30,000 years the selection imposed on dogs by their association with humans has caused evolutionary alterations never experienced in the canine family during the previous 40 million years.

Evolutionary modifications of dogs by humans extend to many other traits as well, including coat color and skeletal modifications. Skull shape variation in domestic dogs exceeds not only that of all other canines combined but also that of all other carnivores—the taxonomic group of families that includes canines, cats, bears, weasels, raccoons, hyenas, civets, seals, and sea lions—combined.

The effects of human selection on dog behavior have been no less dramatic. Most notably, domestic dogs have an evolved capacity to "read" human intentions. For example, they can interpret human gestures, such as pointing, to locate distant food items. Wild wolves cannot do this. In fact, dogs are much better at reading human intentions than are our closest relatives, chimpanzees and gorillas. So in some ways, the social cognition of dogs more closely resembles that of humans than does that of the great apes.

The human influence on the evolution of other domesticated creatures is only slightly less impressive. The placid hyper-uddered Holstein cow does not much resemble its wild ancestor, the noble and fierce auroch, nor do Merino sheep much resemble the mouflon, their wild ancestor, though in both cases the domesticated and wild forms share a common ancestor only 10,000 years ago. That's a lot of evolution in a very short period of time.

Since domestication is an accelerated form of evolution, it is an ideal subject by which to nurture the intuitions and hence the understanding of nonbiologists as to how evolution works. Evolution is a historical process, but most evolution involves timescales that are hard for the uninitiated to digest, much less intuit, given the limitations of the human mind. Domestication, however, occurs over more comprehensible timescales. Dog breeds such as the English bulldog, for example, have evolved extensively in just the last 100 years. It is therefore possible to connect human history, prehistory (5000–30,000 BP), and evolutionary history in a more or less seamless way. This is sometimes referred to as "big history," though I prefer "deep history." This big or deep historical dimension provides the backdrop for the main themes that will concern me throughout.

The predomestication history—the deepest part of the deep history—is, of course, the province of evolutionary biology, specifically the branch of evolutionary biology concerned with reconstructing the genealogical relationships on the tree of life, which is called "phylogenetics." Much of what we know about the period connecting predomestication history and written history comes from the burgeoning field of zooarchaeology, which combines expertise in ancient human cultures, animal biology, and natural history. The more recent history of domestication, for which we have written records, concerns the phase during which humans have assumed the most conscious control. This phase, which is still ongoing, is one of unprecedented change, for good and ill.

While this historical backdrop is, I hope, interesting in and of itself, it is also essential if we are to understand how evolution works. Domesticated creatures are familiar to one and all. As such, their transformations are easier to perceive and appreciate—a prerequisite for my main aim here, which is to consider some recent and current developments in evolutionary biology through the lens of domestication.

We can consider each case of domestication as a sort of natural experiment in evolution. By "natural experiment," I mean a case that is ideally suited for the study of evolution but was not planned as such. We also have natural experiments in reverse domestication, called feralization. The dingo is one such experiment. Transported to Australia approximately 5,000 years ago by proto-Polynesians, dingoes transformed from pets into the apex predator of the outback. In the process, they evolved into more wolflike creatures; indeed, dingoes provide interesting contrasts to both wolves and domestic dogs.

In addition to these natural experiments of past domestication and feralization events, we now have actual ongoing scientific experiments in domestication—that is, attempts to experimentally replicate the domestication process for the sole purpose of illuminating the evolutionary processes at work.

Though each case of domestication involves some interesting and unique peculiarities, some general themes emerge that are equally interesting and significant for evolutionary thought. Among the most noteworthy and revealing of these themes is that domestication has unintended consequences. It seems that, in selecting for one trait, humans invariably inadvertently affect other, seemingly unrelated traits. It turns out that such by-products are a general feature of evolution by natural selection. They can act as a brake on the evolution of a trait under selection and/or create new evolutionary opportunities.

Another theme is that the amount of change in phenotype (which includes behavioral, physiological, and morphological traits) does not correlate closely with changes in the genome (genetic makeup). For example, large physical changes in domesticated animals often involve remarkably few genetic alterations. The genetic distance between dogs and wolves is tiny compared to their physical distance. This is no less true of pig, cow, or horse, which brings us to the third theme: that the human environment has some remarkably consistent evolutionary effects on creatures spanning the vast genomic (and hence evolutionary) distance between horses and dogs.

The paramount theme of this book, though, is the conservative nature of the evolutionary process, even, as in domestication, when it occurs with the most rapidity. In most popular accounts of evolution, the creative side of evolution gets all of the attention. This is part and parcel of the adaptationist program, in which the goal is to demonstrate the diverse—and seemingly boundless—ways in which organisms respond adaptively to environmental challenges. But adaptive change is far from boundless; it is actually quite restricted, channeled by the previous evolutionary history of an organism. In fact, adaptive change is restricted to tinkering at the margins of what has previously evolved. The Pekingese is a tinkered wolf, not redesigned wholesale from its wolf ancestors.

Two recent developments within evolutionary biology in particular have brought the conservative side of evolution to the fore: genomics and evolutionary developmental biology (evo devo). Both will figure prominently in this book.

THE BOOK PLAN

Each chapter will focus on the domestication of either a single species (e.g., dogs, cats, pigs) or two or more related species (e.g., sheep and goats) beginning with the history of their domestication and cultural significance, followed by a brief account of the evolutionary history of the species and family from which the domesticates were derived— including their position in the tree of life—to provide some context for the evolutionary changes wrought by humans. The bulk of each chapter will concern what these human-induced alterations reveal about the evolutionary process in both its creative and conservative aspects. Humans are the focus of the final three chapters. The first two of these will consider the hypothesis that human evolution involved a process of self-domestication—that, in important ways, our evolution parallels, for example, the transition from wolf to dog. The final chapter will consider explanations for the rise of humans from an insignificant component of the African fauna to our current state of dominion. Given that the earth is increasingly man-made, our domesticated animals are in the vanguard of future evolution.

CHAPTER 1

HOUSE FOX

WHAT ARE SOME ADJECTIVES THAT COME TO MIND when you hear the word "dog"? Chances are your list will include "loyal," "friendly," "loving," and "faithful." How about "fox"? That word conjures up a whole different set of adjectival connotations, such as "sly," "wily," "clever," "sneaky," "crafty," or "deceptive." Most of the adjectives we associate with foxes have a derogatory component, at best a grudging respect.

Since most of us have had little direct interaction with foxes, these associations come to us secondhand, often by way of fables such as those collected by Aesop. *The Fox and the Raven* is typical.[1] The raven, upon discovering a nice piece of cheese, retires to a tree branch to dine at its leisure. The fox spies the raven and its prize and covets the latter. But since foxes can't climb trees, it must take an indirect approach. The fox flatters the raven fawningly, praising its satiny plumage and handsome feet. The fox then asks whether the raven has a voice to match. When the raven opens its mouth to caw, it drops the cheese and the sly fox has gained its prize. The raven has been outfoxed.

The photographer and installation artist Sandy Skoagland uses these associations to great effect in her piece called *Fox Games*.[2] The scene is a restaurant, complete with tables, chairs, cutlery, glassware, vases, and breadbaskets—but devoid of humans. Instead, it is populated by 28 life-sized foxes dispersed throughout the tableau and in a multitude of postures and attitudes, all of which suggest they are having quite a good time. Several are in the act of leaping; one leaper is about to come down on top of another fox lying on its back, catlike. In fact, there is a catlike quality to the manner of most of these foxes, most notably the one in the foreground staring insouciantly at the viewer. (See Figure 1.1.)

As in all good surrealist art, there is something emotionally disorienting about the scene. In part that response stems from the lurid whorehouse-red monochromaticity of the restaurant and its accoutrements, which, combined with the peculiar lighting, imparts a hallucinatory illumination that is neither night-like nor day-like. Another disorienting element is, of course, the foxes themselves, which are painted gray, in stark contrast with their background, providing them extra visual pop. (One fox, however, is the color of the background; this mutant adds another disorienting element.) You can't help but focus on the nuances of the foxes' manner and actions, which are cute, engaging, and smile-worthy at first glance. But there is something sinister here as well. These are foxes, after all, with all of the connotations—sly, crafty,

FIGURE 1.1   *Fox Games* 1989 installation © Sandy Skoglund. (Courtesy Denver Art Museum.)

deceiving, and the like—that foxes bring with them to the scene. These are not creatures you can trust; these are not creatures you would want to pet; that playful animal on the table might just bite you if you tried.

Most disorienting is the fact that these foxes are indoors, so comfortably indoors. Foxes are wild animals after all, and wild animals belong outdoors. Wild animals are found indoors only when something has gone wrong. Something must have gone very wrong here. The scene as a whole seems postapocalyptic.

Perhaps, though, it won't require an apocalypse to bring foxes indoors. Perhaps, in the not-too-distant future, *Fox Games* will be viewed quite differently: less disorienting, more natural. The color and illumination will retain their effect, but the foxes themselves will just seem cute, disturbing in this context only because of their sheer number, like the cat-overrun homes of crazy animal hoarders. Much like cats and dogs, foxes might become indoor pets. Such was the goal of a singular Russian scientist, Dmitry Belyaev, who set out to discover how we made cats and dogs—the process of domestication.

A BRIGHT AND BRAVE FELLOW

Belyaev had attained a prestigious position as a geneticist in Moscow in the 1940s but then ran afoul of Soviet authorities. The problem was his refusal to disavow the genetic framework that had been initiated in the middle of the nineteenth century by the Moravian monk Gregor Mendel and that remains the foundation of modern genetics. Mendelism, as it came to be called, was certainly not a problematic or controversial view in most of the world at that time; indeed, it was the orthodox view. But in the Soviet Union, because of a unique combination of circumstances, Mendel had come to be associated with a bourgeois and reactionary worldview, and those who championed it were in real peril. Most retrospective blame for this state of affairs was laid at the feet of an upstart agronomist, Teofrim Lysenko, but it was more complicated than that.[3]

Perhaps the greatest internal problem facing Joseph Stalin during this period was a food shortage, particularly of wheat—a famine that he had created through his forced collectivization of the peasant farmers. Stalin desperately needed a way to quickly improve the food supply, and Lysenko seemed to offer the means. In the standard Mendelian view, improved wheat required highly improbable mutations, which could be gradually made more common through selective breeding. But Lysenko claimed that wheat could be genetically improved much more quickly through manipulation of certain key environmental factors. In essence, he proposed that some kinds of environmental alterations could cause genetic modification in the wheat, and furthermore that these genetic modifications could be directed toward desirable characteristics.[4]

(Lysenko's was a view of evolution derived in part from the great French evolutionist Jean-Baptiste Lamarck and remains a stain on Lamarck's name to this day, even though Lamarck formulated the first theory of evolution and was greatly admired by Darwin himself.)[5]

Once Stalin decided in favor of Lysenko, those who demurred were in a parlous position. The more fortunate were stripped of their academic and research positions; the less fortunate were jailed. Belyaev's older brother Nikalay, who was also a geneticist, was among the less fortunate. He died in a labor camp. So, too, did Nikolai Ivanovich Vavilov, one of the most important geneticists of his generation.[6] Another world-renowned evolutionary biologist, and a godfather of evo devo, Ivan Ivanovich Schmalhausen, lived to see his works destroyed in public, including his classic, *Factors in Evolution*.[7]

Belyaev himself was more fortunate. But he certainly did not escape unscathed; in 1948 he was stripped of his position as chairman of the Department of Animal Breeding in Moscow. He did manage to surreptitiously continue conducting Mendelian research for the next decade, couched as animal physiology. Then, in 1959, he made a life-altering career move.

Though most—including most Russians—would not consider a move to Siberia as career enhancing, it proved as much for Belyaev. In Novosibirsk, beyond the horizon of Moscow's bureaucratic gaze, he helped found the Siberian Department of the Soviet Academy of Sciences; he also became the director of the department's Institute of Cytology and Genetics. Under his leadership the institute was to become one of the world's leading research centers for the study of both classical and molecular genetics.

DOMESTICATION

Belyaev was not just a proponent of Mendelian genetics; he was also a strong advocate for Darwin and the modern synthesis, with some very original ideas as to how evolution works. He was particularly interested in evolution under stressful conditions, such as occurs during rapid environmental alterations or the colonization of a new habitat. For Belyaev, the domestication process was evolution of this sort. The new habitat was one of human creation.

Belyaev had a theory about how domestication occurred, with a focus, initially, on wolf domestication. How can you get a Pekingese from a wolf? Before examining Belyaev's specific hypotheses, let's consider the multifarious ways in which modern dog breeds have diverged from their wolf ancestors.

Size is the most obvious. Chihuahuas are a fraction the size of wolves, while Great Danes, wolfhounds, and bullmastiffs are considerably larger than wolves.

Somewhat less obvious but no less pronounced are the skeletal alterations wrought by humans, above and beyond those related to size. The pelvis and shoulders have been modified in numerous ways. Spines have been lengthened or shortened. One part of the spine, the tail, has proved particularly amenable to human manipulation. Even the most wolflike breeds, such as huskies, elkhounds, and malamutes, have up-curled tails, which are never found in wild wolves. In other breeds, such as German shepherds, the tail is directed unnaturally downward; wolves hold their tails out straight. In many breeds the tail has considerably diminished.

The wolf skull has been modified almost beyond recognition. At one end of the spectrum are the Pekingese and pugs, in which the skull has been shortened to the point of interfering with normal breathing; at the other end, as in collies and Afghans, the skull has become quite elongate and narrow. Skull shape has been altered in a number of other ways as well. In fact, the variation in dog skull shape is even more pronounced than the variation in dog size.

Dog fur has also diverged markedly from the wolf condition. The curly, wiry hair of poodles and some terriers has no precedent in wolves, nor does the long-haired shaggy look of Pekingese and sheepdogs. The hair colors of many dogs are also very unwolflike. Wolves actually vary quite a lot in color, ranging from near white to dark gray. But dogs have added to this palette considerably, notably in the yellow-red-brown range. Even more noteworthy are the color combinations found in various dog breeds, especially the white spotted or piebald patterns. As far as we know, no wild wolf has ever been piebald.

Floppy ears are another signature of domestication. In some breeds, such as Doberman pinschers, in which the wild-type look is considered desirable, the ears must be surgically cropped and pinned to make them wolflike.

The behavioral differences between wolves and Pekingese are at least as pronounced as their physical differences. Wolves don't bark. A wolf wouldn't sit on your lap if it could, seek your approval, or even wag its tail for you. Tail wagging is a distinctively domesticated behavior. Most fundamentally, wolves can't read your intentions like dogs do. Point to an object and a wolf will "understand" you about as much as a cat would, which is to say not at all. Nor will a wolf follow your gaze or detect subtle fluctuations in your emotions or attention, and it certainly won't seek your approval, even if you raise it from birth. Our ability to train dogs to do anything depends crucially on the fact that they recognize our dominance. Wolves, even when hand reared, don't see things that way. And don't waste your time teaching a wolf to fetch. It's not that fetching is beneath a wolf's dignity—though it may very well be—a wolf just wouldn't "get it," both literally and figuratively.

Were each of these physical and behavioral traits altered independently? Or were some of the domesticated traits incidental by-products of alterations in other traits? Belyaev strongly leaned toward the latter view, because of a phenomenon known as "pleiotropy": a single gene affecting multiple traits. Pleiotropy means that selection—whether natural or artificial—for one trait can affect many others. This is why evolutionary biologists make a distinction between selection *for* and selection *of*.[9]

The trait selected *for* is the target of selection; correlated changes in other traits reflect selection *of*. For example, selection *for* red hair in humans will also result in the selection *of* fair skin tone, freckles, and non-brown eyes. It's a package deal. Similarly, Belyaev believed that many dog traits came as a package of by-products as a result of selection for something else. Moreover, he proposed that the target of selection, what was selected *for*, was not a physical trait but a behavioral trait—specifically, tameness.[10]

To test this idea, Belyaev set out to experimentally replicate the domestication process. He chose as his subject another member of the canine family, the silver fox, a color variant of the familiar red fox (*Vulpes vulpes*) native to North America and northern Eurasia, and the model for Skoagland's art piece.

The foxes for this experiment were obtained from a fur farm in Estonia.[11] Thirty males and 100 females were fortunate enough to escape the fur farm with their hides intact. They were not chosen at random but were, rather, the tamest of the thousands of farmed foxes tested. Belyaev and Lyudmila Trut, his colleague and close collaborator from the beginning of this project, selected for one trait and one trait only: the capacity to tolerate human proximity without fear or aggression. Only 5 percent of the tamest males and 20 percent of the tamest females were allowed to breed in each generation.[12]

By the fourth generation, some of the pups began to wag their tails when near their caretakers—an unprecedented behavior for foxes. By the sixth generation, some of the pups were eager for human contact; they not only wagged their tails, but whined and whimpered as well; they would also lick their caretakers' faces in a way that any dog owner could appreciate. This behavior is all the more remarkable when you consider that these pups were reared just as they would have been at the fox farm, with minimal human contact.[13]

The tail-wagging, face-licking pups were assigned to the "elite" category. The proportion of elite cubs increased with each successive generation; by the thirtieth generation it was at 49 percent. By 2005, all of the pups were in the elite category and could be adopted out as pets.[14] In their behavior, the pet foxes are said to be somewhere between a dog and a cat—more independent than a dog, but more responsive to human instruction than a cat. Perhaps most remarkably, these pet foxes can also read human intentions, through gestures and glances.[15] That's roughly 50 years from a wild fox, which, if it ever found itself inside a restaurant would be in extreme distress, to a domesticated fox that could be quite at home there. Though it might take a while for us humans to adjust our deeply ingrained attitudes toward these new additions to the pet menagerie.

At least as noteworthy as the behavioral alterations were the physical changes that accompanied them. First, strange things began to happen to the foxes' hair. Some came to have brown mottling over the standard silver color; others became piebald: white spots of varying size on a black background. The white forehead blaze, so characteristic of domesticated horses, cows, and goats, among others, began to appear with increasing frequency. On some foxes, the hair increased in length as well.[16]

Other physical changes in these foxes were also significant, most rendering the domesticated foxes more doglike. Floppy ears began to appear, and curled tails as well. In later stages the skeleton was affected; leg bones were shortened, as was the tail; the snout also was shortened, while the skull—and hence the face—broadened in a doglike way.[17]

The reproductive physiology was also altered in the tame foxes. Both in nature and at the fur farms, silver foxes breed once a year, when days begin to lengthen (January–February). The breeding season was extended in the tame foxes; some mated twice during a single year.[18]

Remember, all of these seemingly unrelated behavioral, physiological, and anatomical changes occurred as a result of selection for tameness. Belyaev's view that domesticated traits come in packages is amply vindicated. If traits are developmentally linked, you don't need a separate mutation for each one. But in this case there may well have been no mutation at all.

CREATIVE DESTRUCTION

Belyaev, remember, considered domestication an example of evolution under extreme and challenging conditions. We can call these extreme conditions "evolutionary stress." Belyaev proposed that under such stressful circumstances there was a sort of creative destruction caused by what he referred to as "destabilizing selection."[19] As with many ideas in evolution, the idea of destabilizing selection can be traced back to Darwin—in this case, his famous principle of natural selection, which inexorably occurs when:

1. Individuals in a population vary with respect to one or more of the traits that make up their phenotype—the principle of variation.
2. Individuals with the most felicitous phenotypes, given the current conditions, leave more descendants—the principle of differential fitness.
3. The fitness differences in phenotypes are transmitted to the offspring—the principle of heredity.

Wherever these three conditions are met, natural selection inevitably occurs.

The concept of natural selection has been much refined since Darwin, and several distinct forms of natural selection are now recognized. For my purpose here, two are worth mention: purifying (also called normalizing) selection and directional selection. Purifying selection is simply the elimination of mutants with lower fitness—an albino wolf, for example. Directional selection is more interesting; it occurs when environmental conditions promote the sustained alteration of a trait in a particular direction—toward larger or smaller size, for example. Darwin had primarily purifying and directional selection in mind when he formulated his principle of natural selection.

Belyaev's unfortunate compatriot, Schmalhausen, formulated a sort of metaselectionist principle, which he referred to as "stabilizing selection."[20] This idea is easiest to approach through a concept known as the norm of reaction or "reaction norm."[21] Reaction norms are graphically depicted in two-dimensional plots in which the x-axis is an environmental variable and the y-axis is a phenotypic variable (Figure 1.2). In the hypothetical population depicted in Figure 1.2A, the environmental variable is ambient temperature and the phenotypic variable is adult body size. The slope of the line is 45 degrees. Stabilizing selection alters this slope as in Figure 1.2B. Now the reaction norm is flat; that is, no matter what the ambient temperature is, body size stays the same.

Conrad Waddington independently arrived at a similar principle, which he called "canalization."[22] Both stabilizing selection and canalization buffer development not only against environmental perturbation but against genetic alterations (mutations) as well. These mutations are shielded from selection because they don't affect the phenotype. Therefore, these mutations can accumulate as cryptic genetic variation, which, when conditions change, can be exposed for subsequent selection. This is what occurs during the early stages of domestication.

According to Belyaev, the phenotypes of wild foxes had been canalized by eons of stabilizing selection. The new selection regime he imposed resulted in destabilizing selection, one effect of which was to expose the previously cryptic genetic variation that had accumulated

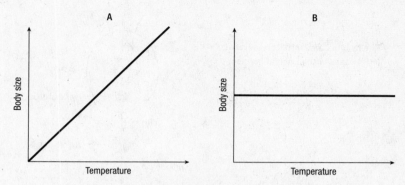

FIGURE 1.2   Norm of reaction before (A) and after (B) stabilizing selection.

through stabilizing selection. This newly exposed genetic variation was the fuel for the rapid response to selection for tameness.

TIMING IS EVERYTHING?

We still need to account for the fact that tameness resulted in all of these correlated changes. In what way are tameness and floppy ears connected? To say simply "pleiotropy" is not a sufficient explanation; this particular form of pleiotropy must itself be explained. To better understand this connection, we need to examine how tameness and floppy ears are developmentally linked.

It is noteworthy that many of these alterations, such as floppy ears and curled tails, are typical traits of fox pups—and wolf pups as well. The same could be said of many of the behavioral alterations. Wild fox pups are more likely to seek human company than are wild fox adults, and not solely for lack of conditioning or learning. Some of the difference between infant and adult foxes can be traced to the physiological maturation of the hypothalamic-pituitary-adrenal (HPA) axis, which underlies the stress response (Figure 1.3). Once the stress response is developed, foxes react to humans—and other foxes—with more fear and aggression.

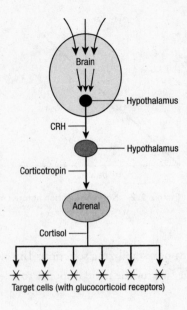

FIGURE 1.3
The hypothalamic-pituitary-adrenal
(HPA) axis, with principal hormones.

The development of the fear response in particular marks the closure of the socialization window.[23] This is true of other canines as well.

Several hormones are involved in the stress response; the levels of many of these hormones were altered by selection for tameness in ways that suggest a general dampening of the stress response in domesticated foxes relative to farm foxes.[24] Stress hormones of one type, the glucocorticoids—a family of hormones that includes cortisol—were especially altered by selection for tameness. Tame foxes had substantially lower cortisol levels than did the farm foxes.[25]

So, many of the adult traits of domestic foxes and dogs may result in changes in the timing of certain key developmental events—a phenomenon known as heterochrony (*hetero* = "different"; *chrony*, as in "chronological" = "time").[26] There are two basic forms of heterochrony; the form observed in the farm fox experiment—which results in the retention into adulthood of traits characteristic of earlier developmental stages— is called "paedomorphosis" ("infant form"). As Figure 1.4 shows, there are three distinct avenues to paedomorphosis:

	Avenues to:	
Paedomorphosis (trait of juvenile ancestor)		**Peramorphosis** (new trait not present in ancestor)
Progenesis (earlier offset)		Hypermorphosis (delayed offset)
Neoteny (reduced rate)		Acceleration (increased rate)
Postdisplacement (delayed onset)		Predisplacement (earlier offset)

Figure 1.4   Paedomorphosis and peramorphosis.

1. Postdisplacement: delaying the onset of the trait's development
2. Neoteny: slowing the rate of the trait's development
3. Progenesis: speeding up sexual maturation relative to the trait's development

At least two may have occurred in the farm fox experiment.

The tame foxes reached sexual maturity about a month before the farm-reared foxes did (progenesis). At the same time, the development of the HPA axis was retarded or decelerated in tame foxes relative to farm foxes (neoteny and/or postdisplacement).[27] A similar retardation occurred in ear, tail, and skeletal development. Even the seemingly uncanny ability to read human intentions may simply manifest another retained juvenile trait: the close attention fox pups pay to their mother's behavior.

It seems, then, that general physical and psychological development of the silver foxes was significantly slowed, and reproductive development accelerated, by selection for tameness. As a result, the adult tame foxes came to resemble the early developmental stage of their untamed ancestors. The genetic changes may have involved just a few genes, regulating a few key hormones that affect developmental rate.

The search for candidate genes commenced only recently.[28] Of particular interest will be genes and nongenic DNA sequences that influence the regulation of stress-related hormones. Glucocorticoids, for example, influence every physiological system in the body, from blood to bones.

The pituitary precursor peptide for corticotropin, which regulates the release of cortisol from the adrenal glands, also affects many types of cells during development, including melanocytes, which are responsible for dark pigmentation.[29] Here, then, is a possible link between tameness and hair color. Lyudmila Trut, who took over supervision of the fox project after Belyaev died, proposed that glucocorticoid levels influence the development of many other traits as well, through their coordinating effects on the timing of multiple developmental processes.[30]

Belyaev had the foresight to create a complementary line of foxes in which he selected for an intolerance of human proximity, the opposite of tameness. Over the years, these foxes became increasingly hostile to humans, growling, snarling, and lunging when approached. In essence, they became wilder than wild foxes. But the wilder-than-wild foxes showed none of the correlated changes in behavior, physiology, and anatomy associated with selection for tameness.[31]

It is worth emphasizing that all of this evolutionary change probably occurred in the absence of new mutations. Rather, successful selection for tameness was achieved solely with what evolutionists call standing genetic variation—that is, the genetic variation already existing in the population of foxes at the outset. But prior to selection, much of this genetic variation was cryptic, invisible to natural selection. In Belyaev's view, this cryptic genetic variation, which accumulated because of previous stabilizing natural selection, was subsequently exposed by destabilizing artificial selection. This newly exposed genetic variation is why selection for tameness was successful absent any mutations.

NEURAL CREST CELLS

There is a new and intriguing hypothesis concerning the package deal that is the domesticated phenotype.[32] While acknowledging a role for heterochrony, the hypothesis's emphasis is on an important population of stem cells called neural crest cells (NCCs), which first appear very early in development. They develop on the crest of the neural tube toward the tail end. These cells then migrate to many parts of the body, where they differentiate into the precursors of many cell types, including those involved in the development of many traits that constitute the domesticated phenotype, such as the cartilage of the tail, cartilage of the ear, tissue of the jaws and teeth, pigment cells (melanocytes), and the adrenal gland. On this view, domestication, through selection for tameness, induces defects in the development, proliferation, and/or migration of NCCs, hence the many correlated changes, such as smaller and floppy ears, shortened tail, shorter face, smaller teeth, depigmentation, and reduced adrenal glands and corticoid production. The authors also implicate abnormal NCC migration with year-round breeding, but I find this evidence much less compelling. No mention is made of early sexual maturation.

Importantly, any one of a large number of genes that affect NCC development and migration, alone or in various combinations, can be responsible for the domesticated phenotype, hence the genetic alterations should vary from species to species. Like the paedomorphic hypothesis, this new hypothesis highlights the role of pleiotropy by virtue of shared developmental pathways. It goes much further, though, in characterizing the possible mechanism. Moreover, it helps us narrow in on the genes that constitute the cryptic genetic variation that is exposed by disruptive selection. Furthermore, the neural crest cell hypothesis is compatible with and perhaps complements the heterochrony hypothesis, but the role of heterochrony is deemphasized.

FROM FOX-DOGS TO WOLF-DOGS

Foxes and wolves come from two different sides of the canine family tree (Figure 1.5). They differ in one particularly salient way with respect to domestication: foxes are largely solitary as adults; wolves are highly social. Social mammals are often presumed to be the easiest to domesticate because they are innately prone to hierarchical interactions in

FIGURE 1.5   Canid phylogeny. (Redrawn from Wayne 1993.)

which human agents can insert themselves at the top.[33] Therefore, fox domestication would seem far less likely to succeed than wolf domestication, which makes the success of Belyaev and his colleagues in creating a fox pet all the more remarkable.

Also noteworthy is the way the fox → fox-dog transition parallels the wolf → wolf-dog transition, especially in the correlated by-products of selection for tameness, from floppy ears to shorter snouts. At bottom, this parallel response reflects shared developmental processes in the fox and the wolf, which you might expect, given their genealogical proximity on the tree of life. But many of these correlated responses occur in other domesticated mammals as well, some quite distantly related to canines. Some are even found in domesticated birds and fishes. So consistent is this suite of changes, in fact, that it has a name: the "domesticated phenotype." And the domesticated phenotype is not just a feature of domesticated animals; we humans exhibit features of the domesticated phenotype as well. The view that humans are "self-domesticated" has become increasingly popular.[34] If indeed this occurred, such self-domestication would not be unique to humans; it is an important aspect of the domestication of most mammals, especially in the early stages.

We will explore the domesticated phenotype and self-domestication more generally in subsequent chapters. For now, though, let's turn to how they figure specifically in the evolution of dogs.

CHAPTER 2

DOGS

THIS MAY COME AS A SURPRISE, BUT THE PEKINGESE is supposed to look like a lion. (They are actually nicknamed "lion dogs.") Their resemblance to lions is not obvious, though, suggesting that the Chinese concept of lion-ness was obtained at some remove from actual lions. During the Han dynasty (206 BCE–220 CE), when sculpted lionlike guardian figures called *shishi* first appeared, lions still ranged throughout much of southern and central Asia, regions made accessible by the newly developed Silk Road. But few if any Chinese had seen an actual lion. This may explain, in part, the rather fanciful features of these lion representations. In any case, it was these *shishi*, not actual lions, that served as models for all subsequent lion representations in China—including the Pekingese—long after actual lions were imported into the imperial courts. (See Figure 2.1.)

The most lionlike feature of *shishi* is the mane. The most unlionlike feature is the pronouncedly squashed, square face. Both the mane and the flattened face feature prominently in Pekingese. And this was enough, in traditional Chinese culture, to embody lion-ness. Other Pekingese traits are neither lionlike nor *shishi*-like: their short, bowed legs, for example, designed to keep them from wandering, and the black face.

The Pekingese breed, which extends back in time at least 2,000 years, is the first for which written standards are available, albeit of a poetical

FIGURE 2.1   *Shishi* sculpture.
(© iStock.com/ThanyaG7.)

form. These are attributed to the Manchu princess Cixi, the de facto ruler of China from 1861 to 1908, and have the strong odor of royal status. According to Cixi, a Pekingese should be dainty, dignified, and evidence self-respect. Among the recommended foods are shark fins and curlew livers. For illness she prescribes "a throstle's [song thrush] egg shell full of the juice of custard apple in which has been dissolved three pinches of shredded rhinoceros horn." These were not items readily available to most peasants. Nor are the "piebald leaches" she deemed most suitable for bleeding.[1]

However distant the resemblance of Pekingese to lions, the resemblance of Pekingese to wolves is even less apparent. Yet the genetic distance between a wolf and a Pekingese is minuscule, far less than that between a wolf and a coyote, which much more closely resemble each other.

How do you get a Pekingese from a wolf? It takes time, of course, but not nearly as much time as you might think. Less time, in fact, than even Darwin suspected. On evolutionary timescales the wolf → Pekingese transition occurred in an eyeblink. And the Pekingese represents but one extreme in what we humans were able to make of wolves by acting as agents of evolution. Great Danes, dachshunds, poodles, greyhounds, Chihuahuas, and pugs are a few of many other evolutionary novelties that we have created from wolves, each as divergent, in its own way, from its canine ancestor as the Pekingese is.

What makes this amazing evolutionary story even more remarkable is that for many thousands of years, all wolf-human interactions were overtly hostile. We competed fiercely for the same prey and probably killed each other at every opportunity. In this respect dogs are unique among domesticated animals. Their evolution by domestication represents, in many ways, a reversal of eons of prior evolution by natural selection. But surprisingly, and notwithstanding the tremendous differences between Pekingese and Great Danes, the prior evolution of wolves has emphatically constrained what humans were able to make of dogs. Wolves, it turns out, were not at all like clay in the hands of the evolutionary potter that is humankind; in many ways dogs came preformed. For this reason, the story of dog domestication opens long before wolves began hanging around the camps of human hunter-gatherers in the Pleistocene.

BEFORE THE BEGINNING

The wolf, like the fox, belongs to an evolutionary family called Canidae. Among carnivores, canines, which first appeared about 35 million years ago, are distinctive in their mobility.[2] They are built to travel long distances in short periods of time. Wolves are quite mobile even by canine standards. They generally run down their prey, often after pursuing them for miles. And wolves are one of only three canine species that hunt in packs.[3] Their capacity for social cooperation and coordination go a long way toward explaining why wolves were once among the most successful predators on earth. An individual wolf is impressive, even imposing, but not nearly as much so as a brown bear or tiger. It is only because they hunt in groups that wolves are able to take prey the size of moose or bison or musk ox.

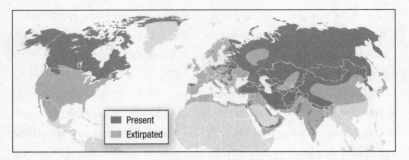

FIGURE 2.2   Geographic distribution of the gray wolf prior to domestication.

Wolf youngsters remain with their parents much longer than is typical of most canids. Foxes, for example, generally leave their parents once they are weaned. Wolf youngsters hang around their parents and often their parent's siblings for extended periods, which is the mechanism for pack formation. This extended period of dependency figured significantly in wolf domestication.

Prior to their domestication, wolves were among the most widespread mammals on earth, ranging from Arctic latitudes to the subtropics throughout the Northern Hemisphere (North America and Eurasia) (Figure 2.1). They also occupied a variety of habitats, from Arctic tundra to thick forests and semideserts. As a species, wolves are quite adaptable. Within this vast range and varied habitats were several genetically distinct wolf populations, but because of their mobility and adaptability there was sufficient gene flow between populations to prevent speciation.[4]

IN THE BEGINNING

The wolf is the only species that was domesticated prior to the agricultural revolution. When and where the process commenced are much disputed. There is some archaeological evidence—based on a skull discovered in the Goyet Cave, Belgium—that some human-associated wolves began differentiating in a doglike way as early as 31,700 BP,[5] but whether these remains and others found in European caves are actually proto-domesticates is disputed.[6]

Genetic evidence for the time and place that dog domestication began is also equivocal. Until quite recently, the two primary candidate regions were East Asia and West Asia (the Middle East). In 2002, a group of researchers led by Peter Savolainen proclaimed eastern China as the region in which dogs were originally domesticated, pinpointing an area south of the Yangtze River.[7] In 2013, this time using whole nuclear genomes, Savolainen's team provided further evidence for a southern Chinese origin and a date of about 32,000 BP.[8] But a rival group, led by Robert Wayne, proposed instead a much later Middle Eastern origin, based on the greater genetic affinity of most domestic breeds to the native wolves of that region.[9] Then, also in 2013, Europe was added to the mix, as a result of evidence from ancient mitochondrial DNA obtained from dog bones dating from 30,000 to 18,000 BP.[10]

It is probably best to keep an open mind on this matter; dogs may well have multiple origins.[11] For our purposes, what matters most is how this wolf domestication occurred. Those who dispute where wolf domestication first occurred seem to agree that it began when wolves began following human hunting parties to scavenge for leftovers.[12] Eventually, some started hanging around human encampments—not around the fireside, but at some distance, in the darkness beyond, waiting for the Paleolithic equivalent of table scraps. No doubt they were unwelcome at first; rocks were thrown, and later, spears, when verbal abuses failed to deter. But these bold pioneers persisted. Wolf domestication was initiated by wolves, and it required that an evolved psychological barrier be surmounted, or at least eased, such that they tolerated closer human proximity than did their forebears.

This process of self-taming was accomplished by standard natural selection. Among the wolves that hung around human encampments, those that better tolerated human proximity got more scraps and hence left more offspring than did their "wilder" cohorts. This naturally selected tameness was the first step toward dogness, and it may have taken thousands of years.

At some point human attitudes toward these wolf-dogs altered; they came to be viewed as at least somewhat beneficial. Wolf-dogs probably first became useful as sentinels. Their superior olfactory and auditory senses came to function as early warning systems. It is significant that dogs don't need to be domesticated much to fulfill this function. Indeed, feral village or pariah dogs continue to serve as inadvertent warning signals throughout much of the world to this day.[13]

Between 15,000 and 12,000 BP, some human societies became more sedentary, which probably accelerated the domestication process and marks the onset of the wolf-dog→dog transition.[14]

The first definitive archaeological evidence for a more intimate relationship between humans and domesticated wolves comes from several late-Paleolithic sites in Europe, in the form of dog burials. One such burial site was unearthed at Bonn-Oberkassel in Germany, and dated to about 14,000 BP.[15] It is significant not only that this dog was buried, but that it was buried with a human—a sign of a more intimate dog-human relationship. Dog burials from around this period of time or earlier have also been found in western Russia and Belgium.[16] Other particularly interesting dog burial sites from about 12,000 BP have been excavated in western Asia.[17] These sites belong to the Natufian culture, a contemporary of the late Magdalenian culture of western Europe. This was a period of massive ecological changes in the Northern Hemisphere, when the Pleistocene ice caps began to melt, and it corresponded with a shift in human hunting technology.

Prior Paleolithic hunters killed their prey primarily with hand axes and thrusted spears, which of course required very close proximity between hunter and prey. By 12,000 BP, there was a major technological advance in both Europe and Asia. Small, sharp rock flakes, called microliths ("tiny rocks"), were attached to spears that could be thrown, greatly reducing the danger to human hunters. It has been suggested that domesticated dogs became particularly useful after this shift in technology and hunting strategy, in tracking and perhaps harrying wounded prey.[18] This new cooperative hunting strategy may partly explain the subsequent rapid incorporation of dogs into human cultures over a wide geographic range. But a more important factor in the range expansion of dogs was the spread of agriculture.[19]

By 8000 BP, dogs could be found throughout the wolf's range in Eurasia[20] and parts of North America.[21] They then began to appear south of the wolf range in agricultural societies: in Mexico, about 5200 BP;[22] sub-Saharan Africa, about 5600 BP;[23] continental Southeast Asia, about 4200 BP; and island Southeast Asia, about 3500 BP. It wasn't until much later—about 1400 BP—that domestic dogs reached South Africa, where cows, sheep, and goats had preceded them.[24] South America first became host to domestic dogs about 1000 BP.[25]

Whatever their role as hunting auxiliaries or pets during this range expansion, most dogs continued to function as sentries, if they served any function at all. They foraged for themselves, largely at human dumps. Foraging at dumps created the conditions for further naturally selected alterations in behavior and anatomy. One of the first physical alterations during early dog domestication was a diminution in size. Initially, this size reduction resulted from a change in the natural selection regime caused by human association.

Outside of human settlements, a small wolf is at a disadvantage, particularly in dealing with other wolves. The human environment alters things in ways that make smaller size less disadvantageous, and perhaps even advantageous. Smaller wolves require less energy, for example. Smaller wolves would have an incentive to stay closer to humans than larger wolves do and hence evolve increased tameness relative to their larger brethren, to the point that there may have been genetic divergence based on size. Once the smaller wolves achieved a certain degree of tameness, they would have garnered increased human attention. In the early agrarian settlements of western Asia, human-dog interactions reached the point of artificial selection for small size. The genetic legacy of this artificial selection is a DNA sequence—called the small-dog haplotype—that is found in all dogs smaller than 22 kilograms, from schnauzers to Chihuahuas.[26]

The human environment also caused physiological alterations in dogs relative to wolves; most noteworthy were alterations in digestion related to a shift to a higher-starch diet.[27] With the advent of agriculture, the human diet shifted toward more starchy foods, which meant more starchy scraps. Village dogs adapted to this high-starch diet in ways that caused them to diverge further from wolves and converge toward humans.

FROM THE GARBAGE HEAP TO BURIAL AND BACK

Given the fraught relationship between wolf and man prior to wolf domestication, it's not surprising that humans have exhibited a somewhat schizoid attitude toward dogs. Even as humans acquired a fondness for wolf-dogs, wolves remained anathema. But even after wolf-dogs were transformed into dogs, humans took a variety of attitudes toward their creations. For example, man's best friend has been, until quite recently, as much eaten as petted. Throughout much of the evolution of dogs, their food function has been quite substantial.

Paleolithic hunter-gatherers, such as the Natufians and Magdalenians, probably ate dogs opportunistically, even while using others as hunting companions. Dog fur was also quite useful. But as the early dog burials indicate, Paleolithic humans could also adopt a quite nonutilitarian view of dogs. We can gain some insight into this state of affairs from more recent hunter-gathers, such as Australian Aborigines and Kung Bushmen. They, too, use dogs both as hunting partners and as food.[28] They seem to make a distinction, though, between food dogs and non–food dogs. (I use the term "non–food dogs" because, among the Aborigines at least, the dogs are not considered pets. They are not fed, for example, but have to forage for themselves, like village dogs.) One way aborigines distinguish food dogs from non–food dogs is by conferring names on the latter. Once it has a name, it is no longer potential food. That naming a dog would put it off limits for dinner is an interesting matter for human psychologists to chew on.

Once humans began to occupy permanent settlements after the agricultural revolution (9000 BP), the nonutilitarian view of dogs intensified. In Egypt, by 4100 BP (Middle Kingdom), dogs were depicted in the tombs along with their human companions; these dogs were named as well, presumably to ensure their continued companionship in the afterlife.[29] Later they were mummified along with their human companions. One famous mummified dog, called Hapi-puppy, was the size of a Jack Russell terrier.[30]

In ancient Greece, dogs were long celebrated for their loyalty. Penelope is justly celebrated for her patience and fidelity to Odysseus, her egomaniacal wandering husband. But the story of their dog, Argos, is even more poignant. He waited steadfastly for his master, well into his dotage. When Odysseus finally made it back home after 20 years of self-indulgent and self-inflicted adventures, Argos was there to greet him, albeit briefly. He could only wag his tail once before he died, having, presumably, stayed that moment by force of will, out of dogged love.[31]

The Romans appreciated their dogs at least as much as their Greek predecessors. In Roman Carthage, a dog named Yasmina was buried at the feet of her human companion, a glass bowl placed carefully at her shoulder. Yasmina was quite old and decrepit when she died. Arthritis, a dislocated hip, and spinal problems would have limited her mobility. And she had few remaining teeth, so she was probably restricted to a diet of soft food.[32] Clearly, Yasmina was well cared for. Also significant was the fact that Yasmina was a toy breed. Toy breeds aren't good for much else than love.

Yet in Roman Britain, toy dogs existed side by side with others much less fortunate. At Calleva Atrebatum, dogs from terrier size to Lab size ended up in the garbage middens, not the burial grounds.[33] Judging by the number of bones, they must have been consumed with great gusto, even as their more diminutive cousins were treated with great sentimentality—

partly, it seems, because they became associated with wealth and status.[34] (The association of toy dogs with status continues to the present day, which may be why the sight of a dog being carried in a purse strikes many as effete.) The distinction between pet dogs and food dogs continued in Europe well into medieval times. Elsewhere in the world things tilted decidedly toward the latter, especially in North America.

It has been suggested that the taste for dogs in the New World was, in part, born of a protein shortage caused by the relative dearth of large mammals.[35] (Large mammals—including mammoths, ground sloths, horses, camels, and various species of bison—had been eliminated during a mass extinction that occurred in the Pleistocene, suspiciously soon after humans first populated the New World.) Whatever the merits of that view, Native Americans, from Greenland to Central America, ate a lot of dogs.[36]

Dog consumption started early in the New World. The oldest bones of a domesticated New World dog were found in a Texas garbage midden.[37] The bones showed clear signs of having passed through a human digestive system. Paleo-Indians probably ate dogs opportunistically, but once city-states had been established in Mexico, dogs were more or less farmed. At the important Olmec site of San Lorenzo (3400–2400 BP), part of the annual tax that farmers owed the ruling elites was paid in fattened dogs that had been fed only maize.[38] These Mexican dogs obviously had the starch-heavy digestive capacities discussed earlier.

Dogs were also a significant component of the Mayan diet.[39] The Chichimecs of central Mexico, ancestors of the Aztecs, were called "dog people"—not because of a predilection for canine pets, but because of their taste for canine flesh. The Aztecs themselves created a hairless breed for consumption at royal feasts.[40] Presumably their hairlessness made them easier to grill. To the northwest, in the states of Nayarit, Colima, and Jalisco, dogs had long been raised to this end, as memorialized in some wonderfully sensitive and naturalistic ceramics depicting these quite cuddly, seemingly small, pudgy dogs in various states of repose and activity. The models for these ceramic sculptures had obvi-

Figure 2.3
Pre-Columbian clay dog
sculpture from Mexico.
(© Irafael/Shutterstock.com.)

ously been fattened. These cute, fat dogs may have been given the run of the house until such time as their services were required in the kitchen. This is the way many pigs of old were raised, though it is beyond the ken of most of us today. Perhaps some of the especially cute ones escaped that fate to become founders of a pet breed. Things may have been quite fluid in that respect.

In Cuauhtémoc, Chiapas, which is on the Pacific side of Mexico at about the same latitude as San Lorenzo, dwelt some contemporaries of the Olmecs. They treated their dogs quite differently, as indicated by dog burials, some of which included humans.[41] But after trade links were established with the Olmecs, things changed dramatically. Dogs began to appear in the garbage middens like other discarded foodstuffs. In precontact Mexico, even a pet dog wouldn't want to get too comfortable. A dog's position on the food–pet continuum was not a fixed one.

But that was probably true throughout most of the New World. The largest settlement north of Mexico was Cahokia, close to St. Louis. This Mississippian culture was renowned for its earthworks in the form of massive mounds. Cahokia also left us much pottery and ample evidence of butchered dogs.[42] By the time Hernando de Soto arrived on the scene in 1540, the Mississippian culture had vanished, but its descendants feted the Spanish conqueror—unwisely, it turned out—with a dog barbecue. According to one member of the expedition, Rodrigo Rangel, the dogs were raised in the homes of the Native Americans, much as they had been centuries earlier in western Mexico.[43] Evidently, De Soto found the dog meat delectable; perhaps, though, the thought of eating the European dogs that had accompanied him on the trip would have made him squeamish. Or perhaps not.

Dogs have long remained on many menus outside of the Americas, including Europe.[44] In East and Southeast Asia, dog meat is prized, as it is in parts of Africa and many Pacific islands. To this day, dogs are still farmed in Korea, the Philippines, and West Africa. Dog consumption is on the rise in Vietnam; currently over a million are eaten there each year, many exported from farms in Thailand.[45]

VILLAGE DOGS

Throughout the bulk of our shared history, most dogs were neither food nor pets, nor were they hunting companions; they lived a life much like their camp-following predecessors, but in more permanent settlements. Today, throughout much of the developing world, we can find dogs very much like the majority of those that existed 8,000 years ago. They are called "village dogs." Like their forebears, village dogs continue to live outside of human habitations, forage for themselves, and, most important, breed with whomever they choose. They are, in a word, feral.

When a village dog in rural Thailand is run over by a car or truck—which is a frequent occurrence, since they have a predilection for sleeping near or on roadways—no one mourns. The dog is simply left for the scavengers, including other village dogs. No Thai would think of letting a village dog into the house. In fact, village dogs seem to be barely tolerated. The same is true of village dogs elsewhere in Asia, as well as those in Africa, South America, and formerly North America (American village dogs fared even worse than their native human associates and are virtually extinct, replaced by European breeds).[46] Village dogs are hardy creatures that don't owe their survival to human affection.[47]

All village dogs, no matter from what continent, share certain characteristics, from which we can infer what the proto–village dogs of 8000 BP looked like and how they behaved. Most were in the intermediate size range (relative to modern dog breeds), with sleek coats of variable color, including piebald. They had shorter legs than wolves and proportionally smaller teeth. Their tails were upturned and their snouts somewhat shorter than those of wolves. Behaviorally, they were wily and somewhat wary, except for the relative few that were human bred and reared.

Perhaps most significant, they were not prone to pack formation; groups of village dogs, when they formed at all, tended to be nonhierarchical; that is, they had lost something of wolf sociality with respect to their behavior toward each other, including the highly ordered dominance hierarchy. This alteration in social behavior had enormous consequences.[48] In a wolf society, only the dominant male and female breed; in a village dog society, everybody breeds. Hence, village dogs have a much greater capacity to multiply than do wolves. Moreover, as a result of the relative ease of breeding, there is an important sense in which both natural selection and sexual selection are relaxed in village dogs relative to wolves, as reflected in, for example, their variable coloration.

Five thousand years ago, village dogs were the most widely distributed mammals on earth, aside from humans. Any species with a wide distribution tends to genetically differentiate along geographic lines, giving rise to distinct subpopulations. The degree of genetic differentiation at the subpopulation level depends on a number of factors, including mobility and geographic barriers such as oceans, deserts, and mountains. Wild wolves, as we have seen, exhibited such genetic variation among subpopulations prior to domestication. The process was accelerated in the even more widely distributed proto–village dogs, as reflected in the genetic differences among village dog populations today. Village dogs from Southeast Asia are quite genetically different from Middle Eastern village dogs.[49] And both Southeast Asian and Middle Eastern village dogs differ markedly from African village dogs.[50]

At a finer geographic scale—such as that of the Turkish highlands, or the Russian tundra, or the Carolina hardwood forests—village dogs evolved adaptations to their particular environments. These locally adapted village dogs are called "landraces," and they are a crucial link between the first village dogs and modern dog breeds.

Landraces are adapted not only to their physical environment—climate, altitude, and such—but to their human environment as well. Landraces in which the human cultural environment looms particularly large, in which artificial selection is primarily for functions deemed desirable in a particular cultural context, provided the raw material for breeds as we know them.

Some of the most ancient remaining landraces belong to the dingo group. Dingoes originated in East Asia in association with the Austronesians, whom they accompanied on their migration southward through peninsular and island Southeast Asia.[51] Along this route were littered various distinct dingo populations, from northern Thailand to Papua New Guinea, the descendants of which include Thai dingoes (village dogs), the Kinatami dogs of Bali, and the New Guinea singing dog. About 3,500 years ago the Austronesians reached the shores of northern Australia,[52] where they did not linger long before embarking for points east, perhaps with the strenuous encouragement of the Aborigines. A few of their domesticated dingoes did disembark, however, and managed quite well. The descendants of these hardy few colonized the entire continent—including rainforests, eucalyptus forests, mountains, and grasslands—with the exception of the driest interior desert.

The aboriginals of Australia—as well as those of New Guinea—who had arrived some 40,000–50,000 years earlier, had no prior experience with any doglike creatures and probably didn't find the new arrivals particularly useful, except perhaps as a new food source. In any case, they did very little by way of domestication; in fact, the Australian dingoes became not just feral, but truly wild wolf equivalents and the apex predator on the island continent. Things began to change somewhat with the arrival of Europeans and their dogs in the eighteenth century. These dogs, naturally, interbred with dingoes to the point that, as of 1995, 80 percent of mainland dingoes were of mixed ancestry.[53]

Despite such hybridization, dingoes provide some interesting contrasts with both dogs and wolves. In behavior and anatomy, dingoes are intermediate between domesticated dogs and wolves. For example, the canine teeth of dingoes are larger than those of any domestic breed but smaller than those of wolves.[54] They retain the upturned tail of their domesticate forebears, as well as their coloration, but they tend to form packs in which only the dominant male and female breed; and they have become seasonal breeders, having lost the domesticated capacity to breed year-round. Dingoes howl a lot and bark very little. At reading human intentions, they are better than wolves but worse than dogs.[55]

In all of these respects the dingoes outside of Australia and New

FIGURE 2.4   Pure dingo. Note the snout width intermediate between that of wolves and domestic dogs. (Photo by author.)

Guinea, such as the Thai dingoes, tend more toward the domesticated end of the spectrum, reflecting their more intimate and long-standing association with humans. And Thai dingoes are thought to most closely approximate what all dingoes were like when the Austronesians briefly came ashore in northern Australia. Australian dingoes represent a reversal of the domestication process. It is important to bear in mind that this reversal occurred at a relatively early stage in the dog domestication process; domesticated dingo ancestors were much less domesticated than most modern breeds, such as the Pekingese.[56]

"ANCIENT BREEDS"

The transition from landrace to proto-breed depends on active human intervention—artificial selection through culling and/or the control of mating. It begins with landrace dogs that are already particularly well adapted to the human environment—not just the human environment in general, but particular human cultural environments. There were three geographically distinct cultural settings in which this transition first occurred. The first encompassed the agricultural and pastoral descendants of the Natufians in West Asia, from Palestine to Afghanistan. Another such transition occurred in eastern Asia, especially in China and Japan. The third cultural setting was northern Asia, among sub-Arctic cultures. Collectively, the breeds produced in these early centers of domestication are commonly referred to as "ancient breeds."[57] But this term is problematic in that the breed concept was a nineteenth-century invention. Moreover, a number of the so-called ancient breeds are recent reconstructions;[58] for others, their seeming ancientness is an artifact of their genetic isolation from European breeds.[59]

The basenji has been long considered the most ancient breed.[60] Though basenjis are currently associated with the Congo basin and West Africa, they ultimately derive from Southeast Asian dingo-like dogs.[61] Basenjis were originally bred by bow hunters to flush game and track it through dense vegetation—tasks for which athleticism was prized over obedience. And to this day they rank at the bottom of the obedience scale (obedience is often erroneously equated with intelligence in the dog community) but are among the most athletic dogs in existence. For whatever reason, they are particularly adept at standing on their hind legs.

Behaviorally, what most distinguishes the basenji is a yodeling vocalization, made possible by an unusual larynx.[62] Basenjis do not bark at all. Their yodeling resembles that of the New Guinea singing dog, a dingo landrace. Among their other dingo-like traits, basenjis have a highly seasonal estrous cycle, once a year.[63]

Two other so-called ancient breeds come from West Asia: the saluki and the Afghan hound. The saluki, which is now largely associated with nomadic tribes of northern Africa, is a sight hound bred for speed and endurance—much like the greyhound—largely to run down and kill gazelles and hares in open habitats. The Afghan is also a sight hound bred for running down and killing similar prey, but in more rocky and mountainous terrain. The status of the Afghan as an ancient breed has recently come into question.[64]

Much farther to the north, three spitz-type dog breeds were first developed to quite different ends. "Spitz type" is an appellation for any breed that has retained or reacquired many wolflike traits. Most spitz breeds and landraces originated in the northern regions of Asia, North America, or Europe. Some spitz-type breeds, such as the Norwegian elkhound, are of recent origin, and their resemblance to wolves comes by way of human artifice. Others, though, such as the Siberian husky, Alaskan malamute, and Samoyed, resemble wolves because of a close genetic resemblance to their wild progenitors. All three breeds were developed to transport humans and their goods in Arctic habitats that were unfavorable for other beasts of burden, such as horses. The Samoyed may have served, in addition, in herding reindeer.

It is generally assumed that the genetic similarity of these three breeds and wolves is long-standing, in which case these breeds would be truly ancient breeds, but it is also possible that the genetic similarity reflects a more recent introduction of wolf DNA. This latter route to wolflike-ness complicates any attempts to construct a family tree of dogs, because wherever wolves and dogs coexisted during the course of domestication, there was undoubtedly some interbreeding, primarily between male wolves and female dogs.[65] This is especially true of breeds from northerly latitudes, where wolves remain relatively abundant.

The majority of so-called ancient breeds derive from East Asia. From Japan come two spitz-type dogs: the Shiba Inu and the much larger Akita. Both were bred to be hunters of game, which for the Akita included boar and bear.[66] The non-spitz-type Lhasa apso comes from Tibet, where it served as a household sentinel for Tibetan royalty. All other putative ancient East Asian breeds originated in China, among them the Shar-Pei, Shih Tzu, Chow Chow, and Pekingese. What is perhaps most notable about these Chinese breeds is that—with the partial exception of the Chow Chow, a midsized spitz-type dog—they were largely bred to be indoor pets.

Many of these breeds share certain personality traits that distinguish them from other dog breeds, including independence or aloofness. The basenji is sometimes described as catlike—not an attribute to which many in the dog community aspire. These older breeds generally do not need as much human society as most dog breeds do, and they also tend to rank low in trainability. In some cases (such as the Akita), these personality traits may reflect a relatively high genetic similarity to wolves. But in other cases, aloofness and independence simply reflect different selection regimes for East Asian dogs relative to common European breeds.

FROM LANDRACES TO MODERN DOG BREEDS

Most existing dog breeds are only recently derived from their landrace ancestors. The landraces from which most modern breeds derive had already been subjected to artificial selection, to various functional ends.

That is, they were more highly domesticated than basenjis or the ancestors of dingoes. These more recent landrace → breed transitions occurred only after the advent of kennel clubs, the first of which was founded in England in 1873.[67] These kennel clubs were to have a massive impact on recent dog evolution.

The kennel clubs were modeled after existing registries for racehorses and other livestock, with the original aim of preserving genetically distinct dog lineages. But these Victorian institutions quickly diverged from their models, driven in large part by the increasing social status attached to particular dog traits at this time, many of which had little relation to the original functions for which the landraces had evolved. Poodles, for example, derive from a landrace that evolved as a water dog to facilitate retrieving in aquatic environments, a task for which their curly hair was particularly well suited. The indignities to which their coats were subsequently subjected were purely status related.

Many European modern dog breeds can be assigned to a number of functional groups, including herding dogs, sight hounds, scent hounds, retrievers, guard dogs, and so on, according to the activities for which they were originally bred. These working dogs were pets only secondarily, if at all, until the nineteenth century. It is possible that dogs bred for a particular function, such as tracking game by sight or scent, arose independently in different settings and cultures, which evolutionary biologists refer to as "convergent evolution." Alternatively, sight hounds may generally derive from one common ancestor, and scent hounds, from another. A recent study showed that, with the exception of the ancient breeds, European dogs in a particular functional category tend to be more closely related to other breeds in that category than to breeds in other functional groups.[68] So, it seems that these functional categories arose less through convergence and more from common ancestry. (Toy breeds, which don't comprise a functional category in this sense, exhibit much more convergence than do utilitarian breeds).

This means that, in principle at least, we should be able to trace dog breeds from a given functional category in a particular geographic region to particular landraces. Such inference is quite difficult in practice because the radiation in dog breeds occurred very rapidly in evolutionary time, so genetic signatures are relatively faint. Fortunately, in some cases, such as herding breeds from the British Isles, we have a decent historical record to augment the genetics.

The border collie was a sheep-herding landrace developed in the border region of northern England and southern Scotland. Similar landraces from the north, such as the Scotch collie, and the south, such as the English sheepdog, evolved around the same time. All of these landraces had a relatively recent common ancestor.[69] The British Kennel Club made of these landraces a variety of breeds, including the border collie, Scotch collie, English shepherd, bearded collie, Old English sheepdog, Shetland sheepdog, smooth collie, and rough collie. (The bearded collie and Old English sheepdog may represent crosses between British herding landraces and herding landraces from elsewhere in Europe.)

The herding breeds described so far are called "headers" because they get in front of the livestock and stare them down. Other herders are called "heelers" because they herd by nipping at the heels of livestock. A distinct landrace of heelers evolved in Wales, primarily to herd cattle. This landrace gave rise to two breeds: the Cardigan Welsh corgi and the Pembroke Welsh corgi. Both corgi breeds have distinctively elongate bodies and relatively short legs, conducive to heeling. Despite these differences, corgis show a greater genetic affinity to other herding dogs from the British Isles than to breeds from other functional groups.[70]

But herding breeds developed elsewhere in Europe, such as the puli from Hungary and the Pyrenean shepherd from southern France, were clearly derived independently from landraces that were genetically distinct from those of the British Isles. So, if we consider herding behavior in a global context, both common ancestry and convergent evolution play a role. This is true not only of other dog traits but also of the evolution of many traits in any species. Convergence complicates the task of reconstructing any part of the tree of life, especially one as tiny as the dog twig.

RECONSTRUCTING THE DOG TREE

Methods for reconstructing evolutionary history constitute an important but often underappreciated subdiscipline in evolutionary biology called "systematics." The goals of systematics are twofold: to sort out the genealogical relationships of existing (and sometimes extinct) species, and to estimate the time since species and groups of species diverged from one another. Traditionally, systematists relied on physical traits, especially stable physical traits such as bones or teeth, from both fossils and extant species. More recently, DNA has become the primary raw material for tree construction.

Within the last few years it has become possible to compare whole nuclear genomes of various dog breeds to sort out their genealogical relationships. The genealogical reconstruction described next is based on genome-wide comparisons.

We can consider the wolf as the root of the dog tree. The first surviving landraces to emerge from the wolf root were the dingoes, including the ancestors of the Australian dingoes, which branched off over 4,000 years ago. The basenji was the next to diverge, probably derived from a West Asian dingo-like landrace, such as the Canaan dog. One tentative reconstruction of the genealogical relationships between the basenji and the other ancient breeds is depicted in Figure 2.5.

From this point in dog evolution the tree begins to get bushy, in part because of the explosive radiation of dog breeds in the nineteenth century and the consequent dearth of genetic divergence, even in the more rapidly evolving genomic sequences. We can distinguish several distinct branches or breed clumps, though. For example, the herding dogs form one clump. Another clump consists of the mastiff-like dogs—which generally functioned as guards, for either livestock or humans. These include the Rottweiler, Saint Bernard, and bulldog. Sight hounds—so called because they were bred to hunt by sight—such as the borzoi, Irish

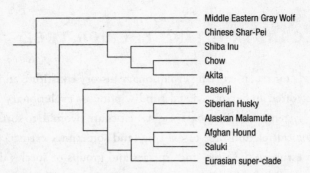

FIGURE 2.5   Genealogical relationships of the oldest dog landraces. (Adapted from VonHoldt et al. 2010.)

wolfhound, and whippet, form another clump, as do scent hounds (e.g., beagle, basset hound, and bloodhound).

There are some suggestive relationships among the breed clumps. For example, spaniels and scent hounds seem to be more closely related to each other than to any other breed groups; the same is true of sight hounds and herding dogs. There is also a surprising lack of close genealogical relationship among some of the small dogs, indicating that small size was developed independently in several functional groups (and "ancient breeds"). Some of the smallest dogs do form a clump of toy breeds, including the Chihuahua and papillon, but the toy poodle is not part of this genealogical clump, nor are the various toy terriers.

It is actually somewhat surprising that systematists can derive any treelike genealogy from dog breeds. Treelike genealogies require vertical ancestor-descendant relationships, in which one common ancestor gives rise to two branches of the tree, from which another common ancestor gives rise to two smaller branches and so on. It is safe to assume vertical relationships in constructing mammalian species trees; it is much less safe to assume so in reconstructing the genealogy of dog breeds. We know that many dog breeds are products of hybridizations of existing breeds or landraces from sometimes distant parts of the genealogical tree. The Newfoundland, for example, was created by the hybridization of the extinct Saint John's water dog and a mastiff-like breed. The golden retriever is a result of multiple hybridizations, including crosses of the flat-coated retriever, the extinct Tweed water spaniel, and the Irish Setter. Many toy breeds were also created in this way. The Labradoodle (Labrador retriever × poodle) and other "designer breeds" are only the most recent examples of a process that has occurred throughout the history of dog breed construction.

Hybridization renders the genealogical relationships more reticular, or, to continue with the botanical metaphor, rhizomatic. With hybridization, one branch fuses to another, as in the rhizomes of a strawberry plant. Evolutionists refer to such nonvertical genealogical relationships as "horizontal." Horizontal genealogical relationships caused by hybridization greatly complicate and compromise reconstructions of the dog tree, rendering them much less robust than species trees.[71]

GENOMICS AND THE EVOLUTION OF DOG TRAITS

Genome-wide comparisons of dog breeds provide much more than genealogical information. We can also learn from these comparisons about some of the genetic alterations that contributed to dog evolution above and beyond traits that can be explained by paedomorphosis and the domesticated phenotype. It will be useful to distinguish five types of genetic alterations ("mutations"), which apply not only to dogs but to all other species that I will consider in this book.

The easiest mutations to isolate, and by far the most studied, are single base changes, called point mutations, or single nucleotide polymorphisms (SNPs), which involve the substitution of one base (C, G, A, or T) for another.[72] The second type of mutation occurs when a base sequence is inserted into or deleted from an existing sequence; these mutations are referred to as indels. The third type of mutation involves the duplication and/or chromosomal rearrangement (termed "transposition") of a gene or large nongenic sequence. These three types of mutations are the most studied in all species, from bacteria to humans, and will be the focus of the following discussion.

Two additional types of mutations that have received increasing attention of late and that I will refer to in subsequent chapters are tandem repeats and copy number variations (CNVs). Tandem repeats are the accretion, often increasing over generations, of sequences of a few bases (such as CCG).[73] This type of mutation is associated with a number of diseases in humans, including Huntington's.[74] In dogs, expansions and contractions occur throughout the genome and with much greater frequency than point mutations. Tandem repeats are therefore potentially a powerful mechanism for rapid morphological evolution.[75] The fifth and final category of mutation, consisting of copy number variations (CNVs), is also a proposed mechanism for rapid evolution in dogs.[76] Like tandem repeats, CNVs involve the expansion of particular base sequences, but CNVs comprise much larger sequences.

To date, most genomic mapping of mutations on phenotypic traits in dogs has been restricted to point mutations, indels, and gene duplications/rearrangements. One point mutation, found in all dogs weighing less than 25 pounds, affects a gene that codes for a growth factor known as IGF1.[77] This mutation does not alter the IGF1 protein itself, but the rate at which it is synthesized. What is particularly interesting about this mutation is that it is located in one of the (noncoding) transposable DNA elements. It has become increasingly evident in recent years that such functional alterations of noncoding DNA play a crucial role in the evolution of all species. Another mutation of a transposable element is associated with the proportionally shortened limbs of basset hounds, dachshunds, and other short-legged breeds.[78] Still another such mutation is associated with a coat trait called "furnishings"—long hair patches in the facial area, such as the eyebrows, mouth (mustache), and chin, which, in some breeds, mimic the appearance of an elderly nineteenth-century gentleman.[79] Well-furnished breeds include the bichon frise and bearded collie.

Other coat quality characters can be traced to mutations in the coding DNA. These mutations alter the proteins involved in hair development. One of these coding-DNA mutations is responsible for long hair in breeds such as the golden retriever. Another such mutation is found in curly-haired breeds, such as the poodle and Portuguese water dog. All three of these coat quality mutations—furnishings, long hair, and curly hair—are completely absent in wolves, which are short haired, or "smooth coated," and hence must have occurred during dog domestication. Various combinations of these three mutations account for the vast majority of coat quality or "pelage" phenotypes in all dog breeds that were tested. The interesting exceptions are the saluki and the Afghan hound, two ancient breeds. These exceptions suggest that in some or all "ancient breeds," coat quality traits such as long hair may have evolved independently of modern European breeds.

Hair color characteristics have a more complicated genetic architecture. Variant versions (called alleles) at seven genetic loci combine in sometimes complicated ways to produce particular coat color patterns, from white to brindled.[80] The fox farm experiments suggest that some of these variant alleles already existed in wild wolves and were developmentally linked to tameness. Many of these variants, though, arose de novo during the course of dog evolution and were made more common through artificial selection. Of particular note is the mutation for black coloration, which arose during domestication and then was transferred to wild wolves through interbreeding with domestic dogs.[81] There were no black wolves prior to domestication; now they are fairly common, especially in North America—another indication of the importance of hybridization in the evolution of dogs. Indeed, the particular trait combinations of, say, hair quality and hair color that are characteristic of modern breeds were generated by human-guided hybridizations to create essentially customized breeds. All modern breeds with curly hair, for example, derive from a common ancestor with curly hair, and breeds with long, curly hair and furnishings, such as the bearded collie, were created through a series of hybridizations among breeds with those characteristics.[82]

Genome-wide mapping studies have also identified the genetic sub-strates for a number of other physical traits, including size of the ears (which are huge in the Pharaoh hound and basset hound),[83] body weight, and snout length (long in the collie, short in the bulldog).[84] All of these traits have been under artificial selection,[85] which greatly intensified with the advent of kennel clubs, to a degree that, in many cases, perverted the process of dog evolution. When humans completely assume the role formerly occupied by natural selection, there are often unintended consequences, some of them quite undesirable.

KENNEL CLUBS AND RECENT DOG EVOLUTION

On a visit to the Natural History Museum in London in the summer of 1988, I happened upon a fascinating exhibit on dogs, which documented changes in various breeds over the previous century. I was astounded at how much evolution had occurred in such a short period of time. In 1888 all of the breeds looked more like generic dogs. Toy breeds were generally larger, and large breeds generally smaller. In general, breed differences in conformation and skeletal structure had become more exaggerated by 1988. But the most striking were the alterations in the short-faced breeds, such as the bulldog. The face of the nineteenth-century version was much less grotesquely (to my mind) squashed. I suspect that if we compared the 2012 version of the bulldog to that of 1988, we would find even more face squashing.

The rapidity of the physical changes documented in the 1988 exhi-bition was unprecedented in the history of dog evolution and can be traced to a single historical event: the founding, in 1874, of the first kennel club in London. The mandate of the kennel club was to "main-tain" breed standards through registries. In this, it utterly failed. Rather, the effect of the kennel club was to massively escalate breed divergence, by means of competitive dog shows, in which the most extreme examples of a given breed type were selectively rewarded and hence selectively bred.

But the selectivity of the selective breeding went beyond anything humans had ever before managed. A single male champion could sire hundreds of offspring, for example. Moreover, these champion studs were routinely mated with their own female offspring. Routine incest of this sort, even in dogs, makes most people a bit queasy. Perhaps the Victorian aristocrats largely responsible for this state of affairs were somewhat desensitized to incestuous relationships, given their own pedigrees.

This kind of artificial selection will rapidly get you large phenotypic changes, but at a cost. First there is the cost of inbreeding and the inevitable accumulation of deleterious mutations. Virtually all purebred dogs have a host of genetic ailments, from narcolepsy to skeletal defects.[86] Cancer is also rampant among purebred dogs, occurring at frequencies that in humans would be considered epidemic. Any account of a breed's characteristics includes the defects, including the particular form of cancer, to which it is prone. This is why purebred dogs have much shorter life spans than outbred dogs (mutts and mongrels) living under the same conditions.

This genetic burden may also help explain why dog breeds are exceptions to what is otherwise the rule among mammals: larger species live longer than smaller species. Elephants live longer than cats, which live longer than mice, and so on. Among dogs, however, large breeds die younger.[87] Irish wolfhounds, Great Danes, and Newfoundlands live only 6–8 years; dogs in the 50-pound range generally live 10–12 years; some toy breeds live 15–20 years.[88]

In part, this reversal of the mammalian trend can be explained by the pace-of-life rule: live fast, die young.[89] Larger dog breeds are larger because they grow faster and hence expend more energy per cell per minute than smaller breeds do. In most mammals, larger species actually grow more slowly than smaller species; they are larger because they keep growing for a longer period. Larger species also expend energy at a lower rate than smaller species do. This trend has been reversed among dog breeds because of artificial selection for size through accelerated growth.[90] So part of the explanation for the life span trend among dogs may be that inherited defects, such as those of the heart and skeleton, have more pronounced effects in breeds that grow fast than in those that grow slowly. Cancer rates are also higher where growth, and hence cell division, is more rapid.

But genetic diseases are only part of the cost in health for breeds like the bulldog. That squashed face, for example, carries a heavy cost in and of itself. It starts with breathing problems. The shortened snout causes the soft palate to bunch up in front of the trachea in a way that impedes airflow. Since panting is the primary way for dogs to cool, bulldogs are also vulnerable to death by overheating. The mouth of a bulldog is also too small to accommodate its teeth, so they are crowded and grow at odd angles, trapping food debris. Gum disease is rampant. A bulldog's eyes don't seat properly in the skull and can pop out, even from straining at a leash. Often the eyelids cannot close completely, resulting in irritation and infection; sometimes the eyelashes rub the eye as well. The excessive skin folds often become infected. Perhaps most telling, the bulldog's head has become too large to pass through the birth canal; most births require Cesarean section. Such are a few of the consequences of an obsession with bloodlines.

Bulldog woes are far from the worst. Under the auspices of the kennel club, the Cavalier King Charles spaniel recently evolved a brain that is too large for its skull—a condition known as syringomyelia. The effects are variable but often involve excruciating pain and ultimately paralysis and death. You would think dogs with this condition would not be bred, but you would be wrong. The problem is that the symptoms often do not become manifest until three or four years of age, and breeders don't wait that long. Several recent dog show champions that died of this disease sired numerous offspring, sometimes through mating with their daughters. This is truly perverse, both evolutionarily and morally.

CONSERVATIVE CREATIVITY

There are limits, though, to what even kennel clubs can make of dogs, and not just ethical ones. These are limits imposed by the previous evolutionary history of wolves, the conservative aspect of evolution. It may seem balmy to speak of evolutionary conservatism with respect to the artificial selection of dogs, given the tremendous variation in dog breeds, but constrained it is.

Consider the skull, the most variable trait complex of dog breeds. The skull consists of a number of bones, evolutionary changes in which are highly correlated. Such correlations cannot be ascribed merely to pleiotropy; they reflect rather a deep developmental integration involving network-like interactions of many coding and noncoding DNA sequences,[91] which evolved long before the wolf did. Indeed, this particular pattern of skull integration is characteristic of the entire canine family, including foxes.[92] These conserved developmental networks would explain, in part, the parallel changes in the skulls of domesticated foxes and domesticated wolves.

Developmental integration is a key concept in evo devo. Another is modularity.[93] In an important sense, the entire development of any creature is tightly integrated. But the total network of interactions that constitute development can be parceled into subnetworks that operate relatively independently. The heart, for example, develops relatively independently of the pancreas. These relatively independent subnetworks are called modules. The wolf skull actually consists of two modules—the neurocranium (or braincase) and the face, including the snout. We will focus on the facial module.

The cephalic index is the ratio of skull width to skull length. Dogs are classified as brachycephalic ("short headed") when the width of the skull is at least 80 percent of the length. Brachycephalic breeds include, in addition to the bulldog, the Pekingese, boxer, and pug. The opposite of brachycephaly is dolichocephaly ("long head"). Notable dolichocephalic breeds include the Afghan hound, saluki, greyhound, and borzoi. The difference in the cephalic ratios of a brachycephalic breed, such as the bulldog, and a dolichocephalic breed, such as the borzoi, exceeds that of all such differences within the entire canine family. Indeed, many of the skull shapes in existing dog breeds represent novel creations, beyond anything previously evolved in any carnivore.

It seems, then, that artificial selection has managed to break through the developmental integration of the skull and the facial module, by restructuring the underlying genetic network previously evolved in wolves and other canines. In fact, however, and despite the massive diversity in dog skull shapes, the genetic networks that underlie the integration of skull development in wolves and other canines remains entirely intact.[94] Breeders have only managed to tinker at the margins of the well-integrated genetic networks inherited from wolves. They have explored the limits of this tinkering, but the modules still abide. A lot more than tinkering—a reorganized module, in fact—would be required to get anything like a cat skull from a dog. But such deep developmental alterations are beyond even the most ambitious breeders.

And this is just as true of all of the other behavioral and morphological traits that breeders have played with. For all of the dramatic interventions of kennel clubs, dogs—from Pekingese to bulldog, from whippet to Chihuahua—very much retain the legacy not only of wolves, but of their evolutionary history as members of the family Canidae in the order Carnivora. In evolution, every creative development results from tinkering with what was already there. For all of its creativity, evolution, even under the extreme degree of artificial selection promoted by kennel clubs, is still fundamentally conservative.

CATS

I HAVE ROUGHLY EQUAL AFFECTION FOR CATS AND DOGS, which puts me in a distinct minority among people of my acquaintance; most are quite partisan in their allegiances. But I am currently living with a cat person, so I have a cat sitting on my lap as I write this. His name is Sylvester because he looks so much like the hapless *Looney Tunes* character with the black tuxedo coat, perpetually tormented by an extremely neotenic bird named Tweetie. Sylvester also shares some of the cartoon character's haplessness: he is ungraceful, if not downright clumsy; once, while soundly sleeping, he actually fell off the top of the couch onto my prone chest, then, in terror, leaped over to an adjacent couch but misjudged and crashed ignominiously on the hardwood floor, from which he strenuously but unsuccessfully sought to gain traction for rapid egress, as if it were an ice surface, and then, upon finally getting his legs, bumped into a cedar chest. (He left me, as a memento, two sets of superficial gouges on my chest and an adrenaline boost.) Sylvester has other problems as well, including a morbid fear of doorbells and the people who subsequently appear. When the doorbell rings, he makes a dash for the closet or the underside of the bed.

Sylvester's sister, Smoke—a gray tuxedo—in stark contrast, is as graceful and athletic a creature as I have ever known. She could do a complete somersault when she was six months old, and she can rest on her hind legs like a meerkat. She is a gray blur when chasing a laser pointer up and down stairs. Sylvester can only watch until she tires. Smoke is also much more "well adjusted." She loves strangers, sidles up, tail high and shivering seductively or plopping on her back for a belly rub.

When we first brought them home as kittens, Smoke exited the carrier immediately and started exploring; Sylvester wouldn't budge. I finally had to dump him out after an hour. When we moved from one part of Brooklyn to another, Smoke mostly took it in stride; Sylvester was a wreck for weeks. If it weren't for Smoke's calming influence, it would have been much longer. For Sylvester is exceedingly fond of his sister and, for the most part, vice versa. When they were respectively spayed and neutered, Smoke and Sylvester were placed in the same recovery cage, primarily for Sylvester's sake. When Smoke had an emergency medical procedure to extract yarn from her intestine, Sylvester was in distress during her absence, incessantly calling for her. They often sleep in a ball so tightly bound that it is difficult to discern their outlines. Their relationship has greatly altered my prior notions of cat sociality. The solitariness of cats is greatly exaggerated.

Sylvester and Smoke were also instructive with respect to a behavioral syndrome that plays an important role in the domestication process—a personality dimension in cats, humans, and many other vertebrates, from goldfish to pigs, called the shyness–boldness continuum. Sylvester is on the shy end, Smoke on the bold end. Since they shared a womb and have spent all but one day of their life together indoors, it is tempting to attribute all of their personality differences to their genes. But Smoke and Sylvester had a rich and formative, albeit brief, existence before we adopted them at 10 weeks of age, which undoubtedly influenced their personalities.[1]

The shared experiences of Smoke and Sylvester—first as kittens handled a lot by humans, and then as coddled indoor cats—also no doubt contributed greatly to their temperament, and particularly their reaction to humans. If Smoke had been born feral, she would not be as human friendly as she is today. And that is true of Sylvester as well; in the big scheme, he is not that shy.

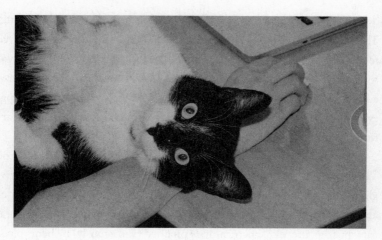

FIGURE 3.1   Meet Sylvester. (Photo by author.)

Our neighbor is part of an innovative program—called trap-neuter-release—to deal with feral and semiferal cats, which abound in this city and elsewhere around the globe. These cats are provisioned and then captured, spayed or neutered, and returned to the empty lots or abandoned buildings from which they came. This method is actually better—and more humane—for controlling the feral cat population than is simple removal.

Prior to her move to New York City, our neighbor worked with feral cats in the Virgin Islands. She adopted three of them. Though they, too, vary in their shyness, compared to these three formerly feral cats Sylvester is quite bold. One of them, also a black tuxedo, called Baby, I have glimpsed only briefly. The difference between Sylvester and Baby is largely due to the different environments in which they grew up. Around humans, the boldest feral cat is shyer than the shyest home-reared one, even if the feral cats are adopted soon after weaning. As with canines, there is a window of socialization, and in cats it seems to close earlier and more firmly than in dogs.

There is an interesting trend among the three feral cats related to the age at which they were adopted. One of them, Pablo, was abandoned between 4 and 6 weeks of age, which is prior to normal weaning. Though still shyer than Sylvester, he is by far the boldest of the three feral adoptees. Lucy, who was adopted at about 12 weeks of age (two to four weeks after cats are usually weaned) is much shyer; and Baby, adopted at a slightly older age, is shyer still. Early human handling, especially during the preweaning period, is crucial to a cat's later reaction to humans.

I use the term "feral" loosely here, for the "alley" cats in both the Virgin Islands and New York City are probably relatively recent arrivals to the street; many are second- or third-generation descendants of abandoned cats and were provisioned to varying degrees. Truly feral cats must secure all of their calories unaided (consciously) by benevolent humans and have bred for many generations under these conditions. It is difficult to determine how many truly feral cats exist in New York City—perhaps a relative few. You are more likely to find them in more rural settings. We would expect a truly feral cat adopted at the same age as Baby to be even less given to human interaction. But even these truly feral cats would seem "friendly" compared to the wild ancestors of all truly feral cats, semiferal cats like Baby, and pampered house cats like Smoke and Sylvester. Because of genetic psychological alterations wrought by the domestication process, even the most feral cats are more human friendly than true wildcats raised under the same conditions.

The domestication of cats, though, has been much less pronounced than that of dogs. Even Sylvester more closely resembles his wild ancestors, both physically and psychologically, than do the most wolflike dog breeds. Generally, there is more wildcat in domestic cats than there is wolf in dogs, because cats took a somewhat different route to domestication than dogs. Yet there are important features common to both. The differences and similarities in the domestication of dogs and cats have a lot to do with the prior evolutionary history of wolves and wildcats, respectively. We explored the evolutionary history (genealogy) of wolves in Chapter 1; let's now consider the evolutionary history of wildcats.

THE CAT FAMILY

Though cats and dogs belong to the same mammalian order, Carnivora, they come from two quite distinct branches (Figure 3.2). The dog branch includes—in addition to other canids—bears, raccoons, otters, skunks, seals, and sea lions. The cat branch includes—in addition to other felids—hyenas, mongooses, and civets. Perhaps what most distinguishes the cat branch from the dog branch is that the members of the cat branch are more exclusively carnivorous; their diet includes very little plant material. The relative specialization of the cat branch is most pronounced in the cat family, Felidae. Felids are considered obligate carnivores, in that they can metabolize only animal protein.

Several distinctive features of felids reflect their meat dependency—most obviously, their teeth. All carnivores have modified premolars and first molars specialized for scissor-like shearing, called carnassials. The size of the carnassials relative to molars provides a good indication as to how much of a carnivore's diet is meat-derived protein. In bears (family Ursidae), the carnassials are small and the molars large, in accordance with their largely vegetarian diet. The size of the molars reaches its extreme in giant pandas, which rely exclusively on bamboo. In dogs

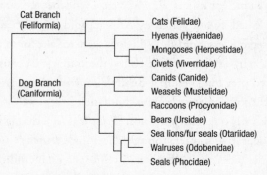

FIGURE 3.2   Carnivore phylogeny with family relationships. (Redrawn from information by Dr. David L. Atkins and by Arnason et al. 2001.)

(family Canidae), both the carnassials and the molars are large, reflecting a diet relatively balanced between meat-based proteins and plant-based carbohydrates. In felids, the carnassials are even larger than those of dogs, and the molars are vestigial. Felid canine teeth are also larger than those of canids, and their incisors sharper. Their dentition is so special-ized for cutting that felids, including domestic cats, can't chew.[2]

Perhaps the greatest evolutionary innovation of the felids is in their claws. Whereas dogs and bears walk on the soles of their feet, cats actually walk on their toes, which provides a mechanical advantage in walking and running, through greater stride length. This mechanical advantage in felids is one reason the fastest saluki or greyhound could never beat a cheetah in a 100-meter dash. But toe walking potentially poses a problem in abrading and hence dulling the claws. That problem was alleviated through the evolution of retractable claws. Each claw is controlled by one muscle above that, when contracted, retracts the claw, and one muscle below that, when contracted, releases the claw. When a cat walks, the claws are retracted, which keeps them sharp for the occa-sions when they need to be deployed.

In contrast to canids, felids are largely ambush predators and kill with a single bite to the skull, jugular, or spine, depending on the size of the prey. The jaw muscles of felids are proportionally larger than those of canids; hence they can generate greater bite force, which also makes them more efficient killers.[3] If you were a wildebeest, you would much rather be killed by lions than by African wild dogs, because death by lion is faster and less painful. Wild dogs seem to take forever to kill large prey—death by a thousand small bites; they often start consuming their beleaguered prey long before it finally dies. Even the most unpracticed lions get the job done much more "humanely."

There is only one feline, the lion (*Panthera leo*), that is a truly social and a cooperative hunter, though cheetahs (*Acinonyx jubatus*) hunt together on occasion.[4] All other 35 felid species, including the ances-tors of the domestic cats, are solitary hunters and live a largely solitary existence.

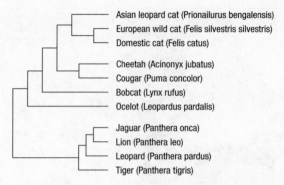

FIGURE 3.3   Felid phylogeny. (Adapted from Johnson et al. 2006.)

The family Felidae first entered the evolutionary stage about 35 million years ago (henceforth abbreviated "mya"), at the very end of the Eocene epoch (of the Cenozoic era). The last common ancestor of all modern felids lived in Eurasia about 11 mya (Miocene epoch), from which eight distinct lineages evolved (Figure 3.3).[5] The first lineage to split off (about 10.8 mya) included the great roaring cats of the genus *Panthera* (tiger, lion, leopard, snow leopard, and jaguar) plus two species of clouded leopards (*Neofelis*). The lineage that includes the domestic cat (genus *Felis*) originated about 6.2 mya (in the late Miocene).

About 2 mya the wildcat, *Felis silvestris*, split from other members of the genus *Felis*. This is the wild ancestor of the domestic cat. Like the wolf, the wildcat has a wide distribution, which includes much of Eurasia and Africa; north to south it extends from Scotland to the Cape region of South Africa; west to east it extends from Iberia to Mongolia. Over this wide range, five distinct subspecies evolved: the European wildcat, *Felis silvestris silvestris*; the central Asian wildcat, *Felis silvestris ornata*; the Near Eastern wildcat, *Felis silvestris lybica*; the Chinese mountain cat, *Felis silvestris bieti*, and the African wildcat, *Felis silvestris cafra*[6] (Figure 3.4). There had long been debate as to which of these subspecies was the ancestor of the domestic cat, which was recently decided in favor of the Near Eastern wildcat (*F. silvestris lybica*).[7]

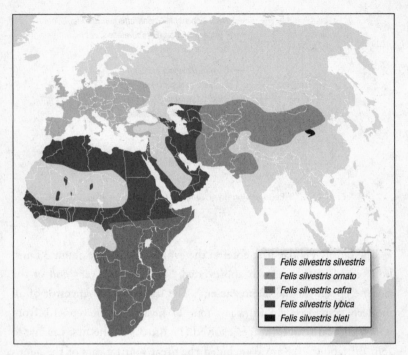

Felis silvestris silvestris
Felis silvestris ornato
Felis silvestris cafra
Felis silvestris lybica
Felis silvestris bieti

FIGURE 3.4 Geographic ranges of wildcat subspecies.

The Near Eastern wildcat is a typical felid in its basic body plan, specialized meat diet, solitary existence, and defense of exclusive territories, in all of which respects it differs from the wolf. These traits also make cats unlikely candidates for domestication and explain in large part the divergent route this process took. For cats are almost entirely self-domesticated through the process of natural selection. Only very recently have a small fraction of the 600 million cats in the world today been subjected to artificial selection for human ends.

Of the five wildcat subspecies, the Near Eastern wildcat is among the more tolerant of people, but that probably wasn't a huge factor in determining why this subspecies alone was domesticated. Contingent historical factors in its interactions with humans figure much more importantly in the domestication of this subspecies.[8]

THE MOUSE CONNECTION

It was long thought that the Near Eastern wildcat was first domesticated in Egypt.[9] But recent archaeological and genetic evidence points
elsewhere. The genetic evidence, which is based on mitochondrial DNA
collected from individuals throughout the range of this subspecies,
indicates that Near Eastern wildcats were first domesticated in the
cradle of agriculture, called the Fertile Crescent, about 10,000 BP.[10] It
was here that humans first began to store grains. These stored grains
proved vulnerable to a recent invader from northern India, called the
house mouse (*Mus musculus*). For the wildcats in the area, these house
mice were a new reliable food source, so some wildcats began hanging
around human settlements. These pioneers in the domestication process did not require human encouragement, but they had to surmount
a significant psychological barrier. The barrier was an evolved fear of
humans and other large predators, including the village dogs. Only
those wildcats that could surmount this fear could effectively exploit
this new resource.

In essence, human agrarian settlements provided the wildcats in the
area a new niche, which required different behavioral dispositions than
the old relatively human-free niche had required. Through natural selection for tameness, a subset of the wildcats was able to increasingly thrive
in this new niche. But in contrast to dogs, which also exploited this
niche, even the more tame wildcats retained their previously evolved
hunting skills and equipment. For example, their canine teeth did not
diminish in size as they did in village dogs.[11] These relatively tame cats
could also still hold their own against the untame wildcats in the area
in any dispute. And this was probably true long after the domestication
process commenced.

The first evidence that the cat-human relationship had progressed
beyond one of mutual convenience is a cat burial in Cyprus 9500 BP.[12]
This cat was buried right next to the grave of a human and oriented in

the same direction. Cats are not native to Cyprus, so they must have been transported there by humans. And since cats are not effective stowaways, like mice or rats, they must have been transported deliberately. Moreover, the cat burial suggests that they were no longer confined to granaries, but were more intimately associated with households. Household cats require both a greater degree of tameness and more active human encouragement.

There is a large gap in the archaeological record after the Cyprus burial, aside from a couple of teeth from Israeli archaeological sites. A small ivory statue from Israel, dated 3700 BP, indicates that the cat-human relationship had become increasingly intimate during the intervening years. It is only after 3600 BP, at the beginning of the New Kingdom period, that cats appear in Egypt.[13] But though domestic cats didn't originate in Egypt, it was there that domestication was taken to the next level.[14]

During this Egyptian "Golden Age," paintings of cats become increasingly common. In some, cats are depicted with collars, sometimes eating out of bowls; so some Egyptian cats, no doubt mostly those of royalty, were spending lots of time indoors.[15] This royal connection led eventually to cat veneration by 2900 BP, and then to deification in the form of the goddess Bastet.[16] When Herodotus visited Bastet's sacred city, Bubastis (2525 BP), the temples to Bastet there were crawling with pampered cats.[17]

Though deification has its perks, there was also a price to be paid. Sacred cats were sacrificed in great numbers for ritual purposes, mummified in the Egyptian fashion, and buried in huge cat cemeteries, which came to contain tons of cat remains.[18] Because of the huge scale of cat burials, it is surmised that the Egyptians must have been actively breeding domestic cats at this time. Whether this breeding was selective—that is, true artificial selection—is not clear, though Patrick Bateson and Dennis Turner hypothesize that the large population of cats in northern Egypt could well have been selected for increased sociability.[19]

Though Egypt had outlawed their export, domestic cats could be found in ancient Greece by 2500 BP.[20] The Romans took control of the Nile Delta and the rest of northern Egypt in 2030 BP, and from their port city of Alexandria domestic cats were transported throughout the empire on grain ships. To protect the grains shipped out of Alexandria from mice, ships were inoculated with domestic cats, some of which probably abandoned ship—in keeping with their independent dispositions—in far-flung ports.[21]

In any case, cat colonies were soon established in many port cities, from which cats worked their way inland. This inland movement was probably also human aided, as, left to their own devices, cats are not prone to move far from where they were born. By 1000 BP, domestic cats were common in Europe.[22] Domestic cats didn't reach the Americas until much later—perhaps as early as the voyages of Christopher Columbus (1492–96), perhaps not until the voyage of the Mayflower (1620); they are thought to have reached Australia by way of European explorers in the seventeenth century.[23]

The eastward movement of domestic cats also commenced during the Roman Empire, along trade routes between Rome and China. By 2000 BP, domestic cats could be found in China and India.[24] From China the domestic cat worked its way through continental Southeast Asia, then much of island Southeast Asia. While the westward movement of domestic cats, through the Mediterranean region to northern Europe, traversed areas with indigenous wildcat populations, there were no such wildcats along much of the eastern route, including most of India, China, and all of Southeast Asia. So, while western domestic cats continued to interbreed to varying degrees with indigenous wildcats, the domestic cats of the Far East evolved in isolation. The result was several distinct landraces, largely distinguishable by distinctive coat color patterns, such as the Siamese, Korat, and Birman.

GENETIC DRIFT AND
NATURAL SELECTION

Just as domestic dogs became differentiated into distinct landraces, so, too, did domestic cats, albeit to a much lesser degree. In the differentiation of these Southeast Asian landraces, the other major evolutionary process, genetic drift, figured more prominently than natural selection.

In genetic drift, populations diverge by means of random genetic alterations that are selectively neutral, meaning that they don't affect fitness. Such is the case, for example, with the genes that underlay color variations of Siamese, Korat, and Birman landraces. Genetic drift is ubiquitous but varies with population size. The smaller the population, the more genetic drift, simply because random influences of any kind are more pronounced in small populations; this is what statisticians call "sampling error."

Evolutionary biologists refer to "effective population size," which is roughly the number of potentially interbreeding individuals. Genetic communication from one population to another, called "gene flow," increases the effective population size, up to that of the combined populations. So, genetically isolated populations, in which there is no such gene flow, are more prone to genetic drift. For domestic landraces it is important to consider not only gene flow between landraces but also gene flow between wild populations and landraces. As long as there is any gene flow between wild populations and domestic landraces, genetic drift is restricted. But when wild populations are absent, as for cats in Southeast Asia, conditions are ideal for genetic drift. The color variations among the Siamese, Korat, and Birman landraces are most likely the result of genetic drift that was later reinforced by artificial selection.

Natural selection, though, was also operating on the Southeast Asian landraces as elsewhere. For example, the short hair common to the Southeast Asian breeds and others, such as the Abyssinian, is likely the result of natural selection for reduced heat retention in a warm climate.

So, too, perhaps, were their characteristically long, slender bodies, which lose heat more rapidly than the stocky body type called "cobb." And the long hair and cobb body type that are characteristic of northern land-races, such as those ancestral to the Maine coon, Norwegian forest cat, and Siberian, are at least partly adaptations to colder climes.

All of the landrace-derived breeds mentioned in the preceding paragraphs are categorized as "natural," in that they were developed with minimal human aid by way of artificial selection and hybridization.[25] As was true of "ancient breeds" in dogs, the term "natural breeds" is problematic. The Abyssinian, Egyptian Mau, and Chartreux, for example, are considered natural breeds but are in fact recent (phenotypic) reconstructions of indigenous landraces.[26] The Japanese Bobtail and American shorthair are, on the other hand, better candidates for the "natural breed" appellation.

Some of the natural breeds are hundreds of years old, but most domestic cat breeds are of a much more recent vintage—the last half of the twentieth century. It was only then that cat fanciers really got to work and cat shows—modeled after dog shows—escaped the limited orbit of aristocrats.[27] A number of "fancy" breeds were created in the 1960s, including the Scottish Fold, the Sphinx, and the Ocicat.[28] In the 1970s came the Singapura and the Australian Mist; the 1980s brought us the York Chocolate, the California Spangled, the Burmilla, the Nebelung, and the Donskoy, among other new creations. The Ragamuffin was created in the 1990s, while the Levkoy, which is shudder-inducingly ugly, is a twenty-first-century creation.[29] By the middle of this century, the number of existing cat breeds—currently more than 60—may well double.

HOW FANCY CATS WERE CREATED

A number of fancy cat breeds began with a mutation—often confined to one individual in a litter—that had an obvious effect on the phenotype. The Scottish Fold, for example, was founded by a barn cat from

Perthshire, Scotland, with peculiarly forward-bending ears.[30] Someone decided it would be a good thing to perpetuate this mutation. The Manx, from the Isle of Man, has a skeletal mutation that causes the tailless condition, among its other effects. In this it somewhat resembles the Japanese Bobtail, a natural breed with a quite different mutation.[31] Munchkin cats have a mutation that causes limb shortening analogous to that of the dachshund.[32]

Polydactyl cats have extra toes and constitute a recognized breed in the United States, called the American Polydactyl.[33] They seem to have originated in southwest England, from where they made the Atlantic crossing by ship to New England, where they are especially abundant. One important reason for their early success was the widespread belief among sailors that they brought good luck—another example of the role of human caprice in the domestication process. The record for polydactyly is 27 toes, set by a Canadian cat.[34] Here's hoping that the record isn't broken.

There is another mutation, called radial hypoplasia (RH), or "hamburger feet," which results in a different form of polydactyly, of a spiraling nature.[35] A creative breeder in Texas sought to build on this deformity in constructing a "Twisty cat" breed, in which the spiraling extends to the bones of the forelimb. Twisty cats also have extremely short forelimbs and relatively long hind limbs, which cause them to sit like a squirrel—hence an alternative name, "squitten." Twisty cats are banned in Europe on humanitarian grounds, but not in the United States; the same is true of the Munchkin. It is time that the United States caught up with the United Kingdom in this regard. The deliberate breeding of skeletally deformed breeds is unconscionable.

Some of the oddest-looking breeds result from a mutation that causes hairlessness. Actually, these cats aren't completely hairless; they just look that way. The first such breed originated in 1966 from a single naked kitten, appropriately named Prune.[36] It is a mystery to me why anyone would want to perpetuate this condition; I suspect it is simple neophilia.

Given the climate there, it is particularly perverse that the Sphinx is a Canadian breed. But then, two other notable hairless breeds, the Donskoy and Levkoy, were created in Russia and Ukraine, respectively. One hopes they are indoor cats. Other cat breeds were founded by less drastic mutations of the coat, including the Cornish Rex (downy hair), Devon Rex (short guard hair), Iowa Rex (dreadlocks), and American wirehair (dense wiry coat).[37]

The other method for generating new cat breeds is hybridization with existing breeds. The Siamese is most commonly used as one part of the cross. For example, the Havana Brown was the result of a cross between Siamese and American shorthair, and the Himalayan represents a cross of Siamese and Persian. Second-, third-, and fourth-order hybridizations begun with Siamese hybrids and other breeds include the Ragamuffin, Ocicat, and California Spangled. Some notable hybrids that lack a Siamese component include the Australian Mist (part Abyssinian), the Nebelung (part Russian Blue), and the Burmilla (part Burmese). The Levkoy is noteworthy not only for its uncomeliness but for the fact that it was created from a cross of two mutant breeds (the ear-challenged Scottish Fold and the hair-challenged Donskoy). The mutant ante can be ever upped.

Some truly creative breeders decided to go outside of the domestic cat box in finding partners for hybridization. The Chausie is a cross between an Abyssinian and a jungle cat (*Felis chaus*). Since the jungle cat is in the same genus (*Felis*) as the wildcat and the domestic cat, it is not surprising that this match worked. But other crosses outside of the genus *Felis* are more ambitious. The Bengal is a cross between a domestic cat and a leopard cat (*Prionailurus bengalensis*). At least the leopard cat is about the same size as a domestic cat; not so two other out-of-genus crosses: the Caracat is a cross between an Abyssinian and a caracal (*Caracal caracal*); and the Savannah is a cross between a domestic cat and a serval (*Caracal serval*).[38] Both caracals and servals are considerably larger than wildcats.

INBREEDING AGAIN

When you start a breed with a single mutant, you have a founder population of two: the mutant and the individual with which it mates. To maintain the mutation at high levels, you must mate close relatives—say, siblings, or mothers and fathers with sons or daughters. Either way, the result is intense inbreeding and the accumulation of deleterious recessive mutations—a phenomenon known as "inbreeding depression." Indeed, inbreeding in some cat breeds begun in this way is as severe as in dog breeds, as reflected in breed-characteristic pathologies.

The opposite occurs when breeds from different species are crossed, as in the Savannah and Caracat. Here the problem is a lack of harmony of various sorts among the genomes—a condition known as "outbreeding depression."[39] Servals and domestic cats, for example, don't have the same number of chromosomes, which creates fundamental problems in partitioning them during the creation of sperm and eggs. More subtly, certain suites of genes that work particularly well with each other are normally inherited more or less as a unit. These "coadapted gene complexes" are broken up with excessive outbreeding.

The optimal condition lies somewhere between these poles, when the porridge is neither too cold nor too hot. The "just right" porridge is called hybrid vigor. This is what you get in mongrel dogs and barn cats. (Sylvester and Smoke are American shorthairs whose mother was the latter.) You would also expect to get hybrid vigor from crossing two distinct cat breeds, such as were used to create the Himalayan (Siamese × Persian). And initially, you do. The problem is that only a relative few offspring of these crosses, which have the desirable characteristics, are used as breeders for the next generation. The intense artificial selection for these characteristics soon results in inbreeding depression again.

The so-called natural breeds were in the "just right" hybrid vigor mode until cat fanciers began to control their breeding in the twentieth century. The effects of these efforts are especially evident in the Siamese, long the most popular of the natural breeds. The Siamese in Europe and North America today are strikingly different from those found in Thailand, as I can attest from personal experience.[40] The Thai Siamese is a larger animal and longer of leg. Though the Thai Siamese has the typical "oriental" lithe body, it is more muscular, and not nearly as thin as that of the western Siamese. In addition, its skull is larger and notably more rounded in shape. These differences reflect the effects of artificial selection in the West.

The first Siamese to arrive in the West—appropriately named Siam— was an 1878 gift to President Rutherford B. Hayes.[41] Six years later the first breeding pair was imported to Britain, followed by several more imports of a small number of these cats. Most Siamese in Britain today may be the descendants of only 11 imported Siamese. This small founder population, with its inherent sampling error relative to the genes of the Thai Siamese, was then prone, by virtue of its small size and isolation, to further random divergence through genetic drift.

The novel Siamese were an immediate hit at cat shows, so they were newly subjected to artificial selection, by means of which they further diverged from the original type. This divergent evolution accelerated in the last half of the twentieth century because judges came to prefer longer, thinner cats with proportionally small heads of a triangular shape, topped by large ears, set wide to emphasize this triangularity—to which end the snout was also thinned and the eyes became more almond shaped. Within a few decades, traditional Siamese had disappeared from cat shows (see Figure 3.5). Some breeders organized to preserve the "traditional" style of Siamese, which is now recognized by TICA (The International Cat Association) as a new breed, called Thai. Such are the inversions of the topsy-turvy world of cat breeders.

The effects of inbreeding have been dire. Siamese have cancer rates rivaling those of Bernese mountain dogs and other cancer-prone dog breeds. They are especially prone to breast cancer. Accordingly, the life span of the Siamese is considerably shorter—with a median length of

FIGURE 3.5   Thai Siamese (left) and European Siamese (right). Note the differences in the skull and face. (Thai Siamese [left]: © iStock.com/Lena Kozlova. European Siamese [right]: © iStock.com/IvonneW.)

10–12 years in one study—than that of the average house cat (15–20 years). Other "natural breeds," such as the Abyssinian, also have shortened life expectancies as a result of inbreeding. Those that live longest are prone to blindness by means of progressive retinal atrophy and other defects of premature aging.

Aside from the Siamese, the Persian and the Himalayan have been the breeds most modified by sustained artificial selection. In addition to their gorgeous long hair, these two breeds are notable for their brachycephalic (squashed) faces, first developed in the Persian and inherited in the Himalayan when it was created through Siamese × Persian crosses. Since creation of the Himalayan, the brachycephaly has been further exaggerated in both breeds, with predictable results. Though neither breed rises to the level of bulldog grotesquerie and its concomitant ailments, they do suffer from breathing problems, chronic sinus infections, and, more generally, abbreviated lives.

In stark contrast, the American shorthair, of which Smoke and

Sylvester are exemplars (OK, just Smoke), is a natural breed that has remained a natural breed. Which means that American shorthairs have long bred with whomever they deemed desirable—and the females often find it desirable to mate with more than one male. They evolved, from a large founding population, by means of natural selection into the perfect domestic cat—robust, athletic, and low maintenance. If properly socialized, they make ideal house cats. As an added bonus, American shorthairs are among the best mousers, right up there with the legendary Egyptian Mau.

There is an attempt under way to create an even better mouser, which would be the first cat breed created for function rather than appearance. The breed is called American Keuda, which is an acronym for "Kitten Evaluation Under Direct Assessment."[42] The breed is being created from American shorthair barn cats. The only criterion for the breeding program is exceptional mousing ability. Inbreeding, which inevitably reduces this ability, is therefore kept to a minimum, as evidenced by the huge variability in coat colors. Interestingly, some Keudas have come to look very much like the Egyptian Mau, a cat breed that perhaps most resembles the ancestral *Felis silvestris lybica*, from which all domestic cats are descended.

CAT GENOMICS

Cat genomics is not nearly as far advanced as dog genomics; it is still in the kitten stage. The first complete cat genome sequence came from an Abyssinian named Cinnamon.[43] Subsequently, 10 other breeds have been partially sequenced. There are clear geographic factors in the genetic similarities of cat breeds. The Southeast Asian breeds, for example, form a distinct cluster; the European and North American breeds form a less distinct cluster; and the Central Asian, West Asian, and North African breeds tend to clump as well. Exceptions, such as the Ragdoll, American Curl, Ocicat, Sphinx, Devon Rex, Cornish

Rex, and Bengal, are generally Western breeds recently created through hybridization or major mutations.

Many of the major mutations affecting body type and coat coloration of domestic cats were identified in the pregenomic age by conventional linkage analysis.[44] Here I will consider a few interesting recent discoveries concerning coat characteristics.

Recall that a mutation (in a gene called *Fgf5*) was responsible for long hair length in many dog breeds. A mutation in the same gene also appears to cause long hair in cats.[45] Actually, four separate mutations in this gene can cause long hair in cats, each different from the mutation that causes long hair in dogs. This phenomenon—same gene, different mutation, similar phenotype—is actually quite common. It occurs when different mutations, causing different amino acid substitutions in the coded protein, disrupt biological activity in similar ways. Since each variant of a gene is called an allele, we can more concisely say that, in this case, different alleles result in the same phenotype.

But it is more often the case that different mutations in the same gene have different developmental effects; that is, different alleles result in different phenotypes. Consider the tyrosine gene (*TYR*), which plays an important role in coat pigmentation. One mutation in this gene is largely responsible for the distinctive coloration of the Siamese: dark extremities, light body.[46] This color pattern is due to the fact that the mutant allele is temperature-sensitive. During development, the extremities are cooler than the rest of the cat and the *TYR* gene is more active; in the more central areas, where the body is warmer, the *TYR* gene is less active, given this mutation. A different mutation in this gene results in an allele that is less temperature-sensitive.[47] The result is the Burmese color pattern, in which the nonextremities are more pigmented than in the Siamese. Different mutations, and hence alleles of a related gene, called *TYRP1*, cause chocolate coloration or albinism.[48]

Like all other genes, *Fgf5*, *TYR*, and *TYRP1* are all coding regions of DNA, in that they code for proteins. But as we saw in the previous chapter, much of the evolutionary action is in noncoding sequences that regulate the activity of genes. One such noncoding mutation is responsible for the polydactyl condition. The gene that it regulates is one of the most storied in all of developmental biology: *sonic hedgehog (shh)*.[49] *Sonic hedgehog* is a master developmental regulatory gene that produces a protein molecule of a sort called a "morphogen," which forms a concentration gradient by diffusion.[50] The effects of this morphogen on the cells of the developing embryo depend on its concentration. In this way, *sonic hedgehog* plays an important role in the development of organs, brain, and limbs. Its activity is regulated by noncoding elements near the gene called "cis-regulatory elements." The limb-specific cis-regulatory element is called ZRS. A mutation in ZRS that causes too much *sonic hedgehog* activity is responsible for the polydactyl condition.[51]

The noncoding polydactyly mutation is an example of a genetic mechanism that also underlies several human developmental abnormalities. And this is but one instance in which breeder-induced cat miseries have served to advance human medicine. For many of the ailments of purebred cats, as for purebred dogs, are also found in humans—a legacy of our shared mammalian evolutionary history. Indeed, these medical applications provided much of the original rationale for both the canine and feline genome projects.[52]

Over 250 hereditary diseases of domestic cats are homologous to human diseases. The goal is to identify the genetic substrates for these diseases in cats, and then look for the homologous genetic substrates in humans. Cat models are especially promising with respect to progressive retinal degeneration, cardiomyopathy, and inherited motor neuron disease.[53] The cat may also prove a useful model for amyotrophic lateral sclerosis (ALS).[54] Cats are already important models for several viral diseases, including HIV-AIDS, which is prevalent in free-ranging cats, as is feline leukemia and the feline equivalent of SARS.[55] Research is under way on cats to determine the DNA variants that make some cats more susceptible to these infections.

Until the wildcat genome is sequenced, feline genomics cannot provide much information about the genetic alterations that facilitated domestication. We can predict, however, that these genetic alterations were more concerned with behavior than with anatomy and physiology. For it is in their behavior that domestic cats most differ from their wild ancestors.

FAR FROM SOLITARY

The vast majority of cats have escaped artificial selection. They are self-domesticated. And except for superficial changes in their coats, domestic cats very closely resemble wildcats. So close is this resemblance, in fact, that wildcats don't make a distinction when it comes to mating. Wherever wildcats—of any subspecies—live in close proximity to feral domestic cats, the two interbreed freely, much more so than wolves and domestic dogs do. This interbreeding threatens extinction by hybridization for several populations of European wildcats (*Felis silvestris silvestris*), notably the Scottish wildcat and the Iberian wildcat.

There is one way to distinguish wildcats from domestic cats without resorting to genetic testing, and that is by their behavior. Even the most feral domestic cats are much more social than wildcats. Wildcats are indeed solitary creatures, with exclusive home ranges. Domestic cats are social creatures, which, in the feral state, often form colonies when food is relatively abundant and localized.[56] When food is less abundant and dispersed, feral cats still interact much more than wildcats.[57] When living in colonies, female feral cats engage in mutual and reciprocal care; they often suckle and protect the young of other colony females, much as lions—the most social of cats—do. Members of a colony also defend their mutual territory against outsiders—again, much as lions do.[58]

Moreover, domestic cats have evolved a novel behavioral signal called "tail up," which they use to signal friendly intentions.[59] This behavior is entirely absent in the much less social wildcats. Lions, however, use the tail-up behavior in the same way as domestic cats.[60] This is a case of convergent evolution, but the shared ancestry of lions and domestic cats may well render this convergence more probable—a part of the felid behavioral repertoire to which only the most social felids have evolutionary access. Put another way, the genetic alteration required for tail up may be minimal—another example of the conservative creativity of evolution.

It has been proposed that neoteny is at work in the sociability of domestic cats.[61] Meowing, purring, and kneading are all kittenish traits retained by adult domestic cats but not adult wildcats. While kneading may just be a nonadaptive (selectively neutral) by-product of infantilization, the purr and meow are important social signals. There is also evidence that the acoustic properties of the cat's meow have been altered to be more audible to the human ear.[62] (If so, Sylvester—whose meow is rather loud and grating—is much better adapted than Smoke, whose meow retains a kittenish volume and softness.) We could even say that Sylvester is in the vanguard of cat evolution.[63]

The domestic cat certainly lends support to Belyaev's hypothesis that in the domestication process, behavioral modifications for tameness come first, physical changes only later. For domestic cats remain only superficially dissimilar to their wild ancestors. In the fox experiment, tameness came by way of intense artificial selection; for the wolf, and especially the wildcat, tameness came by way of natural selection in a human-created environment. For the wildcats, unlike the wolves, there was an additional psychological barrier to overcome: the close proximity of other wildcats, which were congregating around the granaries. Only these more social wildcats, those who were the least stressed by the close proximity of other wildcats, could fully exploit this new resource.

These social wildcats, upon further selection to tolerate humans, were then ready to include us in their social circle. For all of their celebrated independence, cats don't just tolerate us; they enjoy our company. Smoke is asleep on my printer; Sylvester is back on my lap.

CATS IN A DOG WORLD

Human attitudes toward our domesticated animals vary greatly over time and culture. We saw this with dogs, which have been eaten, petted, or ignored, depending on time and place. Indeed, while dogs are taboo foods throughout most of the world today, they are still consumed with gusto in China, Korea, Vietnam, and Polynesia. But, with the notable exception of Islamic cultures, dogs are generally not viewed as unclean, unholy, or otherwise deserving of opprobrium. On average, across human history and cultures, dogs evoke neutral to positive emotional responses. They rank just below horses in positive regard.

Notwithstanding their deification in ancient Egypt, cats are viewed much more ambivalently, though again with great variation over time and space. In western Asia the cat was long associated with female sexuality and fertility, attitudes also found in much of pagan Europe. The spread of Christianity was bad for cats, perhaps as a result of their association with pagan religions; they came to be viewed as the devil's agents and closely associated with witchcraft. During the Middle Ages feast days were particularly dangerous for domestic cats; they were tortured in the most gruesome fashion: boiled or burned alive, slowly roasted on spits, flayed, and maimed. There was a strong current of misogyny in this cat hatred, as in the persecution of putative witches. Fortunately, things have improved for cats since. In our more enlightened age, stray cats are often cared for or adopted. Cats are the most popular pets in the world today.

Yet cats are still viewed much more negatively than dogs, according to a recent survey of Americans. Some cat haters can't abide feline independence; some no doubt are still spooked by hoary associations with witchcraft and paganism generally; and some just see dogs as the paragon of companion animals, by which standard cats are deficient. Dogs are an inappropriate standard by which to measure the merits of cats, yet many persist in this mistaken belief. It was recently explained to me, by way of demonstrating the superiority of dogs, that "a cat won't pull you out of a burning building"—the implication being that it is because of a lack of motivation rather than the manifest physical inability of cats to accomplish such. A cat won't put its life on the line like a dog will.

Maybe, maybe not. There is a widely watched Internet video that suggests otherwise. In it, a four-year-old boy is stalked and severely attacked by a mid-sized dog of uncertain pedigree, when, from off camera, a tabby cat streaks across the screen and flings itself at the much larger attacking dog. The dog, not having a clue what hit it, takes off at high speed. Cats, it seems, can be just as heroic as dogs.

CHAPTER 4

OTHER PREDATORS

ONE NIGHT, WHILE LIVING IN SAN RAMON, CALIFORnia, I awoke around 2:00 a.m. quite thirsty and stumbled to the kitchen for some water. When I flipped the light switch, I was stunned to see a massive raccoon eating food from a bowl meant for our elderly female cat, Misty. He—I assumed the raccoon was a he from its size, not from any closer inspection—briefly looked up when the light came on but quickly recommenced with his meal. I was flabbergasted and not a little disturbed, not only by his nonchalance but by that of Misty as well. She was just sitting on her haunches about 10 feet from the invader, somewhat curious but not particularly concerned. She had obviously seen this before. In retrospect, her *"mi casa es su casa"* attitude was fully in keeping with her pacifist nature—she could kill a fly but that was about it—which served her well throughout her long life of 22 years. If she wouldn't mess with squirrels, she certainly was not going to mess with this brute.

I, however, felt compelled to do something. First I yelled and stomped my foot. That set Misty scampering for cover, but the raccoon only turned toward me, reared up a bit, and emitted a threatening sound, a combination of a growl and a hiss. He then returned to his meal. I then got a broom and waved it at the raccoon menacingly; this behavior did actually seem to impress him somewhat, and I eventually managed to direct his exit through the sliding glass doors to the backyard, albeit in more of a sauntering than a hurried way.

It was only at this point that an obvious question sprang to mind: how did he get in here? It must have been the cat door. But at first I rejected that idea because he was just too damned big. A couple of weeks later, though, I witnessed the feat firsthand. While I was watching TV late one night, the raccoon poked his head through the cat door. He pulled back when he saw that I saw him but then, perhaps unconvinced by our previous encounter, poked his head in again. Then came the rest of the body, slowly, twistingly, and with much obvious effort. At one point he pulled himself with his forelegs while bent awkwardly on his back. It was really quite mesmerizing. But his insouciance grated.

Theretofore I had a phlegmatic attitude toward the raccoons on our property, certainly compared to the attitude of our neighbors. I even tolerated one mother and kits spending about a week each month under the house, though they made quite a lot of commotion in the wee hours. And though they damaged the grape arbor, dug up some of my most prized succulents, and helped themselves to the persimmons, I could not bring myself to call animal control, which would, I knew, mean their doom. The line in the sand was the interior of the house, which the raccoon family, too, eventually trespassed. At that point I knew we had a social transmission issue and therefore a potentially perpetual problem.

I didn't realize this at the time, but our raccoon invaders, as evidenced especially by their cheekiness, were in the vanguard of a new domestication process, one that probably resembles the early stages of cat and dog domestication. The first moves toward the domestication of both cats and dogs were taken by the wildcats and wolves themselves when they began to exploit human resources. This sort of relationship is called commensalism. There are many human commensals, from pigeons and house sparrows to rats and mice. For most, the domestication process never gets beyond that. But there are reasons to believe that for raccoons, it might. The primary reason is that raccoons, as opposed to house sparrows, have undergone significant behavioral changes.

Most raccoons, the ones that live in a somewhat more natural state, won't come near a human, even leaving aside those "coons" that are actively hunted. But they quickly learn how to exploit humans, when we frequent their natural habitats. Raccoons have exasperated me over the years while camping, from Point Reyes National Seashore in California to Belleplain State Forest in New Jersey. These wily creatures steal any unattended food—or insufficiently attended food—outright.

American black bears—and, to a lesser extent, grizzlies—are also omnivores that will facultatively exploit humans in this way. But with the advent of bear-proof food canisters, I have found bears much easier to deal with in this regard. The only raccoon-proof food canister that I have discovered is the automobile, and only when the doors and windows are securely closed.

Raccoons, though, have taken this facultative exploitation of humans from their natural homes in the wild into human-altered environments, such as my former suburban home in California. Judging by their population densities, raccoons actually thrive more in these human-altered environments than in more pristine habitats. I once estimated more than thirty raccoons at a large garbage bin behind an apartment complex in San Ramon. They were crawling over the garbage like rats. Which brings us to the second behavior of raccoons that has been altered in human environments: their toleration of each other.

The domestication of cats is most instructive here. Recall that for wildcats, as for wolves, the first step, the first behavioral change, was the toleration of human proximity, or tameness, as Belyaev experimentally demonstrated with his foxes. But for wildcats, unlike wolves, there was an important second behavioral change: tolerating the proximity of each other. Raccoons are more like wildcats than like wolves with respect to their sociality. And like cats, they seem to be in the process of evolving a tolerance for each other, in order to better exploit the human environment.

AN OMNIVORE'S OMNIVORE

The raccoon, like the wolf and the wildcat, belongs to the order Carnivora. Recall that this order is divided into two suborders: the dog side (Caniformia) and the cat side (Feliformia) (Figure 4.1). Raccoons belong to the family Procyonidae—along with sea lions (Otariidae), weasels (Mustelidae), and bears (Ursidae)—on the dog side of the carnivore tree. Other members of the family Procyonidae include coatimundis, kinkajous, ringtails, and olingos, all of which dwell in the New World, most in Central and South America. Raccoons and other members of the family are most closely related to bears, with which they share much in the way of dentition and therefore an omnivorous diet. In fact, raccoons may have the most varied diet of any North American mammal. One consumed an entire jar of spicy mustard obtained from my campsite.

Members of the raccoon family vary substantially as to sociality. Ringtails and kinkajous are largely solitary, and so, too, are male coatis, but female coatis are quite social. Raccoons were long thought to be solitary, but it is now known that their social behavior is more complex and nuanced. A lot depends on food supplies. Throughout much of their range, both males and females are largely solitary, though they

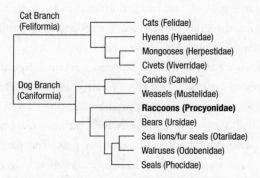

FIGURE 4.1   Carnivore genealogy with family relationships, highlighting the raccoon family (Procyonidae). (See source in Figure 3.2.)

don't seem to actively defend home ranges.[1] Where food is plentiful and population densities are higher, female home ranges often overlap, and females sometimes gather in communal resting spots; where populations are densest, unrelated males are prone to form temporary coalitions to ward off other males.[2]

When Europeans first arrived in North America, raccoons were largely confined to the rich woodlands along river courses in the southeastern United States and south to Panama.[3] Explorers and pioneers settling to the north and west did not observe them. But their range expansion was obvious by the late nineteenth century, aided in no small part by their ability to facultatively exploit the expanding human population. Raccoons were still relatively rare, though, until the 1940s, when their population exploded and their range expanded considerably, to include not only the entire western United States but also the Canadian provinces of British Columbia, Alberta, Saskatchewan, and Manitoba, as well as Ontario and Quebec. In the process, they learned to exploit new habitats, such as prairies, coastal marshes, and mountains.

It probably began with farmhouses but quickly progressed to larger human settlements and ultimately major cities, including Washington, DC, and Chicago. But it is in Toronto that urban raccoons have achieved the apogee of their human exploitation, far exceeding anything I have witnessed in San Ramon, California. Up to about 400 raccoons inhabit each square mile of Toronto real estate.[4] In more natural environments, raccoon densities range from 1 to 5 per square mile.[5] Raccoons have obviously acquired a taste for city living.

Just as raccoons were becoming urbanized, certain misguided souls introduced them to Europe (notably Germany), as well as Japan. The latter introduction was actually prompted by a popular animated cartoon (1977) called *Rascal the Raccoon*, which induced impressionable children to implore their parents for raccoon pets. The parents who acquiesced soon discovered that cute as baby raccoons are, they become surly adults and don't make good pets. Japanese raccoons are now such a scourge that normally pacifist Buddhist monks have been driven to declare war on them in defense of their temples.[6] In Kessel, Germany, raccoon densities are comparable to Toronto's.[7]

Raccoons are no doubt attracted to cities by abundant food supplies, from gardens, garbage cans, and cat bowls. Urban centers also provided unprecedented levels of protection from natural predators. But before raccoons could fully avail themselves of these resources, some behavioral modifications were required. First their fear of humans was ameliorated to some degree, as in the wildcats that invaded the early granaries seeking mice. Those wildcats were then confronted with the unprecedented proximity of each other, and the concomitant stress. Those that got the least stressed out—the mellow elite—were at an advantage in this new environment, not just from avoiding the wear and tear of chronic stress, but also in being able to devote more time to catching mice and less time worrying about each other.

Eventually, the granary cats became genetically distinct from their cohorts with respect to both their tolerance of human proximity and their increased sociality. Has the same thing happened to urban raccoons? At this point we cannot assume so. For raccoons have demonstrated a kind of inherent flexibility that wildcats lack, which may render genetic alterations unnecessary, at least at this early stage of their human exploitation.

PHENOTYPIC PLASTICITY

"Phenotypic plasticity," a key concept in evo devo, refers to the capacity to respond adaptively to the environment without any genetic change.[8] For a given trait, phenotypic plasticity can vary among individuals, populations of individuals, or species. Here we will deal with the species level. The amount of phenotypic plasticity for a given trait (at the species level) is itself an evolved property, whether directly selected or a by-product of other evolutionary processes. For example, omnivores, such as bears, pigs, and raccoons, have more phenotypic plasticity with respect to diet than do carnivores (such as wildcats) or herbivores (such as cows and sheep).

Social behavioral traits, too, can be more or less phenotypically plastic. The social behavior of the wildcat is quite fixed compared to that of raccoons, which can vary from solitary to semisocial, depending on population densities. Moreover, the response of raccoons to humans is more variable than that of many wild carnivores. Those raised in human proximity are more prone to bold behavior than those raised in more natural environments. Therefore, we cannot assume that either their increased boldness toward humans in urban environments or their increased sociality required any genetic alteration. Both behavioral departures from those that are more typical in natural environments may merely be expressions of phenotypic plasticity.

This is probably true of the earliest stages of domestication for any of the carnivores we have considered. As in the evolution of mammals generally, phenotypically plastic behavior often leads the domestication process, and the genetic changes follow. But it is only after there have been genetic alterations that we can call an animal truly domesticated. At this point in time there is no evidence that the domestication process in raccoons has gone beyond what can be gotten through phenotypic plasticity. And the commensal relationships between raccoons and humans may never go beyond this stage. This commensalism does, however, make raccoons vulnerable to further domestication, as we have seen with cats and dogs. Moreover, there is another carnivore, one from a family very closely related to raccoons, that nicely illustrates the transition from a purely phenotypically plastic tameness to a decided genetic predisposition toward tameness. This animal is the ferret.

FERRETS BEFORE DOMESTICATION

Ferrets belong to the family Mustelidae (which also includes polecats, stoats, mink, weasels, martens, badgers, and wolverines), which, like the raccoon family, resides on the dog (caniform) side of the carnivore tree. Mustelids are the most carnivorous members of this side of the carnivore tree and, pound for pound, some of the fiercest carnivores on earth.[9] The genus *Mustela*, to which ferrets belong, also includes polecats, stoats, weasels, sable, and European mink, all of which have long been exploited by humans for their fur. Ferrets are typical members of the genus in their ability to exploit a wide variety of small animals, including insects, fishes, amphibians, birds, a wide variety of rodents, and rabbits. It was for their proficiency in killing rodents and rabbits that ferrets came to be domesticated.

The wild ancestor of the ferret is the European polecat (*Mustela putorius*).[10] The European polecat has a shorter, more compact body than other members of the genus *Mustela*, such as weasels and mink. Because of their shape, polecats are not quite as agile as their congeners, but they have more formidable jaws and teeth.[11] There is some variation in pelage, but most are dark brown with lighter bellies and a raccoon-like face mask. Polecats are generally solitary, but there is evidence that they are somewhat more social than other members of the genus *Mustela*, in that individuals of the same sex sometimes share a home range.[12] Polecats are also somewhat more tolerant of humans than stoats and weasels are; for example, wild-caught polecats readily breed in captivity, in contrast to wild-caught stoats, weasels, and European mink.[13]

Their relative tolerance of each other and humans may have rendered polecats better candidates for domestication than stoats or weasels, which are also magnificent ratters and rabbiters. (Weasels, stoats, and polecats/ferrets are all superior to cats and terriers in their ability to flush out and kill rodents and rabbits, because of their superior burrowing ability, agility, and flexibility.)

It is not known, with any precision, where and when polecats were first domesticated. We do know that the where was probably the Mediterranean region.[14] As for the when, it seems to have been between 2500 and 2000 BP.[15] There are putative references to ferrets in the writings of Aristophanes (2550 BP), and of Aristotle about a hundred years later, but in both cases the subjects could have been undomesticated pole-cats.[16] Better evidence for domestication comes from the Roman period, when, for example, Pliny the Elder described coordinated rabbit hunts with ferrets,[17] and Strabo reported that ferrets were deliberately intro-duced into the Balearic Islands, which at the time were being overrun by previously introduced domestic rabbits.[18]

That rabbits figured prominently in ferret domestication is beyond doubt. The rabbit was domesticated at about the same time as the ferret and in the same part of the world. It was probably at the point when the rabbiting ability of polecats was deemed more important than their fur that they were turned into ferrets. Indeed, the term "ferreting" origi-nally referred to their ability to flush rabbits from their burrows. The ferrets were usually muzzled for this activity, lest they kill the rabbits underground and choose to dine on them at their leisure there. The flushed rabbits were then netted, clubbed, or shot for dinner. Whenever feral rabbit populations reached pest levels, ferrets could be unmuzzled for rabbit control purposes. This latter role became increasingly import-ant as feral rabbit populations exploded in northern Europe.[19]

Ferrets probably moved north along with Roman settlements, even-tually arriving in Germany and England about nine hundred years ago.[20] In England, these exotic rabbit eaters became associated with the aristocracy, such that by 1281 there was an official "Ferreter" position in the royal court.[21] During the apogee of the British Empire, ferrets were deliberately introduced to Australia and New Zealand to remedy the rabbit problem created by a pernicious nostalgia for familiar European animals in these lands of alien creatures. Unfortunately, the introduced ferrets largely eschewed rabbits in favor of native fauna that proved eas-ier pickings, given their evolutionary naïveté for things ferret-like. For-tunately, feral ferret populations never took root in Australia, possibly because of dingo predation, but in New Zealand feral ferrets thrived with disastrous consequences for the native birds, especially those that had evolved flightlessness in the utter absence of ground predators.[22]

Ferrets may have first arrived in the United States via eighteenth-century sailing ships, where they served—like cats—as vermin control. By the early twentieth century, thousands had been released on farms for rodent and rabbit control—a practice that eventually ceased with the advent of better chemical poisons. As in Australia, other predators—and perhaps competition from native mustelids—seem to have largely prevented the spread of feral ferrets in the United States.[23]

FROM POLECAT TO FERRET

The early stages in the domestication of polecats probably resembled that of cats, involving similar behavioral alterations in their response to humans and each other. As with cats and urban raccoons, the first step was to alter their response to humans, as they entered new habitats of human construction. And as with both cats and raccoons—to fully exploit the human environment, with its clumped and concentrated resources—polecats had to become more sociable than their forest cohorts. In this latter regard, there is good evidence that ferrets have moved beyond the mere phenotypic plasticity of raccoons to something more like the evolved changes we saw in cats.

You would not want to stick your finger into the cage of a wild polecat, even one raised from birth by humans. Nor can polecat-ferret hybrids be trusted to pass this test.[24] Pure ferrets, though, will welcome said finger and more than likely rub up against it. Ferrets are also much more likely than polecats to react to one another in a friendly manner. Ferrets are considerably more social than polecats; they are more comfortable in each other's close proximity and therefore can be maintained in relatively large social groups.[25] If polecats were introduced into such conditions, carnage would ensue.

The behavioral changes in the domestication of polecats greatly resemble those that occurred during the domestication of wildcats, another solitary predator. But in one behavioral alteration—their ability to read human intentions—ferrets far exceed domestic cats and approach a doglike state. In one study, ferrets performed as well as dogs did in using human gestures to find hidden food. In contrast, polecat-ferret hybrids reared under the same conditions could not learn to use these cues.[26]

A key factor in reading human intentions in both dogs and ferrets is tolerance of the human gaze. Eye contact is the first step in this sort of interspecific communication. But eye contact is also an aggressive behavior in many mammals. Our female cat, Smoke, uses her stare to quite effectively control the behavior of her larger brother, Sylvester, especially when he plays too rough, but also when he occupies her favorite sleeping spot. I, too, can stare Sylvester into submission, or at least make him damned uncomfortable. And this discomfort is a typical reaction to a stare in many animals, from primates to polecats. The domestication process has obviously ameliorated this discomfort in dogs to the point that they actively seek eye contact from their human companions. Interestingly, the same seems to have occurred in ferrets; they tolerate eye contact much better than do polecats or polecat-ferret hybrids.[27] Moreover, domestic ferrets are much more tolerant of eye contact from their human companions than from strangers.

It is tempting to speculate that this particular sociocognitive convergence in dogs and ferrets may have something to do with the fact that both dogs and ferrets were bred to work in a coordinated fashion with their human keepers, in contrast to cats. But recall that domestic foxes, which were never bred for such activities, also developed a capacity to read human intentions, as a by-product of selection for tameness. So perhaps we shouldn't think of tameness and the capacity to read human intentions as independent traits. The tolerance for eye contact could simply reflect a high degree of tameness, which could then be exploited for interspecific communication when it is beneficial for humans. Cats just never proved useful in that way.

Furthermore, perhaps we should consider tameness a form of interspecific sociality, which may be developmentally linked in some way to the increased intraspecific sociality we observed in cats, raccoons, and ferrets. That is, tameness (as a form of both interspecific and intraspecific sociality) may be a package deal. Again, the stress response may be the common denominator.

While ferrets physically resemble their wild ancestors more than many domesticated species do, they, too, show other elements of the domesticated phenotype. Ferrets are significantly smaller than polecats, for example. But the most marked physical alteration is in the skull, which is broader and shorter than that of polecats. This sort of change in skull shape, as we have seen, along with the behavioral changes, is an indicator of paedomorphosis—the retention of juvenile traits in sexually mature adults. Indeed, the ferret is in many ways a neotenic polecat.[28]

But, as in many other domestic animals, ferrets also exhibit the other side of the paedomorphic coin: progenesis, or accelerated sexual development. Ferrets reach sexual maturity at an earlier age than polecats.[29]

MINK

Like the silver fox, the American mink (*Neovison vison*)—not to be confused with the European mink (*Mustela lutreola*)—has long been farmed for its pelt. Farmed mink are reared in much the same manner as farmed foxes: in individual cages. These farmed mink are considered domesticated by the US Department of Agriculture—a reasonable assessment, given the many generations of artificial selection for coat characteristics. Though they are in a quite early stage of domestication, mink provide some interesting contrasts with the other domesticated carnivores we have considered to this point, because they have never been selected for tameness.

American mink may, however, have experienced a sort of indirect selection for tameness, in that those that were less stressed by captive conditions were more likely to thrive and breed under these conditions. (The same is true of the silver foxes with which Belyaev began his experiments. Those starter foxes may have been genetically altered because of many generations of captive existence.)

So, in farmed mink as in farmed foxes, there may well have been genetic alterations from the wild ancestors. Assuming so would certainly make it easier to interpret the differences between farmed mink and wild mink. But unfortunately, we cannot assume so. Phenotypic plasticity complicates things. It is more than conceivable that all such differences simply result from the different environments in which farmed foxes and wild foxes individually developed. There is ample evidence, from both domesticated and undomesticated mammals, that phenotypes are dramatically altered under captive conditions solely through phenotypic plasticity.[30] Behavioral alterations are the most obvious of such phenotypic changes. A wild wolf, or polecat or mink, reared by humans from birth behaves much differently than a wild wolf raised in the wild. Such phenotypic plasticity is not confined to behavioral traits. Many physical traits, from gross size to limb length, have been known to change under captive conditions as well.[31]

So we cannot assume that any behavioral or physical differences between farmed and wild mink reflect anything more than phenotypic plasticity. Some of these physical differences include the size of hearts and spleens, which are considerably smaller in farmed mink than in wild mink, despite the generally larger body size of farmed mink.[32] The smaller spleens and hearts of farmed mink could be a direct result of being raised in cages, rather than having the normal free-ranging existence. The altered diets of farmed mink could also be a factor.

Farmed mink also have smaller brains than wild mink. But the difference in brain size seems to exist in long-established feral mink populations as well.[33] Such populations generally consist of individuals who have escaped from mink farms, which, unfortunately, is not uncommon, adversely affecting native fauna in Europe, where American mink aren't native. Interestingly, farm escapees seem to adversely affect wild mink populations in North America as well.[34] This effect of feralized mink on wild mink is not due to their size advantage. Quite the contrary:

wild mink dominate farmed and feral mink in competitive situations. The problem seems to be a sort of genetic pollution of the wild populations by farmed-mink genes, similar to what has been demonstrated for farmed salmon.[35]

Natural selection on many traits has been relaxed in farmed mink, such that, over the years, mutations that would have been eliminated in the wild persist, and even increase in frequency through genetic drift. If a bunch of farmed mink suddenly escaped, those mutations would be a burden for any wild mink population in the area.

These adverse genetic effects of escaped farmed mink on wild mink are strong evidence that farmed mink have passed the phenotypic plasticity stage of domestication. Through relaxed natural selection on many phenotypic traits, they have become genetically different creatures. But, as we have seen, it takes more than genetic alterations resulting from relaxed natural selection to create something like a ferret, a cat, or a dog. It takes strong selection as well for particular phenotypic alterations—beyond those that can be achieved through phenotypic plasticity alone.

FROM PHENOTYPIC PLASTICITY TO EVOLUTIONARY DOMESTICATION

During the first few decades of the twentieth century, a number of ideas were proposed concerning ways in which phenotypic plasticity is the vanguard of evolutionary change. I mentioned two of these in Chapter 1: Schmalhausen's and Waddington's. But prior to either Schmalhausen or Waddington, the American psychologist James Mark Baldwin gave an account, later termed the "Baldwin effect," that will provide us the easiest access to the subject.[36] (It is important to note that in the following discussion I will largely ignore the subtle differences in these and other similar proposals,[37] though in fact these differences are quite consequential.)[38]

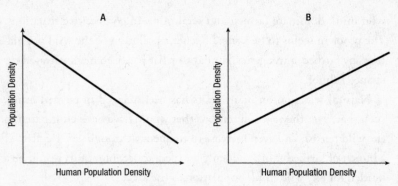

FIGURE 4.2   (A) Norm of reaction before selection. (B) Norm of reaction after selection.

For many, the "Baldwin effect" refers to the evolutionary transformation of a learned behavior to an instinct,[39] but this narrow reading, while
not completely inapt, misses the more general significance of Baldwin's
idea about the role of behavioral or any other form of phenotypic plasticity in evolution.[40]

Baldwin begins with the premise that phenotypic plasticity is a way
for individuals to adapt to their environment within a generation in a
novel environment. Let's say the novel environment is one created by
human occupation. Our hypothetical species has a reaction norm as in
Figure 4.2A, depicting its ability to survive in a range of environments,
from utter human absence to moderate human presence. Though survival in human environments is within the reaction norm, it is at the
extreme end. That is, though at least some members of the species can
survive the human presence, it is far from optimal for this species. The
second part of Baldwin's thesis concerns what happens next, which is a
shift in the reaction norm such that the human environment is closer to
the optimum (Figure 4.2B). This shift is accomplished by natural selection for tameness and increased sociality, and perhaps a change in diet
among those individuals whose phenotypic plasticity was sufficient to
survive among humans. As a result of this selection, a genetic change

occurs in this species with respect to these traits, which renders them better able to survive and even thrive in this new environment.

The Baldwin effect is often misunderstood as further entailing that the reaction norms become flatter—that is, that the amount of phenotypic plasticity is diminished as a result of natural selection for human tolerance. This is the basis for the common reductive interpretation of the Baldwin effect as an evolutionary transition from learning to instinct, or in this hypothetical case a transition from a facultative tolerance for humans to human dependence. In fact, however, this reduction in plasticity is but one possible result of the Baldwin effect. Such a reduction in plasticity does inevitably occur, however, in Waddington's related notion of canalization through genetic assimilation (and in Schmalhausen's stabilizing selection as well).

In a series of experiments on fruit flies, Waddington was able to environmentally induce various morphological changes. For example, some fly larvae exposed to heat shock develop altered wing structure—a state called crossveinless—as adults.[41] Waddington then artificially selected for crossveinless flies for a number of generations. By the fourteenth generation he had produced crossveinless flies without the heat shock. The original environmental inducer was no longer necessary; the reaction norm for crossveinless had become flattened. This flattening of the reaction norm is what Waddington called canalization. Waddington referred to the mechanism by which canalization occurs, the process whereby the crossveinless condition had become canalized, as "genetic assimilation."[42]

The Baldwin effect and genetic assimilation may both figure prominently in the domestication process—sometimes separately, sometimes in conjunction. In both cases, phenotypic plasticity channels subsequent evolution by natural selection. The Baldwin effect, though, is probably the more common bridge from phenotypic plasticity to genetic change in the domestication process, if only because it applies more broadly to evolution in general.[43]

THE ROAD TO DOMESTICATION
FOR PREDATORS

In raccoons, cats, polecats, and dogs, we have various waypoints on the road to domestication. Raccoons may represent the first waypoint, the distance along the road that you can get to through phenotypic plasticity alone, a point through which all domesticated animals must pass. But many animals that reach this point of human commensalism never go beyond it. Whether raccoons proceed further will depend greatly on whether they become more than pests in human eyes. As we will see in Chapter 12, this threshold was only recently surpassed in rats and mice.

Cats, polecats, and dogs have certainly passed the phenotypic plasticity waypoint, as evidenced by behavioral alterations with a large genetic component. The Baldwin effect and canalization are two evolutionary mechanisms involved in the early stages of this transition and, perhaps, beyond. It is often difficult to distinguish the two, but the Baldwin effect alone appears to explain the increased sociality (toward both humans and conspecifics) in cats and polecats. A proper test would require comparisons of the reaction norms of ferrets and polecats, and cats and wildcats, respectively. Dog domestication may involve genetic assimilation as well, since their human dependence is more acute. But even dog sociality may manifest only a shift in the mean values and not a change in the slope of the reaction norm.

The genetic changes resulting from both the Baldwin effect and genetic assimilation are facilitated by disruptive selection and the cryptic genetic variation it unveils, as evidenced in the farmed foxes. Ferrets and cats seem to have undergone a new phase of stabilizing selection for their domesticated reaction norms. For dogs, directional selection, not stabilizing selection, seems to have prevailed to this day. Within breeds, however, social behavior, but emphatically not physical traits, may have reached a new equilibrium through stabilizing selection.

This concludes the section on carnivore domestication. The next group of domesticates come from a different part of the mammal tree and are collectively called ungulates. These include most of the familiar farm mammals. The ungulates took a somewhat different path to domestication than that of the carnivores, partly because they served different human purposes, and partly because of their different predomestication evolutionary histories. The two factors are actually related. The differences between domesticated carnivores and domesticated ungulates are instructive, but so, too, are the similarities, including the role of tameness.

Before we get to the ungulates, though, it's time for an evolutionary interlude.

CHAPTER 5

EVOLUTIONARY INTERLUDE

IT IS CUSTOMARY TO EMPHASIZE THE CREATIVE SIDE OF evolution, the wonderfully myriad ways in which living things manage to meet the environmental challenges that confront them. For a rapidly increasing proportion of creatures on earth, humans constitute the most salient environmental challenges. Our domesticated animals have always been in the vanguard in this respect. At first glance, domesticated animals may seem like putty in human hands, with which we can shape whatever we desire—a Pekingese dog from a wolf, a Holstein cow from an auroch. But evolution has a conservative side as well, an appreciation of which is essential if we are to understand domestication or any other evolutionary process. For all of the readily apparent differences, the Pekingese retains ample evidence of its wolf heritage. More generally, the evolutionary history, or genealogy, of a species provides the essential backdrop against which any changes wrought by selection—natural or artificial—should be viewed.

By way of illustration, let's consider the evolution of an undomesticated creature. This creature is a rather singular fish, called the leafy sea dragon (*Phycodurus eques*), one of the most spectacularly camouflaged animals on earth. Sea dragons belong to the same family (Syngnathidae) as sea horses, which they closely resemble. They have much the same body shape as sea horses, except for an uncoiled tail and longer snout.

FIGURE 5.1   Sea dragon (*Phycodurus eques*). (© iStock.com/kwiktor.)

Sea dragons, like sea horses, have also lost most of their fins—including the tail fin that provides much of the propulsive force for most fishes—so they aren't good swimmers. Camouflage is essential if these small, immobile creatures are to avoid being eaten by the many larger fishes that share their habitat. For sea dragons this means looking like kelp, as they spend their lives drifting with kelp flotsam off the coast of southern Australia. And the kelp-like–ness of leafy sea dragons is truly remarkable. Their coloration is a very good match for the local kelp; they also have elaborate protrusions extending from their bodies at various points that resemble the leaflike structures of the local kelp to an uncanny extent. A sea dragon isolated in an utterly bare aquarium is easily mistaken for a piece of kelp by a casual observer.

Camouflage of this sort provides some of the most compelling evidence for the power of natural selection. It is absolutely clear what selection is *for* in this case—kelp resemblance—and the efficacy of this selection is quite apparent. Richard Dawkins, in *The Ancestor's Tale*, discusses the leafy sea dragon at some length in this regard.[1] But here I want to consider some other factors contributing to the leafy sea dragon's remarkable camouflage—factors related to its evolutionary history, its genealogy, extending back in time long before leafy sea dragons as a species came into existence. These genealogical factors, it turns out, go a long way toward explaining how leafy sea dragons came to be so well camouflaged. In Dawkins's account, it's as if the leafy sea dragon arose de novo, from a generic fish ancestor, but that wasn't the case at all. Selection for kelp resemblance in leafy sea dragons began with a fish that was already more kelp-like than 99.9 percent of the more than 27,000 other fish species. The evolutionary work required of natural selection in bringing about a resemblance to kelp was much less than Dawkins would have you believe.

Dawkins's treatment, as is typical of most popular accounts of evolution, focuses on the power of natural selection to shape organisms in appropriate ways for particular environments. This is certainly an important part of the explanation for how leafy sea dragons came to look so much like kelp, but it is only part of the story. What any creature brings to its environment by virtue of its previous evolutionary history is also crucial in determining how, or even if, it will adaptively respond to its current environment. This internal dimension of evolution is variously referred to as developmental constraint or phylogenetic constraint; a better term is "phylogenetic inertia," which captures not only evolutionary options that are rendered more improbable by genealogical factors but also, as in the leafy sea dragon, evolutionary adaptations that are rendered more probable. Inertia is simply the tendency to move in the direction you have been moving. Phylogenetic inertia does more than constrain evolution; it actively channels evolution. And for leafy sea dragons, this channeling by phylogenetic inertia had been toward camouflage for millions of years before leafy sea dragons ever existed.

Let's begin with the fact that sea dragons and sea horses, along with pipefishes, belong to the family Syngnathidae, which evolved over 50 million years ago. By virtue of common ancestry, syngnathids share certain traits. One of the most interesting syngnathid traits is completely unrelated to camouflage: male pregnancy. Eggs are passed to the males during fertilization, where they develop for six to eight weeks inside a specialized pouch. This trait is not found in any other teleost fish family, but it is present in every single pipefish, sea horse, and sea dragon species.

More relevant to sea dragon camouflage is body shape. As Dawkins emphasized, body shape varies tremendously among teleost fishes as a whole. But body shape is highly conserved at the family level. All syngnathids are quite slender bodied and elongate, making them hard to see even without color camouflage.

Perhaps the most spectacular camouflage elements of sea dragons are their leafy growths, which consist of collagen. Particularly significant is the complete absence of scales in all syngnathids. Scale loss is characteristic of fishes that have become sedentary and no longer require the hydrodynamic advantage that scales provide.[2] The genealogy of leafy sea dragons tells us that scales were lost in an ancestor long predating the evolution of syngnathids, since the trait is shared with related families (Figure 5.2). In syngnathids and related fish families, ringlike segments consisting of dermal bone have replaced scales. The dermal bone and scales share certain developmental pathways but diverge in ways relevant to the formation of the leafy accessories of sea dragons.[3] Both scales and dermal bones begin with nuclei of ossification (bone development) in the dermis, but in scales there is also an epidermal contribution of addi-

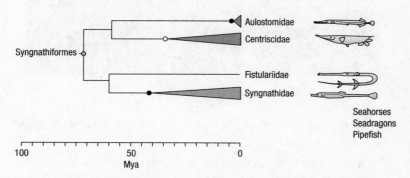

FIGURE 5.2   Evolutionary tree of Syngnathidae and related families. Note that elongated body shape has evolved around 70 mya and has been conserved ever since. (Adapted from Fig. 4 of a PLOS article by Betancur-R et al. 2013.)

FIGURE 5.3
Ghost pipefish.
(© iStock.com/
Trueog.)

FIGURE 5.4
Pygmy sea horse.
(© Ethan Daniels/
Shutterstock.com.)

tional toothlike layers related to dentin and enamel, the latter of which is a particularly hard, highly mineralized substance. Most significant, with respect to kelp resemblance, enamel does not lend itself to the sort of collagenous leafy growths characteristic of sea dragons. Dermal-bone development, however, begins with collagen formation, which is selectively retained and further elaborated, not only in sea dragons but also in many sea horses (especially in the head region), as well as in other dermal-boned relatives, such as ghost pipefishes and pygmy sea horses.

Whereas such collagenous elaborations have never evolved in a scaled fish, they have evolved repeatedly in fishes that have an exterior of dermal bones, where they function mainly in camouflage.[4] The repeated evolution of traits like these filamentous adornments is an example of what evolutionists refer to as "convergent" evolution.[5] The convergence is typically attributed to environmental similarities that result in natural selection for the same traits. But again, this externalist account is only part of the story. As we have seen, much of the explanation for the filamentous growths in sea dragons and other syngnathids, as well as in related fish families, is that they share a common trait: scale loss. Moreover, this trait is shared by virtue of descent from a common ancestor. Traits shared by descent from a common ancestor are called "homologous." Homologies like these can cause even seemingly distantly related species to respond in a similar way to a particular environment.[6]

The discussion throughout this book will be concerned with homologies that predispose all mammals, from dogs to reindeer, to evolve similar traits during the domestication process. This suite of traits, called the "domesticated phenotype" or "domestication syndrome," includes tameness, increased sociality, variable coloration (especially white), reduced size, shortened limbs, shortened faces, floppy ears, reduced brain size, and reduced sex differences. The domesticated phenotype is a form of convergent evolution that occurs in human environments and only in human environments. While the human environment is necessary for the evolution of the domesticated phenotype, the domestic phenotype would not evolve in such diverse species were it not for certain key homologies shared by all mammals through descent from a common ancestor.

Homology has always been an important concept in evolutionary biology. Recently, though, it has come especially to the fore, and with it a renewed appreciation for the conservative side of evolution.[7] This conservative side of evolution has recently become a focal point of evolutionary thinking because of developments in two of the most dynamic fields of evolutionary research: evolutionary developmental biology, more commonly referred to by its contraction, "evo devo"; and genomics. Both research areas present challenges to the received view of evolution, known as the modern synthesis (see Appendix 5A on page 323).

EVO DEVO

Evo devo is based on the premise that all evolutionary change in multicellular organisms results from alterations of existing developmental processes. Much research in evo devo concerns developmental processes that evolved hundreds of millions of years ago and remain highly conserved to this day. Differences in these highly conserved developmental processes are the foundation for the differences in the basic body plans of the major groups of animals that exist today, called phyla—such as mollusks (snails, clams, etc.), arthropods (insects, crustaceans, spiders, etc.) and chordates, the phylum to which fishes, birds, and mammals belong. The celebrated homeobox (Hox) genes figure prominently in these basic body plan differences.[8]

Developmental processes that evolved long ago, such as those that make a chordate a chordate, are highly conserved; they are largely unalterable, simply because all subsequent chordate evolution is predicated on their existence. You don't mess with the foundation of a house once the walls are built. As a general rule of thumb, the more recently evolved a developmental process is, the more vulnerable it is to evolutionary alteration. This book is concerned primarily with highly alterable developmental processes, such as those that get you a Pekingese from a wolf. François Jacob felicitously referred to such superficial evolutionary alterations as "tinkering." To evolve a Pekingese from a wolf requires only some tinkering with the timing of a couple of late-stage developmental events. The basic developmental processes—such as the way bone cells are matured and patterned, limbs are formed, and toes are clawed—do not change at all during the wolf→Pekingese transition.

More generally, evolutionary alterations of development are never wholesale. This constraint follows naturally from the fact that development is incredibly complex and organisms are highly integrated. Therefore, any deep alterations are overwhelmingly likely to be destructive.

GENOMICS

The field of genomics came about because of recent technological advances that made it possible to sequence whole genomes in a relatively short period of time. By "whole genome," I mean the entire DNA sequence of an organism, which consists of billions of base pairs. This base sequence is often referred to as the "genetic code," a term that was coined when most DNA was believed to consist of genes—that is, sequences that specify the construction of proteins. We now know that this view is quite wrong; the vast majority of DNA is noncoding, in this sense. Put another way, most of the genome is not "genic."

Most readers will be somewhat familiar with the Human Genome Project, but the genomes of many other creatures, from yeasts to puffer fishes, have also been sequenced. Because of their economic importance and potential insights into human diseases, many domestic mammals were among the earliest to have their genomes sequenced. This information has proved invaluable in two distinct ways: first, in determining which bits of the genome respond to natural and artificial selection under domestication, and how; and second, in reconstructing the genealogy of domestic breeds and landraces (local varieties of domesticated species), as described in Appendix 5B (page 327).

The implications of genomics with respect to evolution have been enormous. Prior to the advent of genomics, evolutionists could track evolution only by observing changes in particular genes, which is like tracking the evolution of human languages by means of particular words. This practice was informative but limited, not least because, it turns out, most of the biological words being tracked were nouns, while much of the evolutionary action was in the verbs. By "verbs" I mean DNA sequences that control gene activities.[9] Some of these verbs are themselves genes,[10] but many are not. This is one important lesson from genomics.

This realization emerged only gradually, as evolutionary biologists digested the genomic data. The first surprise was the paucity of genes. Most expected gene number to correlate with complexity, such that humans would have the most genes, other mammals somewhat fewer, fishes still fewer, and invertebrates fewer still. That was not the case at all, however. Humans have roughly the same number of genes as puffer fishes have, and not many more than nematode worms have. Even worse for the human psyche, we humans have pretty much the same genes as puffer fishes have, not to mention dogs and cats. That genes, the protein-coding DNA sequences, were so highly conserved was another revelation.

So, why are humans and puffer fishes so different? In evo devo, genes are likened to a tool kit.[11] The basic vertebrate genetic tool kit evolved long ago. There are ways to augment this tool kit, such as gene duplication, and tool kit augmentation is certainly important. But much of evolution is in the various ways in which the genetic tools are deployed, as largely determined by noncoding parts of the genome, all of which were formerly consigned to the category called "junk DNA." We now know that some of this "junk" plays important roles in regulating the activity of genes. The realization that some of the junk isn't junky represents a sea change in evolutionary thought.[12] But I would contend that the most important message from genomics, as well as from evo devo, is evolution's deep conservatism, which is what makes genealogy so informative.

THE TREE OF LIFE

Though Darwin is best known for the concept of natural selection, of equal importance was his metaphor of the tree of life, a very helpful way of representing the ever-branching genealogies of living things.[13] Just as we construct family trees to reflect our genealogy, so, too, do evolutionary biologists, but on a much grander scale. We can think of extant species as individual leaves on the tree of life, their relatedness reflected in their proximity to each other. The distance between leaves reflects their evolutionary history as well, which can be traced backward toward the trunk of the tree.

Systematics is the study of these genealogical relationships, or phylogenies, which are often depicted, in more formal terms, as branching diagrams called "cladograms" (clade = lineage). Clades are hierarchically nested; each level—on the cladogram, not the tree of life—represents a distinct taxonomic category, which ranges from species to domains. Species are the most natural (least arbitrary) of these taxonomic categories, but evolutionary biologists find it useful to recognize a number of others as well. These taxonomic categories are currently under active scrutiny and revision, especially regarding the more basal levels. But here we need consider only the more traditional categories to make the essential point. A "genus" is a group of related species, a "family" is a group of related genera, an "order" is a group of related families, a "class" is a group of related orders, and so on through "phylum," "kingdom" and "domain," in that order.

So what do we know, in advance, about the ancestors of our domesticated mammals? All of them belong to the same kingdom (Animalia), phylum (Chordata), and class (Mammalia). That is, they all share, by virtue of descent from a common ancestor, contingently evolved features that are common to all animals (as opposed to plants, etc.); and, more particularly, to all chordates (as opposed to arthropods and mollusks,

etc.); and, more particularly still, to all vertebrates (as opposed to tuni-
cates, etc.); and, even more particularly, to all mammals (as opposed to
birds and fishes, etc.).

Let's consider some homologies that distinguish mammals from other
branches (lineages) on the tree of life. Mammals, which first entered the
evolutionary arena about 230 million years ago (middle Triassic of the
Mesozoic era), are distinguished from all other living things by a couple of evolutionary innovations. One innovation is hair, which was one
key to elevated body temperatures and, hence, increased activity levels,
which in turn facilitated ways of life unavailable to amphibians and reptiles. Elevated body temperatures were also a prerequisite for the brain
enlargement that is characteristic of mammals.[14] The second key innovation was milk, which resulted in greater maternal investment in each
offspring and hence fewer of them. One aspect of the increased maternal
investment was an extension, beyond the reptile condition, of the dura-
tion of maternal care, which opened up unprecedented opportunities
for the social transmission of behavior, especially learned behavior.

More directly relevant to their domestication, all mammals share not
only the same hormones, but also entire endocrine systems, from cell
types to receptors to feedback relationships. One of these endocrine systems figures particularly prominently in the domestication process: the
hypothalamic-pituitary-adrenal (HPA) axis, which regulates the stress
response.

Mammals also share some key brain homologies, including the limbic
system—a number of interconnected subcortical nuclei that constitute
the neural substrates for our emotions. Fear and aggression are the two
emotions central to domestication; their dampening is at the root of
what we call "tameness."

CHAPTER 6

PIGS

PIGS GET LITTLE RESPECT IN OUR CULTURE, EVEN though they may well be the most intelligent of our domesticated fauna. They have long been associated with filth, greed, and gluttony. Pig abhorrence is not at all universal, however. In much of Asia, pigs have always been esteemed creatures. And formerly, in Europe, pigs were viewed much more positively.[1] Odysseus was a proud pig farmer, for which his image didn't suffer in the minds of his contemporaries. Indeed, in Mycenaean Greece (Greek Bronze Age: 1600–1100 BCE) pigs were associated with fertility, the moon, and the female, as symbolized in the goddesses Demeter and Persephone, and later the Roman goddess Ceres. After the masculine transition to the solar-centered Olympian deities during the Doric (Archaic) period (760–490 BCE), pigs still remained sacred to Artemis and Aphrodite, and boars were admired as both wily and ferocious adversaries by which hunters demonstrated their worth, as in the famous Caledonian boar hunt.

Pigs were also admired elsewhere in pagan Europe, and they figured prominently in Celtic, Teutonic, and Norse religions. The Norse God Frey, who embodied phallic fertility, rode as his vehicle a boar; his sister, Freya, was transported on a boar-powered carriage. From this high point the pig's image has declined considerably throughout Europe. Perhaps, as with cats, this decline reflects their association with paganism in a Christianized Europe. Perhaps it is also related to more intensive pig-farming practices, in which garbage-fed pigs were maintained in filthy conditions.

But elsewhere in the world, the image of pigs never suffered, no matter how much garbage they consumed. Pigs are held in particularly high regard throughout much of Asia. Vishnu's third avatar was the boar Varaha, to which a number of temples are dedicated throughout the subcontinent. Varaha outdid Noah with respect to flooded conditions in that he rescued the entire earth, not just the creatures inhabiting it, by digging it up from the bottom of the cosmic ocean and bearing it to the surface with his tusks.

The Chinese proudly claim for themselves priority in pig domestication, recognizing as they do the pig's importance in the development of Chinese civilization. The ideograph for "home" is a character for "pig" beneath a character for "roof."[2] The pig is no less prized in Southeast Asia, with the notable exceptions of the Islamic populations.

It is in the islands of the tropical Pacific, though, that pig appreciation reaches its apogee. Throughout much of Polynesia, pigs are considered food for the gods, to whom they are sacrificed in great numbers. Pigs, and particularly boar tusks, are associated with fertility and other good things, such as bravery. Elsewhere in the South Pacific, the boar's formidable strength and aggression are emphasized in the decorative arts. The Iatmul people of the middle Sepik region of New Guinea create some particularly marvelous masks, of a variety called mai, in which boar tusks figure prominently.[3] (See Figure 6.1.) And for most New Guineans, wealth is judged primarily by the number of pigs owned.[4] While doing research at a field station in the Madang Province of Papua New Guinea, I was told to be especially careful not to run over any of the free-ranging, frequently road-crossing pigs; far better to kill a man's dog.

The geographic nadir for pigs is the Near East, where pork is taboo. There, again, the problem seems to be an association of swine with filth and impurity. This attitude was born in ancient Egypt, but it was not always so. Predynastic Egyptians consumed many pigs, and in the north,

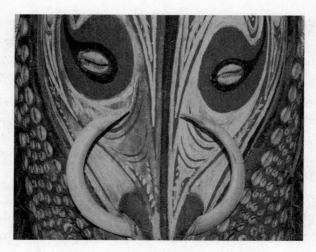

FIGURE 6.1   A mai mask. (Photo by author.)

pigs became associated with the god Seth. There seems to have been a steady reduction in pig consumption, until by the New Kingdom (1567–1085 BCE), there were indications of opprobrium,[5] especially among members of the upper class. By the time Herodotus visited Egypt near the end of the Late Dynastic Period (1038–332 BCE), pigs were considered unclean, even to the touch. While cats were worshipped, pigs were shunned.

Some interpret the shift in attitude to the conquest of the Seth-worshipping pig eaters of the north by the Osiris worshippers of the south.[6] Indeed, after the conquest Seth was transformed into the evil slayer of Osiris, whose death was ultimately avenged by Horus—a central story in Egyptian mythology.[7] Some believe that the Jewish taboo derives from the Egyptian taboo, to which Moses was exposed during his life in the court of Ramses II.[8] The fact that Jews were pastoral nomads, unsuited to pig farming, may have also been a factor.[9]

The stated reason for the pork taboo comes from Leviticus—written about 450 BCE (2460 BP)—and it does indeed seem somewhat arbitrary. The problem with pigs, according to the author of Leviticus, boils down to this: pigs have cleft feet, like cattle, sheep, and goats, but do not chew their cud like cattle, sheep, or goats.[10] According to this religious taxonomy, that combination of traits is deemed self-evidently unnatural, even sinister. Pigs are indeed similar to cattle, sheep, and goats in some ways but quite different in others. They occupy an interesting taxonomic position, but one that, from an evolutionary perspective, is perfectly natural.

THE EVOLUTION OF PIGS

With pigs we enter a new mammalian order, called Artiodactyla (*artio* = "even"; *dactyl* = "toe or finger"), because they have an even number of digits (two or four). There are several families of artiodactyls (Figure 6.2). Pigs and other wild swine, such as warthogs, belong to the family Suidae. Other important artiodactyl families include Bovidae (cattle, sheep, goats, antelope), Camelidae (camels, llamas), Cervidae (deer), and Giraffidae (giraffes).

The first artiodactyls to appear on the evolutionary stage (about 55 mya) were quite diminutive, about the size of a hare.[11] They were distinguished from all other mammals in possessing an even number of toes (two or four), in which the axis of symmetry passed between the two middle digits. The key evolutionary innovation of artiodactyls is a double-pulley system of tendons in the legs, which greatly restricts rotation but makes forward movement more efficient, and hence facilitates long-distance travel. All artiodactyls—with the exception of camels— also have keratinized (fingernail-like) hooves covering their toes.

FIGURE 6.2   Artiodactyl phylogeny. (Adapted from Price et al. 2005.)

Pigs and other members of the family Suidae retain many of the primitive characteristics of early artiodactyls. For instance, pigs have the full complement of mammalian teeth (incisors, canines, premolars, and molars); the incisors and canines are greatly reduced or absent in most artiodactyls. This altered dentition reflects the fact that the majority of artiodactyls are strictly vegetarian, which is also manifest in stomach specializations related to cud chewing. Pigs, however, are highly omnivorous; their diet includes—in addition to leafy matter—fruits, vegetables, insects, fungi, and even small mammals, snakes, and lizards. Hence they lack the dental and gastrointestinal specializations of most artiodactyls, as a result of which, as rabbis observed, they don't chew cud. But pigs are far more efficient converters of food into body mass than are more specialized artiodactyls, such as cattle, sheep, and goats. This trait played a key role in pig domestication.

Another key feature of suids, which distinguishes them from most other artiodactyls, is their canine tusks, which grow continuously throughout life.[12] The lower tusks play a particularly important role in defense against predators; rapid upward head thrusts maximize the bodily harm that these tusks can inflict. These lower tusks are also deployed in fights between males for dominance and access to females. The upper tusks function primarily as sexual adornments to attract females, most spectacularly in the babirusa (*Babyrousa babyrussa*) of Sulawesi. (See Figure 6.3.)

The genus to which pigs belong, *Sus*, originated in Southeast Asia about 3.5 mya.[13] The wild ancestors of domestic pigs (wild boars) belong to the species *Sus scrofa*, which originated in island Southeast Asia,[14] from which they migrated north to the mainland via the Isthmus of Kra when sea levels were much lower; they then colonized Southeast Asia and the Indian subcontinent, followed by western and northeastern Asia, and finally Europe. During this range expansion, wild pigs differentiated into about 25 subspecies. These subspecies can be roughly divided into two genealogical clumps, one in the eastern part of the *Sus scrofa* range and one in the western part.

FIGURE 6.3   Babirusa skull. (Drawing by E. W. Robinson.)

PIG DOMESTICATION

There were probably two distinct routes to pig domestication. The first route resembled that of dogs and cats, in that pigs were attracted to human habitations and refuse, and initiated the domestication process through self-taming. Melinda Zeder calls this the "commensal" route.[15] The second route to pig domestication was through human management of wild populations, by means of first something like herding and eventually complete captivity. This is the process by which cattle, sheep, goats, and horses were domesticated, and it is more human initiated.[16] The two routes to domestication are not mutually exclusive. Some degree of commensalism may have developed prior to herding.

Sus scrofa is unique among large domesticated mammals in that, to this day, large viable wild populations remain in most of its former range. Hence, throughout the process of domestication, introgression of wild genes into domestic stocks has been common. One upshot is that, in many parts of the world today, there is something of a genetic continuum among domestic, feral, and wild pig populations.

While introgression sometimes makes it difficult to differentiate wild, feral, and domestic pigs, the continued existence of wild populations offers a huge benefit for those seeking to trace the genealogy of domestic pigs worldwide. Hence, we know a lot more about the details of pig domestication and subsequent dispersal than we do for dog and cat domestication.

From archaeological evidence, it was long suspected that the pig had been domesticated independently in western Asia and China.[17] Genomic evidence lends support to the archaeological findings but also points to a number of other potential sites of independent domestication.[18] Pigs may have been domesticated twice in China—once among the millet cultivators of central China (Yellow River), and once among the rice cultivators of the south (Yangtze River), by about 8000 BP.[19]

Somewhat ironically, it is in western Asia, where pigs are most anathematized, that the first steps toward pig domestication occurred, perhaps commencing as early as 11,000 years ago.[20] West Asian domestic pigs were dispersed, along with farming, northward and then westward to Europe. Though the first domestic pigs in Europe came from the Near East, their genetic legacy has been obscured through interbreeding with local wild boars, or possibly, independently domesticated pigs from central Europe.[21] Italian pigs seem to be derived from local wild boars, suggesting an additional independent domestication in this part of southern Europe. Another independent domestication event has been claimed for the Iberian Peninsula, because of the genetic affinities of Iberian breeds and local wild boars.[22]

SUBSEQUENT DISPERSAL

The range of the already widespread wild boar was greatly extended following domestication. Actually, this human dispersal may have commenced prior to domestication. Jean-Denis Vigne, who discovered the buried cats on Cyprus, also found pig bones at Akrotiri, which was one of the earliest sites of human habitation on this island.[23] Pigs are not native to Cyprus, so they must have been transported by humans, probably from Turkey. Vigne dates these bones to 11,400–11,000 BP, about a thousand years before the earliest morphological changes associated with domestication. He further claims that the Cyprus pigs point to a long period of human management of wild populations beginning around the end of the Pleistocene (14,000 BP). Whatever the merits of Vigne's view, most human-mediated dispersal of pigs to areas beyond their normal range occurred after domestication.

Perhaps the most interesting of such dispersals occurred through Oceania, the islands of the vast tropical Pacific Ocean, from the Philippines and New Guinea in the west, to Hawaii and French Polynesia in the east, where pig appreciation reached its zenith. The consensus is that this pig dispersal came by way of the same Austronesian peoples and associated Lapita culture that brought dingoes to New Guinea and Australia. The peopling of Polynesia, in particular, was the final leg of the epic human migration out of Africa that commenced 60,000 years ago. There is dispute, though, as to the path taken by the Austronesians from Asia to Polynesia. The most popular hypothesis is the "Express Train from Taiwan to Polynesia," according to which the Austronesians originated in Taiwan 3,000 years ago, headed south to the Philippines and then New Guinea, and then (after off-loading dingoes in northern Australia) headed east until finally arriving in Hawaii and Rapa Nui (Easter Island).[24] Pigs, however, pose a problem for this account.

The problem is this: the pigs of Polynesia, as well as those of the Philippines and New Guinea, are genetically unrelated to the pigs of Taiwan, both modern and ancient.[25] Instead, all of these widely dispersed domesticated pigs are descended from a population of Southeast Asian pigs called the Pacific Clade.[26] From New Guinea eastward, there were never any wild pigs; the only pigs to be found are domesticates of the Pacific Clade. The so-called wild pigs of New Guinea, which figure so prominently in tribal cultures there, are actually feral descendants of domesticated pigs brought (along with dingoes) by Lapita farmers about

3,000 years ago. Since many of New Guinea's ethnic groups have been on the island for 30,000–40,000 years, pigs were a relatively late cultural acquisition. In Polynesia, though, pigs were an important part of the Lapita cultural package from the beginning. They—along with dogs and chickens—were presumably aboard the ships of the first Austronesian colonizers.

But if the pigs were, in fact, distributed throughout this part of the Pacific by Lapita peoples, the train out of Taiwan was a local, not an express, with stops in mainland Southeast Asia, where Pacific Clade pigs (and perhaps dingoes) were first acquired—Java, Sumatra, New Guinea, and the Solomon Islands—prior to any Polynesian destination. It is also noteworthy that the domestic pigs that you find in Taiwan, as well as those of the Philippines, came by way of China, obviously with human aid.[27] These Chinese-derived pigs were then transported to Micronesia in the northwestern tropical Pacific. This second dispersal more resembles an express train out of Taiwan, but of much more limited scope.

Debates about the origin and migrations of the Austronesian Lapita culture continue unabated. While pigs certainly do not provide the last word on the subject, their human-mediated peregrinations, as recently reconstructed from genomic analyses, should certainly inform the conversation.

FROM WILD BOARS TO LANDRACES AND BREEDS

Given the multiple centers of domestication from quite distinct subspecies, domestic pigs were highly genetically differentiated from the outset—much more so than dogs or any other domesticated creature. Pigs probably became even more genetically differentiated as a result of domestication, as they adapted to local habitats and cultures, and through genetic drift. Hence, from the early days there were many distinct landraces of pigs within the natural range of wild boars. Many more landraces were created when humans transported pigs to places where wild boars had never been present, including the islands of the Mediterranean, New Guinea, northeastern Australia and numerous Pacific Islands, later Africa, and still later the Americas and New Zealand. While many of these human-dispersed pigs became feral, others served as the foundation for domestic landraces.

As humans took increasing control of the breeding process and eventually constructed what we now think of as distinct breeds from these landraces, the genetic differentiation of pigs became even more pronounced, because until relatively recently, the construction of pig breeds was extremely local.[28] Historically, there was little interbreeding among village pigs separated by more than hundreds of miles.

One important force countered this tendency toward genetic differentiation: genetic introgression from adjacent wild populations. Genetic introgression varied on a regional scale. In China, for instance, where pigs were most intensively farmed, there was very little introgression from adjacent wild populations.[29] In Europe, by contrast, domestic pigs were freer to roam, and introgression from wild populations was more pronounced, which, as already mentioned, eventually caused the complete elimination of the Near Eastern genetic legacy. In populations that were dispersed to places without wild pigs, there was no such countervailing force.

It was probably in China that the first steps toward the transition from landraces to proto-breeds occurred. And it was in China that some of the most marked alterations of pigs occurred, such as the evolution of squash-faced (brachycephalic) forms analogous to bulldogs. China produced well over a hundred distinct pig breeds by the end of the nineteenth century, when the "breed" concept was developed.[30]

In China, pig breeding remained local throughout most of the twentieth century,[31] as is reflected in the geography-based genetic differences among Chinese pig breeds that still exist today.[32] In Europe, things were more geographically fluid from the beginning of the seventeenth century, especially in the north. Then, late in the eighteenth century, the genetic mixing went to another level, when pigs from northern China were imported to Europe and interbred with existing stock to create new, "improved" breeds.[33] Most existing European breeds reflect some degree of Chinese ancestry, though it varies. Common breeds, such as the Large White, Hampshire, and Berkshire, have lots of Chinese DNA. Of the common breeds, the red-coated Duroc is believed to have the least Chinese influence, but it originated in the United States.[34]

Many of the older European breeds that derive from local landraces without any Chinese component are now rare or extinct. They comprise a large portion of the "heritage breeds," which have garnered much attention of late from the culinary community. These include the Gloucester Old Spot and Tamworth from the United Kingdom, the Mangalista from Hungary (originally the Balkan Peninsula), the Basque Pig (a.k.a. Ass Black Limousin) from France, the Casertana and Calabrese from Italy, and the Negro Iberico from Spain. Heritage breeds, with or without Chinese "blood," are threatened primarily because they are not suitable for the ultraintensive farming practices that predominate in the West today. Generally, they don't grow fast enough for factory farming, or they require too much space and resources. The Iberico, for example, needs about an acre of oak woodland (called dehesa forest) per pig to supply the acorns for its famous hams.[35]

Another reason for heritage status in some breeds is a change in the way pig flesh is used. Pigs, alone among domesticated artiodactyls, are prized solely for their flesh; but in Europe, many breeds became specialized as to the type of flesh they provide. Eventually, distinctions were made between bacon pigs and lard pigs. Lard-type breeds, such as the Mangalista, Guinea Hog, and Large Black, tend to have compact bodies and a high proportion of fat; bacon-type pigs, such as the Yorkshire and Tamworth, are long-bodied and lower fat. Intermediate breeds, such as the Chester White, Hampshire, and Large White, are sometimes referred to as "meat-type" because they are now used primarily for hams, loins, and such, but they were originally more all-purpose.

The Berkshire was originally a lard-type breed that was genetically leaned out to a meat-type pig after World War II. That was a good thing for Berkshires, because other lard-type breeds fell out of favor as lard was replaced by shortening for cooking purposes. Most lard-type pigs now have heritage status.[36]

Other so-called heritage breeds are actually landraces, often of feral populations. These include the Kunekune of New Zealand, which was originally deposited there by European sailors during the eighteenth century.[37] There are several such heritage "breeds" in the United States. Spanish explorers to the Gulf Coast introduced the Mulefoot—so called because of its fused toes—in the sixteenth century. The Ossabaw Island Hog (Georgia) also came by way of Spanish explorers, while the Red Wattle, originally from New Caledonia (Melanesia), was a gift to Louisiana from the French.[38] All of these American "breeds" have become culinary stars. So, too, have some of the old European and American lard-type breeds, such as the Mangalista, Guinea Hog, Negro Iberico, and Large Black, because their high fat content makes their meat tastier. Moreover, the fact that foodies now prefer lard over shortening enhances the prospects for lard-type breeds.[39]

SIGNATURES OF DOMESTICATION

Perhaps the first structural change in domesticated pigs was the shortening of the snout. Snout shortening is especially pronounced in some Chinese breeds of the Taihu group, such as the Meishan, which are the pig equivalent of bulldogs. Also as in dogs, there was a general shortening of the limbs and curling of the tail. Wild boars actually have fairly straight tails, while the tails of domestic pigs tend to curl and twist and, in some cases, corkscrew. Feral pigs tend to be intermediate in all of these respects.[40]

Then there is the matter of the brain, or relative lack thereof. Domesticated mammals, from dogs to yaks, generally have smaller brains than their wild forebears, and pigs are no exception.[41] It is noteworthy that feral populations, even those that have been feral for hundreds of years, more closely resemble domestic pigs than wild boars.[42] This signature of domestication seems more indelible than other structural alterations. Whether it translates into a reduction in intelligence is far from clear, though. Much of the reduction appears to occur in parts of the brain related to motor control and sensory (visual, acoustic, and olfactory) processes.[43] The parts of the brain related to olfaction are particularly adversely affected by domestication in pigs. Feral pig populations retain the olfactory deficits of domestic pigs.[44]

Other components of the domesticated phenotype, such as floppy ears, are also well represented among domestic pigs.[45] But the most obvious and universal signature of domestication in pigs is hair color. Wild boars have the most common mammalian hair color: mostly reddish brown, but with black at the tip and base. This coloration is called the "agouti" pattern, and it provides camouflage in a wide range of environments.[46] Domestic pigs, though, are a relative riot of colors, from black to white, with various shades of yellow, brown, and red in between. And coloration has become an important breed-defining trait.

White is the color most divergent from the wild type and most characteristic of domestication in mammals.[47] All-white pig breeds include the Chester White (UK), Large White (UK), British Lop (UK), Yorkshire (UK), and Landrace (Denmark).[48] The Pietrain (France) is mostly white, while the Hampshire (US), Essex (UK), and Berkshire (UK) have large white belts on otherwise brown-black bodies. Breeds with small areas of white include the Hereford (US) and Poland China (US).

In general, white coloration is more common in European breeds than in Chinese breeds. The characteristic color of Chinese breeds is black. Virtually all Chinese breeds have some black, and many, such as the Meishan, Tibetan, and Xiang are all-black.[49] The only all-black European breed, the Large Black (UK) derives from a Chinese cross.[50] Feral pigs, though tending to revert to wild-type coloration, do so only very slowly. Many existing feral populations continue to exhibit some coloration influence of the domestic population from which they derived.[51]

NATURAL, ARTIFICIAL, AND SEXUAL SELECTION

The trait in which feral pigs most resemble wild boars and most diverge from domestic breeds is the tusks of the males. These tusks originally evolved in the context of strong sexual selection in which boars competed fiercely to dominate other males (intrasexual selection) and attract females (intersexual selection). One hallmark of domestication in pigs is the reduction of these tusks.[52] There were probably a number of factors involved. First and foremost, tusks were no doubt deemed undesirable by human pig handlers. Second, when humans took control of the breeding process, they eliminated much male-male competition, and chose males using criteria quite different from those that a female pig would use. The net result is reduced sexual selection, as well as reduced sex differences.

This reduction in sex differences is not confined to the tusks; the body size of males and females also tends to converge under domestication.[53] When pigs become feral, they revert to the old sexual selection regime, and these sex differences reemerge. The famous razorbacks of the southeastern United States exemplify this process quite nicely. Razorbacks are the feral descendants of domestic pigs brought by Spanish explorers, beginning with De Soto.[54] The name "razorback" refers to a ridge of stiff, erectable hair along the spine; when erected, it is a sign to other males of an aggressive disposition. If that doesn't deter, male razorbacks have tusks that any wild boar would respect.

Neoteny born of selection for tameness may also play a role in the reduction in tusks and other sex differences among domestic breeds. Tusks are a late-developing character, so if development is slowed or truncated, they will be smaller or absent altogether.[55] The reacquisition of tusks in feral pigs may, in part, be explained by the reversion to a wild-type developmental trajectory.

The shortened faces of pigs, as well as the floppy ears found in many breeds, also suggest neoteny. If so, we might expect to find correlations among snout length, ear floppiness, and perhaps tusk size in domestic breeds. To my knowledge, such correlations have not been studied.

Some changes in hair color, though, probably reflect trait-specific selection. First there was a reduction of purifying selection for the cryptic coloration during early domestication.[56] As a result, genetic variants that exist at low frequencies in wild populations became more common in domestic populations. Furthermore, new color mutations were not eliminated in domestic pigs, as they would be in wild pigs. Once this color variation became manifest, two processes—genetic drift and artificial selection—came to the fore. Local color variants could increase through drift in relatively isolated populations—that is, populations isolated from other domestic pigs, feral pigs, and wild boars. More important, human breeders could select from among the color variants.

One of the more interesting stories in pig evolution during domestication concerns the culture-specific nature of this artificial selection, most notably the black coloration characteristic of Chinese breeds. Blackness in Chinese pigs seems to reflect a cultural preference that may date back to the early days of pig domestication. It was certainly present in the early Shang Dynasty (Chinese Bronze Age: 3000–1500 BCE). Pigs figured prominently in the sacrifices so central to Shang religion, and black was the color the gods preferred. Since the preferences of the gods were to be honored, there was intense artificial selection for blackness in China.[57] The genetic legacy of Shang pig sacrifices continues to be evident in the coloration of Chinese pigs today.

More universal examples of trait-specific artificial selection figured prominently in the increased productivity of pig farming, including early maturation, year-round breeding, increased growth rate, and increased litter size.[58] As we saw in the domestic foxes, early maturation and the loss of seasonal breeding can be, to some extent at least, by-products of selection for tameness. But artificial selection by pig breeders no doubt accentuated both traits. The accelerated growth of domestic pigs relative to wild and feral pigs is almost certainly due to artificial selection of ever-increasing intensity.[59] (One reason that heritage breeds have lost out in the pig market is that they do not grow as rapidly as "improved" breeds. Their continued existence rests precariously in the taste buds of culinary sophisticates.) Finally, artificial selection on litter size has increased that number from about six in wild pigs to twelve or more in domesticated breeds.[60] We have managed to create pigs that give birth to more offspring than they can suckle.

DOMESTICATION AND THE PIG GENOME

Pig genomics is in an earlier stage of development than cat genomics—far earlier than dog genomics. Some preliminary findings of note concern the footprints of artificial selection. Let's first consider hair color. Several genetic loci contribute to hair color, each with multiple variants or alleles. Of the loci contributing to hair color in pigs, we know the most about *MC1R*, which codes for a melanocortin receptor, MC1R. Melanocortin plays an essential role in pigmentation, and the receptor MC1R mediates the effects of melanocortin. When MC1R is present, the melanocytes (the color cells in the hair) become brown to black; when it is not present, the melanocytes become yellow to red. Various mutations have produced variant alleles of *MC1R* that vary in activity and are associated with specific coat colors in dogs and cats. In pigs there are six different alleles, each associated with a different color.[61] One of these mutations is associated with black color in Chinese pigs. A recent genomic study demonstrated that the high frequency of this allele in Chinese domestic pigs was the result of artificial selection for black coloration.[62] So we can now link an ancient Chinese cultural practice—pig sacrifices—to a specific genetic alteration, one that has no influence on pig productivity.

Mutations in a gene (called *KIT*) that encodes a receptor for a cell growth factor are associated with white coloration in European breeds.[63] Interestingly, the Rongchang, one of the few white Chinese breeds, lacks this mutation; hence, its white color came by a different genetic route.[64] Also of note, this breed has the same *MC1R* mutation found in black breeds, which must somehow have been counteracted. Both the *MC1R* and *KIT* mutations speak to the fact that Chinese breeds evolved independently of European breeds.

Other mutations have been linked to muscular growth,[65] fat deposition,[66] brain development, olfaction, and immunity,[67] among other traits. Most of these mutations have not been traced to particular genes, and many, no doubt, have occurred in noncoding DNA regulatory sequences. Fat deposition, for example, is influenced by several noncoding DNA mutations.[68] Of particular note is a study conducted by scientists from Belyaev's Institute of Cytology and Genetics at Novosibirsk on porcine endogenous retroviruses, which happen to have one of the more delightful acronyms in genomics: PERVs.[69]

Endogenous retroviruses are genomic elements that ultimately derive from viral infections, after which the viruses are incorporated into the genome. Over time, they can be "domesticated"; that is, their ability to multiply and move becomes restricted by alterations in the rest of the genome. Once domesticated, they can actually become regulatory elements, affecting the expression of nearby genes. Usually ERVs have adverse affects, throwing fine-tuned regulatory networks out of whack; but like mutations, every once in a while they create a selective advantage in the carrier. In domestic animals, the selective advantage is the trait desired by the human breeders. When ERVs confer a selective advantage, they of course increase in frequency.

The Russian researchers were able to distinguish four distinct clusters of European pigs based on the frequencies of various PERVs: cluster 1 consisted of wild boars; cluster 2, of bacon-type pigs; cluster 3, of lard and all-purpose types; and cluster 4, of miniature pigs. Wild boars have the fewest PERVs, while miniature pigs have by far the most. In lard-type pigs the Russian researchers found PERVs associated with genes involved in fat deposition, and in bacon-type pigs they found PERVs that influenced muscle development.

Genomic studies are also useful for determining genealogical relationships on the pig twig of the tree of life. Chinese breeds are geographically clustered on the genealogical tree.[70] This is what you would expect, given that until quite recently, each breed was confined to the area in which it evolved, through long-standing traditions of government control over pig transport. In contrast, the more freewheeling pig husbandry practices in Europe have caused a breakdown in geographic-genealogical associations there.

The deliberate breeding of local populations with Chinese pigs, beginning late in the eighteenth century, further scrambled things geographically. The most recent example is a so-called synthetic line called the Tia Meslan.[71] This breed was created through crosses of Chinese boars (Meishan × Jiaxing) with European sows. Not surprisingly, it falls in the middle of the genealogical tree, halfway between European and Chinese breeds.

The degree of Chinese genetic introgression into European breeds varies greatly, though, and is generally much less than in Tia Meslan pigs. The Duroc evidences little Chinese influence, as do a number of heritage breeds. Generally, there is a north–south trend in European breeds with respect to Chinese influence: more in the north than in the south. Southern European breeds, especially those from Italy and the Iberian Peninsula do tend to form geographic clusters on the pig tree, as do some northern breeds, but much less so than Chinese breeds do. All of these results showing genealogical relationships should be considered extremely preliminary, since they are based on only small parts of the genome.

THE CONSERVATIVE PIG—
A BEAUTIFUL THING

Despite the best efforts of human pig designers, the domestic pig has remained remarkably true to the heritage of its wild ancestors. It is tamer, of course, and males have sacrificed some of their tusks. But the structural alterations are mostly modest or superficial. The modest modifications include snout and leg shortening; the superficial modifications include hair color and obesity. Behaviorally, the domestic pig retains the adaptable improvisational ways of the omnivorous wild ancestors from which it evolved. Maternal behavior is unchanged, as is the struggle for dominance and teat order among piglets, which has lasting effects on their subsequent development.[72] If anything, piglet competition may be even more intense in domestic pigs, given the increased litter size and, in some cases, teat shortage.

It is only quite recently that we have managed to develop pigs that would not be able to survive in a feral state. Throughout most of the history of pig domestication, certainly in Europe, escaped porkers could manage well without human provisions. And if humans were to vanish from the face of the earth, pigs would do just fine—much better, in fact, than most of their more "advanced" artiodactyl kin. In part, this ability to survive without humans is testimony to their intelligence, which is considerable.[73] But it also reflects the success of a basic body plan that evolved quite early in the history of artiodactyls and remained viable even long after the cud chewers appeared on the scene. The pig, both wild and domestic, is quite an evolutionary success story, and it deserves much more admiration than it is generally accorded.

An antidote for anyone still suffering from anti-pig prejudices is a wonderful painting by Jamie Wyeth, located in the Brandywine Museum, Chadds Ford, Pennsylvania, appropriately titled *Portrait of a Pig*. And Wyeth did truly accord his subject—an adult pinkish white sow—the respect and close observation characteristic of the finest realist human portraits. She—I will call her Emily—is seen from the side at about eye level, to maximize her exposure, from somewhat crumpled snout and robust forward-pointing ears to daintily curled tail. Her prominent teats announce her femininity. Emily is perhaps not noble, but she is certainly a captivating presence with a forceful personality. And she is strangely beautiful. I have visited the museum three times; with each visit I become more captivated by Emily. (See Figure 6.4.)

FIGURE 6.4   *Portrait of Pig* by Jamie Wyeth. (© Jamie Wyeth [born 1946], *Portrait of a Pig*, 1970, oil on canvas; collection Brandywine River Museum of Art, gift of Mrs. Andrew Wyeth, 1984.)

I hope the following anecdote doesn't undermine my attempt to reha-bilitate the public perception of pigs. While painting Emily, Wyeth was asked by the farmer who owned the pig to help with a chore. During Wyeth's absence, Emily managed to consume 17 tubes of highly toxic oil paint. He returned to find her face covered with cerulean blue, and Wyeth feared the worst. Fortunately, the only effect of Emily's undis-cerning palate was the unprecedented palette of her excretions. Hardi-ness is another admirable pig trait.

CHAPTER 7

CATTLE

PLACID DAIRY COWS GRAZING IN A BUCOLIC SETTING evoke, for most city dwellers, a comforting wistfulness for the simpler pleasures of rural life. Artists from Cuyp to Constable to Katz have deployed this motif to create a particular mood or emotion, one that is pleasant but not ecstatic, warm but not hot—a mood of comforting beauty. That, at least, is my reaction. But the thought of milking one of those beasts at 5:00 a.m. in midwinter is enough to snap me out of my reverie. Those simple pleasures were hard-won through heavy toil.

Indeed, when you consider their source, dairy cows themselves were hard-won. Their ancestors, the aurochs, were not only the largest but also the most formidable creatures we humans have managed to domesticate. In retrospect, those placid dairy cows are unlikely creations—not as unlikely, perhaps, as a Pekingese, but not anything that an intelligent alien would have predicted 18,000 years ago when humans first began depicting aurochs on cave walls. Those paintings were more religious than purely aesthetic creations, an expression of something like awe. That is not an emotion we tend to associate with dairy cows. But then, even female aurochs were very un-cow-like; they were powerful and dangerous yet tempting prey.

The wild auroch is extinct, so to begin to understand what inspired those cave paintings, we need to consider some of their living relatives:

FIGURE 7.1   Wild gaur. (Photo by author.)

the African buffalo, the gaur and water buffalo of Asia, and the bison of Europe and North America—formidable creatures all. A full-grown gaur, for example, will give pause to even the largest tiger; so, too, a water buffalo. Gaurs generally aren't worth the very considerable risk. Only a well-coordinated and well-conditioned pride of lions can bring down a male African buffalo, and not a few die trying. There is much reward for success, but much risk in the attempt. Such was also the case for Paleolithic humans who hunted aurochs. This combination of danger or dread and magnetic appeal evokes the sense of the sublime so famously described by Edmund Burke.[1]

The sublime is a casualty of domestication in any creature, and not just because familiarity creates the mundane. The dread factor evoked by creatures as diverse as tigers, wolves, great white sharks, and aurochs is also dissipated with the advent of tameness and docility. And cattle, along with sheep, are among the most docile of our domesticates. There is, though, a wide range in this trait among domesticated cattle. Dairy cattle are at the extreme docile end; beef cattle tend to be less so, and free-ranging breeds, such as the Texas Longhorn, much less so.

In some cases, humans have either halted or reversed the descent from the sublime. Bull riders engage some very nondocile cattle that are also extremely athletic and powerful. But the closest contemporary approximation of an auroch hunter is that anachronism known as the matador.

My first impressions of bullfighting came by way of a hilarious Bugs Bunny cartoon, called "Bully for Bugs," in which, on his way to the Coachella Valley "and the carrot festival therein," Bugs takes a wrong turn in Albuquerque (pronounced AL-buh-KOY-kee) and pops up in a Spanish bullring. He then takes over after the premature departure of a cowardly matador upon encountering a particularly formidable bull. The ebb and flow in the conflict between Bugs and Bull is among Chuck Jones's best work. Though I later read *Death in the Afternoon*, the Looney Tunes account continued to inform my view of bullfighting more than Hemingway's did. Then, one September day in 1993, I found myself with some free time in Seville; on a whim I made my way to the Plaza de Toros de la Real Maestranza, the oldest bullring in Spain.

Despite my skepticism, I was infected with the enthusiasm of the crowd well before any bull appeared, somewhat like the excited anticipation characteristic of a big sporting event. Aficionados energetically reject the label "sport" for bullfighting, though; some view it as more like an "art." But that term seemed inapt as well, notwithstanding the theatrical and balletic elements. The proceedings were more carefully ritualized than a Catholic mass, but the atmosphere was like that of a Pentecostal church service. In many ways the bullfight seemed like a strange combination of the two forms of worship, and by the time the preening matador actually engaged the weakened and bloodied bull at close range, I had the same sense of alienation I get in any church. As was the case while sitting in a pew in my youth, suffering the droning inanities from the pulpit, I stopped paying attention and let my mind drift.

In retrospect, I think that this reaction, of someone so out of sympathy with religious ritual, is a clue to the origin of bullfighting—in auroch worship, as counterintuitive as this may seem. For we can trace much of the spirit of bullfighting to the spirits that absorbed the minds of Paleolithic hunters, for whom aurochs were a source of both veneration and meat.

We humans have a penchant for ritualizing our sense of awe—perhaps to tame it, perhaps to channel it toward another end. In any case, wild aurochs were certainly awe-inspiring, and the subject of ritualized religious behavior. The cave paintings at Lascaux (about 17,300 BP) are not art for art's sake, nor simply a reflection of magical thinking about the hunt. (It is noteworthy that the most important meat source at that time, reindeer, was not depicted at all.) These paintings are, rather, an expression of veneration. Some of the most impressive of these paintings, and the largest, are of aurochs.[2]

Once the domestication process commenced, cattle veneration was increasingly confined to the bulls. From Çatalhöyük, Turkey, near where aurochs were first domesticated, come some fine depictions of domesticated bulls. These are dated to about 9400–8000 BP, near the onset of the domestication process.[3] This focus on auroch masculinity, as represented by the skull and especially the horns, became more pronounced over time in western Asian and Egyptian deities, including Apis (a deific intermediary between humans and Ptah, and later Osiris), Moloch and Baal of Canaan, and the Minotaur of Crete. Yahweh was also initially depicted as a bull by the Baal- and Moloch-hating Israelites.[4] As is often the case in the sacred realm, objects of worship become objects of sacrifice. As deities evolve, their iconic animal representations often become the appropriate animal sacrifice. Bulls were prominently sacrificed to Yahweh, Zeus/Jupiter, and Mithras of Persia.

Later, other symbolic rituals replaced out-and-out sacrifice. One of the more intriguing of these rituals was "bull leaping," which was especially well developed in Minoan Crete. An almost mind-boggling feat of athleticism and bravery, bull leaping was also a religious rite.[5] The athlete, or acrobat, literally grabbed a bull by the horns; when, as a natural reaction, the bull rapidly raised its head, the acrobat was somersaulted into a backflip, the goal being to land on his feet either behind the bull or on the bull's back. Failure to properly execute this leap meant a severe and probably fatal goring. The archaeologist, Arthur Evans, who first discovered evidence of this activity when he unearthed the so-called *Toreador Fresco* at Knossos was the man who coined the term "Minoan" for this culture, after the mythical king Minos, who employed Daedalus to construct the famous labyrinth to contain the half bull–half human god, Minotaur, worshipped on the island.

Bull leaping was not confined to Crete; it was also practiced throughout much of West Asia and perhaps Egypt. Noteworthy variants occurred in India as well, which survive to this day in the southern regions of the subcontinent. In the Indian version, called Jallikattu, the objective is to hang on to the hump or horns for as long as possible.[6] Like bullfighting bulls, Jallikattu bulls are specially bred for this activity; unlike bullfighting bulls, Jallikattu bulls emerge unscathed. The same cannot be said about Jallikattu riders.

Another contemporary variant of bull leaping is popular in parts of southern France, albeit with the much less dangerous cows.[7] Bullfighting is about as degenerate a form of bull leaping as is cow leaping, and all of these strange rituals may derive from the sense of awe experienced by Paleolithic hunters when they beheld one of evolution's more spectacular creations.

EVOLUTION OF THE AUROCH

Like wild boars, aurochs are members of the mammalian order Artiodactyla, and hence they possess an even number of toes and double-pulley foot tendons. Aurochs, though, belong to a quite different branch of the artiodactyl tree, one that includes other types of wild cattle, such as yaks, bison, water buffalo, bantengs, and gaurs, as well as sheep and goats. All are members of the family Bovidae (Figure 7.2). Bovids, along with species from other families on this branch, including deer (Cervidae) and giraffes (Giraffidae), are collectively referred to as "ruminants."

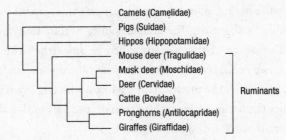

FIGURE 7.2   Artiodactyl phylogeny, showing the genealogical relationships of the main families, including Bovidae. (Adapted from Price et al. 2005.)

Ruminants are much more specialized than suids in their eating habits—largely restricted, in fact, to leafy vegetable matter. This specialization for such a low-nutrient diet is reflected in the digestive system of ruminants, beginning with their teeth. Whereas pigs have the full mammalian complement of teeth, most bovids and other ruminants have lost all or most of their canines and incisors while their molars became more elaborate.

The key adaptation in ruminants, though, is a complex digestive process called "rumination" (literally, "to chew again," from which the term "ruminate," as in "mentally chew over," derives by way of figurative analogy). In the ruminant digestive system, the stomach is elaborated from one to four chambers: rumen, reticulum, omasum, and abomasum (Figure 7.3).

After an initial chew, the food enters the first chamber (rumen), where it mixes with saliva. At this point the ingested material begins to separate into liquid and solid (the bolus) layers. This processing continues in the second chamber (reticulum), after which the bolus is regurgitated as cud, which is then chewed a second time. The cud reenters the rumen (first chamber) and is further digested there and in the reticulum (second chamber), at which point it has become largely liquid. It then enters the third chamber (omasum), where water and minerals are absorbed into the bloodstream. The digesta then finally pass into the fourth chamber

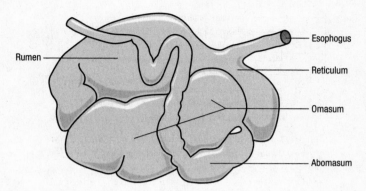

FIGURE 7.3   Cow stomach, showing the four chambers.

(abomasum), which is the equivalent of the human stomach, and food is processed there much as it is in our stomachs, before passing into the small intestine, where most nutrients are absorbed. The entire process depends critically on a complex bacterial fauna, without which even four chambers would not be enough to digest cellulose, one of nature's more indestructible biochemicals.[8]

Cellulose is prevalent in leafy matter, including grasses. Grassland ecosystems greatly expanded about 25–30 mya.[9] That rumination is a particularly successful way to deal with cellulose is evidenced by the fact that bovids and other ruminants largely replaced other large mammalian plant eaters that employed different digestive methods—such as horses, titanotheres, rhinos, and tapirs—at about this time.[10]

The single feature that distinguishes members of the family Bovidae from all other artiodactyl species—including other ruminants—is the horns, specifically permanent hollow horns. These horns function not only in defense against predators but also in male-male competition for status and mates, where they are deployed in ritualized but sometimes fatal combat. As a consequence of this sexual selection, the horns of male bovids are much larger than those of females.

Bovids (Figure 7.4) first arrived on the scene about 20 mya, though estimates based on molecular data vary fairly widely.[11] The first bovid

FIGURE 7.4   Bovid phylogeny, showing two main branches (tribes). (Adapted from Hernandez-Fernandez and Vrba 2005.)

fossil, *Eotragus*, dates from about 18.3 mya.[12] Bovids experienced an explosion of speciation over the next 5 million years and continue to be the dominant herbivores to this day. (Over half of all artiodactyls belong to this single family.) Partly because of their rapid speciation, it is difficult to resolve precisely the fine structure (points of genus and species divergence) of the bovid branch.[13] The main branches, though, are not in dispute.

There are two main branches within Bovidae: one that includes sheep, goats, some antelopes and so-called goat-antelopes;[14] and another that includes wild cattle, including aurochs, gaurs, water buffalo, yaks, and bison, as well as large spiral-horned antelopes such as kudus and elands. On the branch to which wild cattle belong (Figure 7.5), the first to split off were the ancestors of kudus and elands (genus *Tragelaphus* and *Taurotragus*, respectively). Next to go its own way was the water buffalo (*Bubalus*), followed by bison (*Bison*). The remaining bovids all belong to the genus *Bos*, including the auroch (*Bos primigenius*), the gaur (*Bos gaurus*), and the yak (*Bos grunniens*).[15]

The auroch first evolved in India 1.5–2 mya during the Pleistocene epoch (Ice Age).[16] From there they spread west to western Asia. From western Asia, some populations spread southward to North Africa via Egypt. Other populations moved north and west along the north shore of the Mediterranean, reaching Spain about 700,000 years ago. The southern European populations spread northeastward, during warm

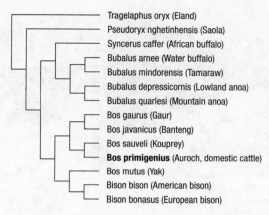

Tragelaphus oryx (Eland)
Pseudoryx nghetinhensis (Saola)
Syncerus caffer (African buffalo)
Bubalus arnee (Water buffalo)
Bubalus mindorensis (Tamaraw)
Bubalus depressicornis (Lowland anoa)
Bubalus quarlesi (Mountain anoa)
Bos gaurus (Gaur)
Bos javanicus (Banteng)
Bos sauveli (Kouprey)
Bos primigenius (Auroch, domestic cattle)
Bos mutus (Yak)
Bison bison (American bison)
Bison bonasus (European bison)

FIGURE 7.5   Phylogenetic tree of the tribe Bovini, to
which all wild cattle species (genus *Bos*) belong, along
with bison, water buffalo, African buffalo, and yaks.
(Adapted from Fernandez and Vrba 2005, 286, fig. 4.)

periods, following the retreating ice, reaching Germany about 275,000 years ago. They continued to spread eastward, eventually occupying most of the temperate forests of Eurasia. By the time of their domestication, aurochs had differentiated into three distinct subspecies (Figure 7.6): one in South Asia (*Bos primigenius namadicus*), one in North Africa (*Bos primigenius africanus*), and one that occupied all of northern Eurasia (*Bos primigenius primigenius*).[17]

From the auroch's first contact with humans, its environment began to deteriorate, first from hunting and ultimately from loss of forest habitat, which accelerated with the advent of farming. The historical pattern of auroch disappearance is closely correlated to human population density. Aurochs were extirpated first in the Near East, then in India, and then in southern Europe. Ironically, they persisted longest in the places they colonized last, which, not coincidentally, had the lowest human population densities. During the Roman period, aurochs were still common in France (Gaul) and other parts of western and central Europe (though extinct in Italy). Julius Caesar saw wild aurochs for the first time during his invasion of Gaul, and he was besotted. He somewhat

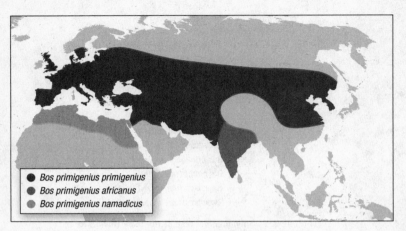

FIGURE 7.6   Range map of the three wild subspecies of the auroch, *Bos primigenius*, prior to domestication. Adapted from C. Van Vuure, *Retracing the Aurochs: History, Morphology and Ecology of an Extinct Wild Ox* (Sofia, Bulgaria: Pensoft Publishers, 2005).

hyperbolically claimed, "They are a little less than elephants in size . . . Their strength and speed are extraordinary. They spare neither man nor wild beast that they have espied. They cannot be brought to endure the sight of men, nor be tamed, even when taken young".[18]

But the French aurochs, like aurochs elsewhere in Europe, were doomed by overhunting and forest destruction for agriculture. The final remnants of the species made their last stand in a medieval wilderness in eastern Poland. The last of these, the last of its kind, was a female that died in 1627.

ASCENT OF THE DOMESTICATED AUROCH

Obviously, humans were bad for wild aurochs, despite their veneration. The notable exceptions were those aurochs that managed to make themselves useful. As wild aurochs declined, their domesticated descendants proliferated; by the time wild aurochs became extinct, domesticated aurochs had come to be among the most numerous large mammals on earth.

The fateful domestication process commenced in two places, independently: in the fertile crescent of western Asia, and in the upper Indus valley,[19] involving two distinct subspecies: *Bos primigenius primigenius* and *B. p. indicus*. The domesticated descendants of *B. p. indicus* are called "zebu" cattle; the domesticated descendants of *B. p. primigenius* are referred to as " taurine" cattle. Some believe there was another independent domestication in North Africa, involving the third subspecies, *B. p. africanus*,[20] but the current consensus is that the first domestic cattle in North Africa were taurine cattle from the Middle East.[21]

Domestication first occurred in the Near East, probably in the upper Tigris region of northern Iraq and southeastern Turkey.[22] From there, domesticated taurine cattle expanded in all directions, but mainly to the west, and then to the south and north, eventually arriving in Europe and North Africa. Domestic cattle first appeared in Europe about 9000 BP, when West Asian farmers migrated through Greece and the Balkans.[23] They then took two distinct routes to the rest of Europe: the Danubian (northern) route and the Mediterranean (southern) route.[24]

Cattle movement along the northern route, from the Balkans to central Europe and finally northern Europe, was facilitated by the migration of farming peoples into areas formerly inhabited by hunter-gatherers, who were gradually displaced.[25] Breeds descended from the northern pioneers are genetically distinct from breeds descended from those who came by way of the Mediterranean route, who predominate in southern Europe and arrived mainly by sea.[26]

Zebu cattle differ in obvious ways from taurine breeds; they have a prominent hump on the back of the neck, and a complementary dewlap hanging from the throat. Zebu cattle also have physiological adaptations for heat and drought that are lacking in taurine breeds.

The domestication of zebu cattle took a somewhat different course than that of taurine cattle. Of particular note is the fact that zebu cat-

tle are more genetically diverse than taurine cattle.[27] There are probably two reasons for this difference. First, *Bos primigenius primigenius*,
the auroch subspecies from which taurine cattle derived, experienced
a significant population bottleneck caused by Pleistocene glaciation;
B. p. indicus, which inhabited more southern, unglaciated regions, never
experienced this bottleneck. Second, interbreeding with wild aurochs
continued long after domestication commenced, especially in southern
India, where zebu aurochs survived longest.[28] Therefore, the sort of bottleneck effect typically associated with domestication in all species was
somewhat muted.

Zebu domestication began around 8000 BP in what is now Pakistan;[29]
there is some evidence for a secondary domestication in southern India
as well.[30] Domestic zebu soon dispersed throughout the subcontinent,
and then eastward throughout Southeast Asia and southern China.
Some traveled quite far north, eventually reaching southern Siberia
and Korea.[31] Others dispersed westward, eventually displacing taurine
breeds from most of their original homeland throughout western Asia,
probably aided by climate change toward more arid conditions. Zebu
first arrived on the African continent around 5000 BP, probably by sea.
They seem to have first alighted at the Horn of Africa, from which they
dispersed north, south, and west, in the company of nomadic pastoralists.[32] At some point, perhaps in Egypt, they encountered taurine cattle
and things took a novel turn, genetically.

Long-horned taurine cattle were first depicted in Egyptian tombs
about 6000–5000 BP.[33] Today they survive mainly in West Africa.[34]
Short-horned taurines, which reflect more intensive domestication, were
first depicted around 4500 BP; they still predominate in Mediterranean
Africa.[35] The first Egyptian depictions of long-horned zebu come from
the Twelfth Dynasty (about 4000–3800 BP).[36] These cattle began interbreeding with both long-horned and short-horned taurines from the
time members of the two subspecies first sniffed each other, resulting in
a uniquely African cattle type called Sanga cattle. A final genetic complication was the introduction, about 1400 BP (670 CE) of a new type

FIGURE 7.7   Ankole bull.

of zebu, with short horns, courtesy of Arab sea traders.[37] These, too, were incorporated into the Sanga stew.[38]

Unsurprisingly, Sanga are quite diverse, reflecting not only varying proportions of taurine and zebu genes, but also the extremely variable tribal cultural and ecological conditions for which they were selected.[39] Sanga cattle usually have a neck hump, but it is much smaller than that of most zebus; their horns vary markedly in size, from the polled Mashona cattle of Zimbabwe to the spectacularly adorned Ankole-Watusi, pride of the Tutsi people of Rwanda-Burundi. (See Figure 7.7.)

FROM THE SUBLIME TO THE DOMESTICATED

It was as a meat source that aurochs first became useful. At some point human hunters, perhaps as they became increasingly sedentary, must have adopted something like conservation tactics to ensure a steady supply of auroch meat. This meant some degree of control over wild aurochs, perhaps by creating conditions that tempted aurochs to remain nearby.

At this stage, not much was required for human management, except limiting cattle movements in the forest somewhat. This period of loose management may have extended for a thousand years or more.[40] Any changes in these proto-domesticates from the wild-type condition would have been subtle. Eventually, though, humans took a greater management role, increasingly confining and/or directing the movements of these large and still dangerous creatures. It was only after significant human control that this meat source could be used in other ways, such as for dairying, carting, and plowing. How long did this take?

On the traditional view, first proposed by Andrew Sherratt, these additional uses evolved only near the end of the Neolithic about 5000 BP, in the Near East, in what amounted to the "secondary products revolution."[41] This hypothesis has recently been challenged because of evidence of milk residue in pottery dating to as early as 10,000 BP, very near the time cattle domestication was thought to have commenced.[42] That date seems implausible, though, because no matter how tame these early domesticates were, by auroch standards, you would still need to be a lot braver than a bull leaper to push their calves aside and pull on their teats. Moreover, the meager yield from those teats wouldn't have justified the effort or risk, except perhaps for its symbolic value.[43]

But dairying did become important in some areas well before 5000 BP. There is good evidence for the processing and storage of milk in northwestern Turkey by 8000 BP.[44] Within 500 years, dairying was also present in southeastern Europe.[45] From there, dairying gradually developed along the Danubian route until reaching northwestern Europe, where it was later developed to an unprecedented degree.[46] The early use of dairy products is noteworthy because dairying requires intensive human management, and a degree of tameness above and beyond what is required of beef cattle.

And as is true of all domesticated mammals, it was tameness—the ability to tolerate close human proximity—that was the initial target of first natural selection and then artificial selection. Perhaps the first physical alteration was a reduction in overall size. Wild aurochs exceeded in size even the magnificent gaur—the largest remaining species of wild cattle. Males could weigh over 3,300 pounds (three times the size of a Spanish fighting bull); females were about three-quarters that size, which is still considerably larger than all but the largest domestic bulls living today. Aurochs were also much longer legged than their domestic descendants, almost as high at the shoulder as their trunk was long. They were particularly well muscled in the neck and shoulders, especially the males. Their skulls were substantially larger and longer than those of domestic cattle, and they carried much more massive horns.

Auroch horns also had a characteristic and complex shape with three curves (Figure 7.8); from the base they grew upward and outward, then forward and inward, and then upward. Only a few breeds retain anything like this horn shape, one of which is the Spanish Fighting Bull. But the horn size of Spanish Fighting Bull cattle is much reduced from the auroch state, even after accounting for the overall size reduction. The apogee of horn reduction was achieved in "polled" breeds, in which all individuals, male and female, lack horns altogether.

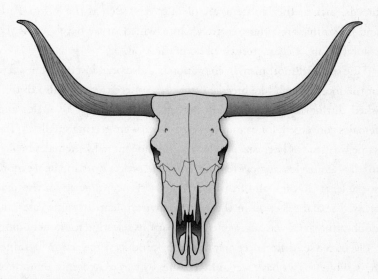

FIGURE 7.8   Auroch horns have three characteristic curves.

In mammals generally, from foxes to pigs, one of the first changes under domestication is the loss of wild-type coloration, primarily because of relaxed natural selection. We would not expect the auroch to be an exception. This trait may have helped early herders separate their cattle from wild individuals. Under the circumstances, they were remarkably successful in preventing genetic introgression, even where wild aurochs were common and domestic cattle relatively free-ranging.[47] The main danger, in that regard, would have been the wild males. Whereas wild female aurochs would no doubt prefer wild males over the relatively scrawny domestic bulls, wild bulls, like male wild boars, would be less discriminating. It was particularly important to prevent these wild bulls from introducing their genes into the domestic pool, entailing perhaps lots of culling in the early stages of domestication.

Aurochs had a distinctive coloration, resembling that of gaurs. Females and young males were a rich reddish brown; older males became much darker, almost black, with a buff "eel stripe" running down the spine. Again, few domestic cattle exhibit this particular color pattern, despite the great variation of coloration among existing breeds. Many breeds, such as the Holstein, are pied; others, such as the White Park and Chianina, are almost entirely white, which, as we have seen, is the coloration most characteristic of domestication.

The sexual dimorphism in coloration has also been lost in all but a few of the most auroch-like breeds. Other sex differences were also diminished during the domestication process. The body size of males and females converged, for example, even as both were getting smaller.[48] The same was true of horn size. Males also became more like females behaviorally—much less aggressive than their wild progenitors and hence more sociable. As in pigs, all of this sexual convergence came about through relaxed sexual selection in the human environment. In some cases, no doubt, this trend was augmented by conscious selection for tamer bulls.

Reduced sex differences may also be by-products of selection for tameness, which, as we have seen, often comes by way of neotenic alterations in development. Males are the slower-developing sex, maturing two to three years later than females; hence neoteny would tend to affect them more, especially their late-developing traits, such as horns. The general reduction of horn size in both sexes may also be a neotenic feature, as well as the shortened snout and legs, which are part of the domestic phenotype package. The evolution of short-horned and polled breeds may reflect a more specific heterochronic alteration in horn development.

Many features of domestic cattle, though, are clearly the result of artificial selection for specific traits—none more so than milk production. Consider that the udders of wild female aurochs were not visible from a distance, in stark contrast to the grotesquely tumescent milk chambers of a modern Holstein. A lot of artificial selection went into the enlargement of those udders and the associated increase in milk production.

Milk benefited the human diet in several ways. It is a good all-around food source, as any mammal baby can attest. Milk is also rich in calcium, which is especially important for the sun-deprived populations of northern latitudes, where low levels of ultraviolet light result in vitamin D deficiency and hence low calcium levels.[49] The water in milk can also be important during periods of drought.[50] All three benefits, in varying degrees, may have factored into the development of dairying, not only in Europe but in northern India and East Africa, as well as the Middle East.[51]

There is a problem to be overcome, though, before dairy can become an important dietary component: lactose, the most common sugar in milk. Lactose is not a problem when we are young, because we have naturally high levels of the enzyme lactase, which breaks it down. As we mature, though, lactase levels drop in most human populations and lactose intolerance develops, symptoms of which include gastric distress. So, in order for dairying to develop beyond a rudimentary stage, a biological alteration in dairying human populations was required in the form of lactase persistence into adulthood. The putative mutation conferring lactase persistence was first identified in northern Europeans.[52] The frequency of this allele declines toward southern Europe, where cow milk is a much less important part of the diet.

It is noteworthy that the same mutation conferring lactase persistence to northern Europeans has also been identified in dairying populations of northern India.[53] It is unlikely that this mutation originated independently; there was probably a common ancestral population, but where it was located is not clear. It may have been in northwestern Turkey, where dairying first developed, but not necessarily.[54] The first dairy consumers were probably lactose intolerant.[55] Wherever the mutation conferring lactase persistence first became common, a few migrants must have made their way southeastward, while others headed northwest.[56] In India as in Europe, there is a decline in this mutation toward the south, where dairying is much less important.[57]

This mutation is not present in pastoralist dairying peoples of East Africa, though.[58] Nor is it present among pastoral Bedouin Arabs of Saudi Arabia and the Sinai, who also rely on dairy products for many of their calories. Instead, they evidence several other mutations that confer lactase persistence. African pastoralists are particularly instructive, as they often live adjacent to tribes that consume little or no dairy and that lack the mutations.[59] Similarly, urban Arab populations, such as Palestinians, have low levels of the mutation that is common among pastoral Bedouins living nearby.[60] Lactase persistence in Europe and Africa is a case of convergent evolution in humans caused by similar environmental conditions—in this case, independently developed cultural practices.

Of course, human cultural practices and their diffusion through human migrations were a crucial factor in the development of modern cattle breeds as well (see Appendix 7 on page 330).

CATTLE GENOMICS

Much has been learned, genetically, about the history of cattle domestication, using remarkably small DNA sequences, including microsatellites, mitochondrial DNA, and Y-chromosome DNA. But now we have entered the new era of genomics. The entire cattle genome was first sequenced in 2009; it came from a Hereford, a taurine breed.[61] Several more breeds have been sequenced since.[62] As in dogs and pigs, the cattle genomes were soon scanned for mutations that might have functional significance.[63] As always, the initial focus was point mutations (so-called single nucleotide polymorphisms, or SNPs), by means of which the footprints of natural and artificial selection can be detected in both coding and noncoding DNA.[64] Among the traits evidencing strong selection are milk production[65] and growth rate,[66] which is unsurprising, but also immune response,[67] which is less expected.

Other sorts of mutations also figure prominently in functional traits and breed differences. Copy number variation (CNV) is particularly well studied in cattle.[68] One notable study looked for CNV differences between the Holstein (a milking breed) and the Angus (a meat breed).[69] The Holstein was particularly enriched for CNVs related to lactation, suggesting that the positive results of intense selection for increased milk production are due in part to DNA amplification, not just point mutations. In fact, there are reasons to believe that the genome can respond more rapidly to strong selection of any kind through copy number variation than through point mutations.[70]

FROM AUROCH TO HOLSTEIN AND BACK?

Within 10,000 years—an evolutionary eyeblink—humans have made of wild aurochs more than 700 distinct cattle breeds.[71] The wild aurochs themselves were a casualty of this process. Now, many of these breeds are about to become casualties of further developments in the domestication process—primarily the ever-increasing mechanization and globalization of agriculture, which demand certain forms of efficiency, for which breeds developed under more localized and unmechanized conditions are poorly suited.

When I was growing up in the Central Valley of California in the 1960s and early 1970s, every dairy herd was composed of at least three breeds: Holstein, Jersey, and Guernsey. Brown Swiss and Ayrshires were not uncommon. Now there are only the high-producing Holsteins; all other breeds seem to have vanished. In fact, as I was shocked to learn while researching this book, Guernseys are now considered a heritage breed, which means they are in some danger of extinction. Their distinctive golden milk, so rich in beta-carotene, is almost impossible to find now.[72] You will also have to look long and hard to find the high-butterfat Jersey milk.

This loss of genetic diversity extends to individual breeds, most dramatically in the Holstein. Through artificial insemination, a few bulls are responsible for the vast majority of Holstein pregnancies. Two bulls, a father and son, are responsible for 7 percent of the genomes of the entire American population of Holsteins; genetic diversity be damned.[73]

Things are no different on the beef side of the equation. In the United States today, over 85 percent of beef cattle are Angus, Hereford, or Simmental, or some combination thereof. While the United States, as always, seems to be in the vanguard of diversity loss, similar trends can be observed throughout the world, from England to Brazil. As with pigs, the best hope for some heritage beef breeds—such as the Belted Galloway, Dexter, White Park, and Lincoln Red—lies in their appeal to the discerning palates and novelty-seeking propensities of the foodies. The Slow Food movement also seems to be benefiting both beef and dairy breeds.

Some endangered breeds that were used primarily as draft animals have also benefited from the culinary community, including two quite auroch-like breeds: Maremmana and Pajuna. Another draft breed, the Sayaguesa (a.k.a. Zamarona), though not known for its flavor, may also have a future because of its resemblance to the extinct auroch. For scientists are now attempting to genetically resurrect these creatures.

In the early decades of the twentieth century, some Europeans experienced remorse over the extinction of wild aurochs. Remedies were sought. In Germany, Hermann Goering sponsored a Nazi effort to conjure aurochs, which was executed by the brothers Heck: Heinz and Lutz. Heinz, in Munich, crossed Hungarian Grey cattle and other primitive-looking breeds from the Podolian steppe region, with Scottish Highland and German Friesian cattle, among others. Lutz, in Berlin, took a completely different tack, focusing on southern European breeds, such as Camargue cattle and Spanish Fighting Bull cattle. Both claimed success in generating neo-aurochs within a remarkably few generations. Both were wrong.

These neo-aurochs were a heck of a lot smaller than true aurochs, and they retained the truncated domestic body shape. Their horns, while longer than those of the average cattle breed, were generally not auroch shaped. Some of the neo-aurochs did have auroch coloration, including the eel stripe, but of the herd I recently viewed at the Edinburgh Zoo, I could find only one such bull; most looked like Scottish Highland cattle complete with shaggy long hair. In retrospect, these failed experiments were based on superficial resemblances and the shoddy genetics typical of the Nazis.

Recently, a more sophisticated attempt at auroch resurrection was initiated in the Netherlands, which holds more promise. Called the TaurOs Programme, it combines state-of-the-art genomics with zooarchaeology, history, and ecology. This team has decided to use as founding breeds primarily southern European auroch-like cattle, including the Sayaguesa, Pajuna, Maronesa, Tudanca, and Spanish Fighting Bull. The goal is to eventually repopulate what remains of wild Europe with increasingly closer approximations of aurochs. It is much too early to judge the results; indeed, if things are done right, the process should take hundreds of years, with most of the mate selection being made by the cattle—not humans— before anything resembling the sublime wild aurochs emerge. And it will be a brave matador indeed who dares engage one in a bullfight.

CHAPTER 8

SHEEP AND GOATS

THE DEEP RESEMBLANCE OF SHEEP AND GOATS reflects their proximity to each other on the tree of life—two adjacent twigs, whose leaves can rub against each other with the slightest breeze. Because of their evolutionary history, it is possible to maintain sheep and goats in mixed herds, which proved a particularly efficient way to utilize the land for much of human history. Yet in Western culture, we have chosen to emphasize their differences. Usually this works to the detriment of goats, which have come to have all sorts of negative connotations, including downright evil. Satan is often depicted as a human-goat chimera.

Perhaps it is through their satanic associations that goats are also considered paragons of hypersexuality. In fact, though, male goats (bucks) are no hornier in this metaphoric sense than male sheep (rams). Both are highly promiscuous, mounting any remotely receptive female, yet rams get a pass. Consider the quaint British expression "randy as a goat." In British English, "randy" means lustful or lascivious; as such, "randy as a ram" would be more apt, and alliterative as well; yet it is goats—both male and female—that are pilloried for their sexual behavior. It seems goats are being scapegoated.

The term "scapegoat" derives from Jewish mythology. Each year on the Day of Atonement (Yom Kippur), the priest would select one ox and two goats for sacrifice.[1] The ox was sacrificed for the sins of the high priest (sons of Aaron); one of the goats was sacrificed for the sins of the community. The second goat was the scapegoat and experienced a far worse fate—from the Jewish perspective—than that of the slaughtered goat. It was exiled, sent into the wilderness, also to expiate the sins of the community. Lest it return—as it inevitably would—and thereby reinstall the community's sins, someone was usually sent out to discreetly push the scapegoat off a cliff or otherwise ensure its demise.[2]

But the Jews had no particular animus for goats. In fact, their use as sacrificial animals indicates quite the opposite. It is in the Christian tradition that the reputation of goats took a plunge, perhaps by way of distinguishing Christians from other Jews, as in a famous passage from the book of Matthew in which Christ proclaims that on Judgment Day the sheep will be separated from the goats[3]—the sheep to join the heavenly flock and sit on Jesus's right hand, the left-leaning goats to become the eternal property of Satan, the evil goatherd. (It is noteworthy that the high incidence of male homosexuality among sheep—more than 8 percent[4]—does not count against them on Judgment Day, and that goats do not benefit from their sexual conventionality).

From even a cursory reading of the New Testament, it is clear that all good Christians aspire to be like sheep (except sexually, presumably) and the Lord—who was himself the lamb, not the kid, of God—their shepherd. There is some logic to this sheep-goat dichotomy that any pastoralist would recognize. Sheep are far more docile and tractable than goats. A single well-trained sheepdog (header or heeler) can keep hundreds of sheep in line. For goats you need more dogs. Goats are social but not natural followers. They are much more individualistic than sheep.

Perhaps it is because I am left-handed, but my sympathies lie with the goats. There are more objective reasons to admire goats as well: Goats are far more curious and playful than sheep. Moreover, while sheep are not nearly as stupid as they seem,[5] goats have much more in the way of practical smarts. Goats—but not sheep—monitor each other's gaze, lest they miss something important.[6] And goats are much more adept at locating any weak part of a fence. When the weather deteriorates, goats head for the nearest shelter. Sheep, on the other hand, just clump up, often in areas of greatest snow accumulation, where they sometimes freeze to death in unison.

Goats are not only mentally superior to sheep; they are athletically superior as well. Consider the case of a goat born without front legs—call him "Pan"—that came into the possession of a Dutch veterinarian named Otto Slijper when he was but a few months old.[7] Remarkably, Pan functioned quite well with just two hind legs, holding his own with his four-legged cohorts. He could forage for himself without any special human aid, and Pan remained in good health until an unfortunate accident befell him. The nature of the accident was not divulged by Slijper, which raises suspicions that Pan's death may have been assisted like those of the scapegoats of old. The good doctor certainly took advantage of the opportunity to dissect the resourceful creature, and what he found was remarkable. The bones and musculature of Pan's hip region were highly modified by his bipedal existence such that they resembled those of kangaroos and humans, which are naturally bipedal.

The two-legged goat is a striking example of phenotypic plasticity—in this case, how skeletal muscular and nervous systems developmentally interact to accommodate what would seem an insurmountable perturbation. Here I want to focus on a potential difference between sheep and goats in this particular form of phenotypic plasticity and this particular developmental challenge. Goats may more readily meet this challenge than sheep do, by virtue of previous selection for a related form of athleticism. To understand why, we will need to refer back to the evolutionary histories of sheep and goats that diverged prior to their domestication. First, though, let's look further back in evolutionary time to explain their considerable similarities.

EVOLUTION OF WILD SHEEP AND GOATS

Goats and sheep belong to the same family as cattle (Bovidae), which is in the large ruminant branch of artiodactyls (Figure 8.1). Therefore, they have the same general adaptations to a high-cellulose diet, including a four-chambered stomach. Like all other bovids, sheep and goats also have permanent bony horns. The bovid family is composed of eight subfamilies (Figure 8.2), one of which is Caprinae, to which sheep and goats belong. This subfamily diverged from other bovids during the Miocene, about 15 mya, and subsequently came to occupy the mountainous habitats of the Northern Hemisphere, with a center of distribution in Asia.[8]

The caprines can be further subdivided into four branches, or tribes (Figure 8.3), to one of which—Caprini—both sheep and goats belong. The Caprini tribe diverged from the other caprine tribes about 7.1 mya. It is only when we get to the genus level, the taxonomic level just above species, that we can distinguish sheep from goats. Sheep belong to the

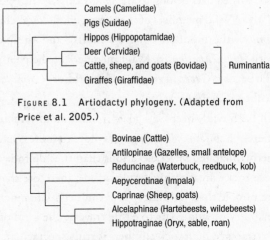

FIGURE 8.1   Artiodactyl phylogeny. (Adapted from Price et al. 2005.)

FIGURE 8.2   The major bovid lineages. (Adapted from Bibi et al. 2009, 3, fig. 1.)

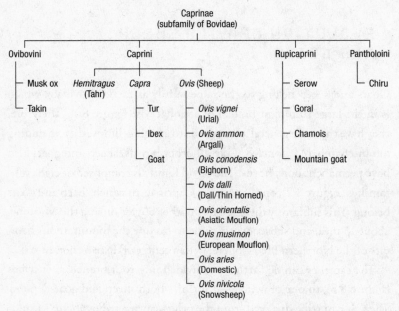

FIGURE 8.3   Major lineages of the subfamily Caprinae. (Adapted from Kopecna et al. 2004.)

genus *Ovis*, while goats belong to the genus *Capra*; the two genera split about 5.7 mya.[9] Wild members of the sheep genus *Ovis* include the argali (*Ovis ammon*), the mouflon (*Ovis orientalis*), and the North American bighorn (*Ovis canadensis*). Wild members of the goat genus *Capra* include the bezoar (*Capra aegragus*), the markhor (*Capra falconeri*), and the alpine ibex (*Capra ibex*).[10]

Though both wild sheep and goats inhabit mountainous areas, they prefer different habitats. Wild sheep gravitate to more open grassy areas and formerly inhabited lower elevations than they do today. Wild goats prefer more rocky areas at higher elevations; therefore they need to be even more nimble than wild sheep and can negotiate much steeper terrain. The ability of wild goats to easily traverse the kind of vertical terrain that attracts human rock climbers is truly spectacular.

Sheep are dyed-in-the-wool grazers; goats graze as well, but they also browse shrubs and trees. To the latter end, they often stand on their hind legs for extended periods in order to reach the more tender leaves. This propensity of wild goats for occasional bipedalism—in contrast to the firmly planted feet of sheep—goes a long way toward explaining why Pan could survive without front legs. He merely perfected the partial capacity for bipedalism found in all goats. This capacity for bipedalism no doubt contributed to the Greek conception of satyrs, the sexually promiscuous companions of the libertine god Dionysius. Greek satyrs may have contributed to the Christian conception of Satan.

Of the candidate wild ancestors of domestic sheep, genetic evidence implicates the mouflon (*Ovis orientalis*), which inhabits mountainous areas from the Caucasus south through southeastern Europe and south-western Asia (Figure 8.4).[11] Today mouflon are largely confined to the Caucasus, northern Iraq, and northwestern Iran. The wild ancestor of the domestic goat, again based on genetic evidence, is the bezoar (*Capra aegragus*), which formerly exhibited a distribution quite similar to that of the mouflon: from the Caucasus throughout much of Southwest Asia (Figure 8.5). Today bezoars occupy more of their former range than mouflon, but both species are rapidly sliding toward extinction.

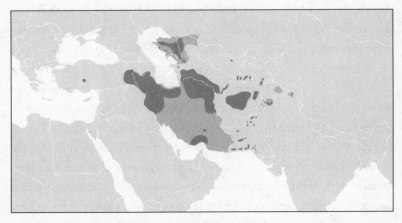

FIGURE 8.4   Current geographic distribution of mouflon. (Courtesy of the IUCN Red List of Threatened Species.)

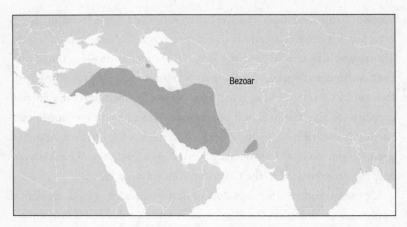

FIGURE 8.5   Current geographic distribution of bezoars. (Courtesy of the
IUCN Red List of Threatened Species.)

DOMESTICATION OF THE MOUFLON

Female mouflon—like all wild sheep and many other bovids—live in
groups consisting of other females and juveniles of both sexes. Adult
males generally live in bachelor herds, interacting with females only
during the breeding season, and then only after establishing their cre-
dentials through highly ritualized but physically demanding contests.
Actually, most contests are decided through visual assessment, especially of an individual's condition and horn size. If posturing fails to
deter, the horns are deployed in violent collisions that can be heard for
miles, echoing through the mountain valleys. The animals run at each
other from far enough away to achieve considerable velocity before
impact; these are the original battering rams. Understandably, the
combatants often seem a bit dazed after such an encounter, but they
quickly shake it off and renew hostilities again and again, until one ram
suddenly gives up and makes a rapid egress. The winners of these contests get the vast majority of intimate female interactions, though the
losers sometimes sneak in a few copulations when conditions permit—

that is, when the dominant ram is distracted and a ewe is compliant. (See Figure 8.6.)

Archaeological evidence points to two distinct regions of Turkey as the sites of initial sheep domestication: the Upper Euphrates region of eastern Turkey; and central Turkey, the general region where cattle were also first domesticated.[12] Genetic evidence, primarily from mitochondrial DNA, indicates three other possible sites where domestication may have independently commenced in western Asia, probably in the Taurus and Zagros mountain regions.[13]

It is clear that sheep were first domesticated as a reliable meat source and that their domestication came by way of the prey route, as in cattle, not the commensal route, as in dogs and cats.[14] Initially, this process entailed an extensive period in which wild populations were managed to varying degrees. As in pigs and cattle, young males were preferentially culled, since only a few males were required for breeding purposes.[15] The original meat sheep closely resembled the mouflon but were of somewhat smaller stature.[16] They were probably humanly dispersed along trade routes to Africa, Pakistan, India, and China, as well as Europe.[17] As was true of cattle, the European migration occurred along two dis-

FIGURE 8.6   Mouflon. (© iStock.com/LeitnerR.)

tinct routes: the Mediterranean route (primarily ship transport) and the Danube corridor through the Balkans northward.[18]

Most of the primitive meat sheep landraces are extinct; the survivors are primarily feral populations, some of which were long mistaken for native wild sheep. These include the "mouflon" of Cyprus and Sardinia, which are in fact feral descendants of primitive meat sheep that escaped human control early in the domestication process.[19] Other feral populations derived from primitive meat sheep include the Orkney, Soay, and Faroes sheep from islands of the North Atlantic, as well as Icelandic and other Nordic landraces. These primitive northern landraces had a longer history of human control than those from the Mediterranean islands, before they became feral; hence, they don't resemble the mouflon as much. Nevertheless, they retain such primitive mouflon traits as horns in both sexes, seasonal molt, dark coloration, and coarse hair.[20]

It wasn't until thousands of years after they were first domesticated for their meat that humans began to systematically develop the fiber potential in sheep. Archaeological evidence for this shift is somewhat slim,[21] but it seems to have occurred first in Southwest Asia about 5,000 years ago.[22] There ensued a second wave of migrations of these wool sheep. One route led to North Africa, another to China by way of Pakistan.[23] Wool sheep appeared in southern Europe about 4000 BP, probably via Phoenician traders along the Mediterranean route. Northern European wool sheep probably arrived later, via the Danubian route.[24] The Vikings eventually dispersed these northern wool sheep throughout Scandinavia, Iceland, and the Faroe Islands.[25] British wool sheep seem to have come primarily through the northern route.[26]

Once the wool—or more accurately, wool/meat—sheep arrived in Europe, they soon displaced the more primitive meat landraces, such as the Soay and Orkney, which survive now only as feral populations in marginal parts of Europe where sheep farming is tenuous at best. The wool/meat sheep differentiated into landraces in Europe as elsewhere in the world, and it was from these wool/meat sheep landraces that most of the hundreds of existing sheep breeds were eventually constructed.[27]

DOMESTICATION OF THE BEZOAR

The bezoar dwells in more formidable mountain habitats than the mouflon. It is roughly the same size as its sheep cousin but longer legged. Bezoar coloration varies among populations but is generally a shade of gray or brown with darker, sometimes black, areas on the muzzle, chest, and legs. As in most wild goats, both sexes have a beard of longer hair under the chin. Aside from the chin hair, the most obvious physical difference between the bezoar and the mouflon is the horns (Figure 8.8). Those of the bezoar are much slimmer and longer, curving gently backward in a scimitar shape. Bezoar horns are also punctuated by ridges at regular intervals and, uniquely among wild goats, have a sharp frontal edge.

The social structure of the bezoar, as well as its social and sexual behavior, closely resembles that of the mouflon. Like their mouflon cousins, male bezoars engage in ritualized contests for dominance and access to females. After side-by-side comparisons in the assessment phase, unin-

FIGURE 8.7   Illustration of a bezoar. (© iStock.com/ilbusca.)

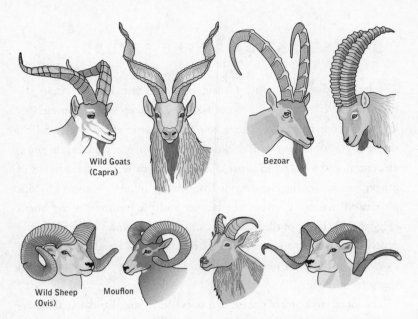

Wild Goats
(Capra)

Bezoar

Wild Sheep
(Ovis)

Mouflon

FIGURE 8.8   Diverse horn shapes of wild goats, including the bezoar. (Adapted from Figure 5 of Article 8.1.11 of *Palaeontologia Electronica* © Society of Vertebrate Paleontology, May 2005.)

timidated males engage in head-to-head combat resembling that of the mouflon. The primary difference is that, instead of running at each other from a distance, bezoars engage their bipedal capacity and simultaneously rear up on their hind legs, butting horns on the way down. Though seemingly less likely to addle the brain, bezoar head butts are still quite forceful; bezoar neck muscles are impressive. As in mouflon, bezoar males live separately from the females and young for most of the year.

The earliest known site of goat domestication, based on archaeological evidence, was in the southern Zagros Mountains of western Iran about 10,000 years ago.[28] Genetic evidence, again based on mitochondrial DNA, points to possible independent origins in other highland areas of western Asia, one in eastern Turkey.[29] The genetic evidence suggests that the vast majority of domestic goats alive today are descendants of wild bezoars that lived in eastern Turkey, not the southern Zagros.[30]

It wasn't until 9500–9000 BP that bezoars began to appear outside their normal range,[31] in more lowland settlements of western Asia, from which they were humanly dispersed in all directions: Central Asia to the north and east, Europe to the north and west, India to the east, and Africa to the south.[32] Like sheep, goats are much more portable than cattle, so especially for early-stage domesticates, their spread was quicker. Perhaps, partly as a consequence, domestic goat dispersal is less well documented archaeologically. Evidence for early dispersal to Europe along the Mediterranean route comes from island goats—such as those from Cyprus, Crete, and Ionia; they so resemble wild bezoars that they were long thought to be part of the native fauna.[33] In fact, though, they are feral descendants of bezoar escapees along the Mediterranean route.

As in sheep, the hair of goats was always an important element of their domestication. In contrast to sheep, though, there was no secondary products revolution in goats for their hair fiber. Instead, goats have always been valued primarily for their meat. It is perhaps for this reason that they have diverged less from their wild ancestors than have sheep (and cows), retaining more of their behavioral and physical attributes.

THE EARLY DOMESTICATION PROCESS

The early stages of sheep and goat domestication were quite similar but are best documented in goats, so I will focus on the latter. The goats of the Zagros Mountains had been hunted by humans for thousands of years—first by Neanderthals, and later by anatomically modern humans.[34] This subsistence hunting was indiscriminate with respect to the goats' age or sex. That began to change about 10,000 years ago, with the harvest becoming increasingly selective of young males.[35] Selective culling alone, without any other form of herd management, represents the onset of the domestication process; this culling amounts to a considerable alteration of the former natural selection regime, since the males are culled without respect to their fitness. We wouldn't expect much anatomical change at this stage, except for diminished sex differences in body size and horn size.

The convergence of males and females with respect to body size and horn size under these conditions is due entirely to reductions in the males. The culling of young males reduces male-male competition in adults, and the result is less intense sexual selection, which in turn results in reductions of body size and horn size in males. A decrease in sexual size dimorphism is often one of the earliest signs of domestication in any species with sheep- and goatlike mating systems, including pigs, cattle, and horses.[36]

Domesticated goats (and sheep) probably remained in this state for hundreds of years. But between 9500 and 9000 BP goats begin to appear outside their normal range, in settlements at lower elevations.[37] This shift had the effect of cutting off genetic communication between wild and domestic bezoars, greatly accelerating the domestication process. For obvious reasons, it was the more tame goats in the native population that were transported, and tameness, in turn, became increasingly advantageous—to the goats being domesticated—because the goats became more dependent on human provisions in this new environment.

These transported animals began to show altered horn shapes, limb shortening, and smaller size, in addition to more decreases in sexual dimorphism.[38] We can infer—but not observe—that coloration began to diverge from the wild type and became more variable, as natural selection for wild-type coloration was diminished in the goats' new human-dominated habitat. Floppy ears and other elements of the domesticated phenotype probably began to crop up in some individuals—partly because of relaxed natural selection for the wild phenotype, and partly because of their developmental links to tame-ness, which was under strong positive selection. Genetic drift would also be a factor in, for example, fixing some color variants in these small populations.

FERAL POPULATIONS FROM PRIMITIVE LANDRACES

During the dispersal of sheep and goats through Eurasia and North Africa, a number of feral populations were left in the wake, some very hard to distinguish from their wild ancestors. This similarity could reflect the fact that the domesticates from which these wild-like pop-ulations descended had not diverged much from their wild ancestors; that is, they became feral very early in the domestication process. Alter-natively, the resemblance could represent a reversion to the wild pheno-type of domesticated sheep and goats under renewed natural and sexual selection. It's probably a combination of both. The longer the history of domestication, though, the less likely the complete return to the wild type, because some alterations wrought by domestication are hard to reverse. It is instructive to compare the feral "mouflon" of Cyprus to Soay sheep in this regard. Both are descendants of domestic meat breeds, but despite their primitive characteristics, no one would mistake a Soay sheep for a wild mouflon.

As it happens, the Soay sheep has been subjected to an excellent long-term study, the results of which have important implications with respect to ecology and evolution in general, and domestication in particular.[39]

Consider the horns. Both male and female mouflon have horns. Though the horns of the female are considerably smaller, they still func-tion in the establishment of dominance for food resources.[40] The horns of the male Cyprus sheep closely approximate those of mouflon; so, too, do those of the females. The situation is quite different on the island of

Hirta (in the Saint Kilda archipelago), where Soay sheep are studied. Male Soay sheep tend to have much larger horns than wool/meat breeds, and the horns have the wild-type curving shape, but they are much smaller than mouflon horns. Moreover, some males have only vestigial horns, called "scarps," and many females are entirely hornless (polled).[41] Why, after these thousands of years with no human interference, have the Soay not fully reverted to wild-type horns?

The explanation provided by Soay researchers is that, though the bigger-horned individuals (both male and female) sire more offspring, there is a genetic complication called heterozygote advantage, which maintains the smaller horn condition.[42] We can refer to heterozygote advantage colloquially as the "Goldilocks Principle." In essence, the rams with large horns (genotype LL) have a lot of reproductive success but a short life; the rams with vestigial horns (ll) don't attract many mates but live a long time. The rams with intermediate-sized horns (Ll) have their porridge just right: a good long life with plenty of sex.

This is fine as far as it goes, but it raises the question as to why the heterozygote advantage exists in this feral population but not wild populations. To understand that, we need to consider the history of the feral populations before they became feral. Soay sheep are descendants of domesticated sheep. As such, they experienced a genetic event, common to all domesticated animals, that is known as the "bottleneck effect," which refers to the fact that any domestic population has but a fraction of the genetic variation found in wild populations. Moreover, this reduced genetic variation is usually a biased sample—that is, not representative of the wild population. Therefore, domesticates begin with all manner of atypical (for wild populations) genetic architecture. This was true for both Soay sheep and Cyprus mouflon.

This atypical genetic architecture born of sampling error is further altered during the domestication process by genetic drift and altered selection, especially sexual selection as already described. Cyprus "mouflon" escaped this domestication process earlier than Soay sheep and hence experienced fewer domestication-related alterations. It is their long history as domesticates that stands in the way of Soay sheep re-evolving wild-type horns.

History, not just current ecological conditions, matters in any evolutionary process, including domestication, because history is highly contingent and therefore not readily reversed. Their history as mammals was important in channeling the evolution of artiodactyls; their history as artiodactyls was important in channeling the evolution of bovids; their history as bovids was important in channeling the evolution of wild sheep and goats; their history as wild sheep and goats was important in channeling the evolution of domestic sheep and goats; and their history as domestic sheep was important in channeling the evolution of feral Soay sheep and Cyprus mouflon. The weight of history is one aspect of the conservative side of evolution.

Feral sheep and goats that derive from extensively domestication-altered landraces or breeds are even less likely to revert to the wild phenotype. This is one reason why there are many more feral goat populations than feral sheep populations: domestic goats have not been as highly modified by domestication as sheep have, because they never underwent the secondary products revolution. But goats are also just generally more adaptable than sheep, as exemplified by Pan. This difference in adaptability of sheep and goats is the legacy of their wild forebears.

You can find feral goats on some of the most desolate and remote islands, including Saint Helena (where Napoléon was exiled), the Galápagos, and the Juan Fernández Islands (home of Robinson Crusoe). Seventeenth- and eighteenth-century sailors planted many of these feral goat populations, in order to provision future trips. (They often left sheep as well, but feral sheep survived only on relatively lush islands with plenty of grass.) These feral goats represent a huge environmental problem; they are responsible for the extinction of many island birds and mammals, and their extermination is now a high conservation priority. But it takes a concerted effort, involving helicopters, marksmen, dogs, and poison to eliminate goats from even relatively small islands.[43] The fact that the eradication of feral goats from small areas requires so much effort provides us some appreciation for how much human pressure their truly wild counterparts are under.

FROM LANDRACES TO BREEDS

As in all domesticated mammals, from dogs to cattle, domestic sheep and goats differentiated into locally adapted landraces as they dispersed worldwide. Goats largely remain at this landrace stage today. There are few true goat breeds subjected to intensive artificial selection.[44] Most of those breeds, such as the Saanen, Alpine, and Nigerian Dwarf, were selected for their milk products; some, such as the French Rove, are quite recent creations. A few breeds have been selected for fiber, notably the Turkish Angora (originally from Tibet) and the cashmere breeds of China. The so-called Boer goat, now widely distributed, was originally a landrace selected by the Khoisan peoples of South Africa for their meat.[45] Another landrace, the Baladi, was selected for both milk and meat in the Middle East.

Most other landraces, even those that are referred to as breeds, are multipurpose (milk, fiber, meat, with an emphasis on the latter) and not under systematic selection. They often forage for themselves and are free to choose their mates. These "village goats" comprise the vast majority of the world's rapidly expanding goat population. You will see them wherever you travel in the developing world. They are almost as ubiquitous as village dogs.

Sheep have been domesticated well beyond the landrace state; most living sheep can be classified into distinct breeds or combinations thereof. Though originally domesticated for meat, domestic sheep have long been an important source of fiber and, to a lesser degree, milk. After the secondary products revolution, most domestic sheep were at least dual-purpose meat and fiber providers, and an occasional milk source. Some specialization occurred before true breed development, in the landrace state; and though the trend continues, most sheep breeds remained dual-purpose until quite recently, and many still are. As such, the genealogical relationships among breeds do not correspond to function.[46]

The family tree of sheep breeds has a mouflon trunk and then a single branch of primitive meat landraces, such as the Soay sheep. Though stout, this branch is quite short, as if struck by lightning. But before that happened, it gave rise to another longer branch from which the vast majority of current sheep breeds have leafed. The branching structure that supports these leaves is difficult to resolve, partly because this part of the tree grew so rapidly but also because sheep, like goats, are easily transported over large distances, and hence both landrace and breed development are extremely sensitive to the vagaries of contingent human history.[47]

The Merino breeds are a case in point. Originally a Spanish landrace, Merinos were bred for their fine fleece from as far back as the twelfth century.[48] Their export was banned by royal decree until the eighteenth century, but they began to trickle out in the form of royal gifts from an increasingly globalized royal family, eventually becoming widespread in Europe. The trickle became a flood, and ultimately, Merino breeds were dispersed to North America (e.g., Rambouillet), New Zealand, and Australia.

Because of their portability as well as hybridization, sheep breeds show poor geographic-genealogical associations—much weaker than those of cattle, goats, and especially pigs. There is a detectable phylogeographic signal, however. For example, preliminary studies indicate a fairly deep bifurcation separating European sheep from both Asian and African sheep,[49] and then a bifurcation of Asian from African sheep.[50] Among European sheep there is a genealogical axis extending from southeastern Europe to northwestern Europe.[51] As expected, breeds from southeastern Europe show a closer genealogical affinity to Middle Eastern breeds.[52] The southeastern breeds also evidence greater genetic diversity, which is to be expected, given their proximity to the region where domestication commenced.[53] At a finer scale, alpine breeds tend to form a distinct cluster, as do Iberian breeds.[54]

The Jacob breed of England represents perhaps the most mysterious breakdown in genealogy-geography associations. Jacob sheep are more closely related to West Asian breeds than to other English, or even European, breeds.[55] But this is only one of their peculiarities. They are also piebald, which is uncommon in sheep, and multihorned (polycerate) as well. Their somewhat tendril-like horns often grow in seemingly random directions, presenting what is to me a much more satanic countenance than any goat has. I find it ironic, therefore, that this breed's name derives from the biblical figure Jacob, who struck a deal with his father-in-law, Laban, according to which he received every "spotted" sheep born in the flock. In yet another irony—this time with respect to biblical literalism— by creating this purebred flock of spotted sheep, Jacob represents the first historical example of applied Darwinism. (See Figure 8.9.)

There is some genealogical clustering along phenotypic lines—that is, among closely related breeds that share a particular trait. Perhaps

FIGURE 8.9   Jacob sheep. (Photo by author.)

the most notable of these breed clusters are "Fat-Tail" breeds, so called because they store fat in their rumps.[56] ("Fat-assed" would be more apt.) These breeds, including the Karakul of Central Asia, the Awassi of West Asia, and the Afrikaner of South Africa, are particularly hardy in arid environments, where they are prized for their meat.

Making up another phenotypically distinct genealogical group are the Zackel sheep, including the distinctively spiral-horned Racka breed of Hungary. Zackel sheep are quite goatlike, with respect to not only their horns but also their build. Members of this breed group are now distributed widely, but they are still most abundant near their center of origin in southeastern Europe.[57] (See Figure 8.10.)

The existence of Merino, Fat-Tail, and Zackel genealogical clusters notwithstanding, current phenotypic classifications of sheep breeds do not at all correlate with genealogy. Sheep are often classified into six or seven function-defined breed clusters—for example, meat breeds (such

FIGURE 8.10
Racka sheep.

as Cheviot, Dorset, Dorper, Suffolk, Texel, and Southdown); long-wool breeds (Coopworth, Cotswold, Scottish Blackface, and Romney); medium-wool/dual-purpose breeds (such as Corriedale, East Friesian, and Finnsheep); double-coated breeds (Navajo-Churro and Romanov); and hair breeds (Katahdin, Barbados Blackbelly, Wiltshire Horn, and St. Croix). But this classification bears little relation to the genealogy of these breeds.

SHEEP AND GOAT GENOMICS

There has been much recent progress in sheep genomics, and considerable progress in goat genomics as well. Mutations in both coding and noncoding DNA regions have been identified in sheep that are related to coat color, body size and shape, reproductive traits, and growth rates.[58] The strongest footprint of selection is a gene that promotes the hornless, or polled, condition.[59]

The goat genome project has also yielded some extremely useful information. Of the 44 genes that are most rapidly evolving under directional selection, seven have immune function and three are related to pituitary hormones.[60] Given Belyaev's thesis, the latter, in particular, warrant further investigation.

In addition to evolution by point mutations (single nucleotide polymorphisms, or SNPs), both sheep and goats also evidence evolution by copy number variation (CNV). As in cattle, some CNVs are associated with functionally important traits such as coloration and size.[61] Sheep, for example, have one to several copies of the growth hormone gene, and there is an association between the copy number of this gene and growth rate.[62] It remains to be established how artificial selection or genetic drift has influenced CNV in this gene.

In sheep and goats, CNVs also seems to play a role in the white coloration so characteristic of the domestic phenotype.[63] Formerly, the white coloration was assumed to be the result of a dominant point mutation at the *Agouti* locus.[64] Recently, though, it was discovered that some all-white goat breeds, such as the Saanen, also have CNV at this locus, which would explain why this mutation doesn't behave like a simple Mendelian trait.[65]

Transposable elements (TEs) are another important genomic component (see Appendix 5B on page 327). These so-called jumping genes have the capacity not only to move about the genome, but also to increase in number (repeats) by exploiting the genome's machinery for repair and duplication. The degree of expansion of particular TE repeats illuminates some interesting features of goat evolution, both before and after domestication. The expansion of one type of TE is shared with cows, sheep, and all other ruminants.[66] But another type of TE has expanded in a goat-specific way.[67] The degree of expansion of this TE may prove very useful in reconstructing goat domestication, as well as the genealogies of goat landraces and breeds.

Another kind of genomic element, which I alluded to in the pig chapter, is the result of past viral infections, specifically retroviral infections. Retroviruses, which include HIV, among numerous other disease agents, have the capacity to incorporate themselves into the genomes of host cells through a process called "reverse transcription."[68] If they incorporate themselves into the sperm or egg cells of the host, they are transmitted to future generations and eventually can become a species-wide genomic component called endogenous retroviruses (ERVs). This can be particularly problematic, in an evolutionary sense, when ERVs incorporate in or near regulatory elements, and thereby mess up gene regulation. Over evolutionary time, successful genomes come to surmount the ERV problem—and in many cases turn them into new regulatory elements—through a process often called "genomic domestication."

ERVs have become an important tool for constructing genealogical trees, because each ERV in the genome has a unique relative date of incorporation. Therefore, specific ERVs can be used as markers to distinguish lineages. Analysis of ERVs is what enabled investigators to distinguish the lineage of primitive meat breeds of sheep, such as the Soay, from the lineage that includes all of the breeds and landraces that evolved after the secondary products revolution.[69] Eventually, ERV markers will prove useful in constructing more fine-grained genealogies of existing breeds.

THE DIVERGENT FATES OF SIMILAR SPECIES

Sheep and goats are closely related; are similar in size, morphology, and behavior; and were originally domesticated in the same geographic region in much the same way. Yet, for reasons that have more to do with human cultural practices and contingent human history, their evolutionary trajectories have diverged considerably under domestication. In fact, it is fair to say that domestic sheep and goats differ more from each other than did their wild ancestors, at least on the surface.

The primary difference between sheep and goats (both wild and domestic) is that the latter have much greater phenotypic plasticity; goats are therefore more adaptable than sheep and can potentially thrive in a wider range of habitats, as is apparent in feral populations. Yet sheep vastly outnumber goats in the world today, and this is especially true in the parts of the world most dominated by western European cultural traditions. In other words, goats are an underutilized resource—perhaps the most underutilized of all the barnyard mammals.

Goat milk, especially, is an underexploited resource. Goat milk is much more like human milk than is cow milk.[70] Hence, goat milk is more nutritious for humans than is cow milk. As a bonus, goat milk is much lower in lactose than cow milk.[71] Many of the world's population who are lactose intolerant tolerate goat milk and goat milk products, and could therefore enjoy the considerable nutritional benefits of milk. Goat milk could be a huge boon to health in the developing world.

Artificial selection programs for traits related to goat productivity—whether for hair, milk, or meat—lag far behind those for sheep. As I indicated earlier, most goats alive today cannot be assigned to distinct breeds. Modest investments in systematic artificial selection for milk or meat production could yield huge dividends. At the same time, all goat populations, domestic and feral, require careful management, given their adaptability and hardiness. Goat overgrazing presents a substantial ecological cost, up to and including ecosystem collapse, as has occurred on some oceanic islands and large swaths of the regions where goats were first domesticated. Wild goats (bezoars), as well as wild sheep (mouflon), require careful management for a completely different reason: they are about to vanish forever.

CHAPTER 9

REINDEER

CHRISTMAS, AS CELEBRATED IN AMERICA, IS A GUMBO of traditions befitting a country of immigrants. One outspoken segment of the Christian community would like to make of it a more pure celebration of the birth of Jesus, shedding the pagan elements, the unholy symbols and rituals such as Christmas trees, Santa Claus, and gift giving. A common mantra-like refrain is "Jesus is the reason for the season." But this attitude is quite parochial. The "reason for the season" is the sun—more specifically, its having reached an annual nadir in the Northern Hemisphere, after which things start to change each year for the better for humans and other living things. This transition has been celebrated in the higher latitudes since humans were able to accurately track the sun's annual course.

It was not until centuries after the death of Jesus that, by fiat, the Catholic Church fixed the date of Jesus's birth, on which the New Testament is conspicuously silent. The pope's choice of December 25 had some precedent in the exegeses of a few early church luminaries, but it was also a savvy ploy to appeal to the pagan sensibilities of northern Europeans, for whom the first days of the solar calendar had long been designated for bacchanalian revelry. And the Catholic Church, wisely, did not attempt to discourage these earlier forms of celebration as the pagans became Christianized. Rather, the celebration of Jesus's birth was simply accreted to them; hence the odd juxtaposition of crèches and Christmas trees.

But where do reindeer fit into this odd mix? And what about their capacity to fly? In their excellent history of New York, *Gotham*, Edwin Burrows and Mike Wallace claim that Christmas as we (Americans) know it was actually created in the Big Apple.[1] That is, the particular mixture of legends and rituals that enchant American children so—including the ornamented tree, the hung stockings, and the jolly, white-bearded, fat man bearing gifts, the efficient distribution of which is greatly facilitated by reindeer with the gift of flight—was created by a few New Yorkers in the early nineteenth century. Before then, Christmas in America was no big deal, far less important than the celebration of the New Year. In fact, the celebration of Christmas had been banished by the Puritans as a papist plot to paganize.

Washington Irving, ever droll and mischievous, seems to have initiated the change in tide in his *Knickerbocker's History of New York*,[2] a satire on the local aristocracy, particularly the Dutch component. According to Irving, Saint Nicholas was the patron saint of New Amsterdam, the original Dutch settlement on Manhattan Island. Saint Nicholas had a long-standing reputation in both the Catholic and the Orthodox traditions as a secret gift giver, but Irving made of him the jolly old man who slides down chimneys bearing gifts for sleeping children, which he deposits in large hanging stockings. (Presents under Christmas trees were not yet known; Christmas trees only entered the American tradition in the 1830s by way of German Brooklynites.) And it was Irving who supplied the nickname "Sancte Claus," probably a brutal botch of the Dutch contraction of the saint's name—Sinterklaas. But notably absent from Irving's account of Sancte Claus was any mention of reindeer.

It was another New Yorker, Clement Clark Moore, who introduced reindeer into the Christmas casserole in his famous 1822 verse, *The Night before Christmas* (originally, *A Visit from St. Nicholas*), written for his own children.[3] The fact that these reindeer could fly certainly made

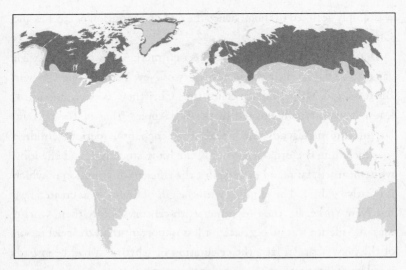

Figure 9.1　Geographic distribution of caribou and reindeer. (© 2015 Defenders of Wildlife.)

Santa's labors more efficient and explained his ready rooftop access, but from what part of the cultural ether did Moore's imagination extract the reindeer, flying or not?

Reindeer (in North America, caribou) were then, as they are now, associated with northern climes. Reindeer have what is known as a circumpolar distribution, from Alaska, throughout much of Canada, and eastward through Greenland, Scandinavia, and Siberian Russia all the way to the Kamchatka Peninsula (Figure 9.1). At the time, only those from Scandinavia and Siberia had ever pulled sleighs. It is doubtful, though, whether anyone who had actually ridden a reindeer-borne sleigh had emigrated to America by 1822—perhaps a Scandinavian or two. The Scandinavian community was the most likely source for the reindeer-sleigh motif.

Explanations as to how Santa's reindeer acquired their power of flight vary. One hypothesis leans on the Norse god Thor, which would really rankle Christian purists. Evidence (though that's probably too strong a word) for the Thor hypothesis is of two kinds. First are the names of two of Santa's reindeer: Donner (Donder) and Blitzen. Donner (which means "thunder") and Blitzen (which means "lightning") are two key attributes of Thor. Second, Thor was notoriously flight prone but not himself a flier. He was conveyed through the sky on his special chariot by goats with the gift of flight. Substitute reindeer for goats, sleigh for chariot, and Santa Claus for Thor, and there you have it. But that's a lot of substitution, and the substitution of reindeer for goats is particularly suspect. Thor was a god of agriculturalists, for whom goats were import-ant, both practically and mythologically. The Thor worshippers were not reindeer people. The reindeer people were pastoralists who lived much farther north, and their animist religions did not include a Zeus-like sky god such as Thor, just a variety of local spirits and the shamans who communicated with them.

Shamans figure prominently in another theory of reindeer flight, but the main actor is the mushroom *Amanita muscaria*, which contains the hallucinogen psilocybin.[4] In broad outline, the idea is that Siberian sha-mans, who use this magic mushroom to commune with the spirit world, experienced visions of reindeer flight while under the influence. Their vivid descriptions of flying reindeer were passed down, orally, through the generations.

Some advocates of the "shroom theory" go so far as to claim that Santa himself was a shaman;[5] they point to his rubicund nose in sup-port, which in their view resembles the living mushroom: red with white spots, which also happen to be the exact colors of Santa's attire. I cannot do justice to the shroom theory here, as it has many elaborations. In my view, however, the shroom theory is likely to be found compelling in direct proportion to shroom consumption.

Finally, there are the Bronze Age megaliths of Mongolia and south-ern Siberia, called deer stones for their characteristic stylized deer depic-tions.[6] The deer on the deer stones are clearly flying, their legs extended horizontally fore and aft. They also have extremely elaborate winglike antlers (some with feathered tines), and some have birds incorporated into the horns. The bird and feather motifs are especially apparent on some mummified humans from this period.[7] Their tattoos, which are excellently preserved, show in much more detail the same reindeer depicted on the weathered stones. It is perhaps significant that the megalith region is thought to be where reindeer were first domesticated. Could this be the ultimate source of Santa's flying reindeer? Some think so.[8] But there is quite a cultural distance, both spatial and temporal, between Bronze Age Mongolia and nineteenth-century New York.

It is safest to conclude that there are no compelling explanations for Santa's flying reindeer. Moore clearly accessed some Germanic (in the broad linguistic sense) tales that were current in New York at the time, but the source of the tales themselves remains a mystery, and given that cultural evolution is adept at obscuring its past, perhaps a mystery it will remain. Fortunately, because of its conservatism, biological evolution leaves better traces, which we can access to at least understand how reindeer ever came to be domesticated to the extent that they could pull sleighs.

PRESLEIGH REINDEER

Reindeer belong to the family Cervidae, which includes most mammals we know as deer, including the red deer and roe deer of Europe; the wapiti (American elk), white-tailed deer, and mule deer of North America; the chital, sambar, muntjac (barking deer), thamin and barasingha of Asia; and the pudu and brocket of South America. The moose (*Alces alces*), called elk in Europe, is also a member of this family, and the largest. The various species of musk deer (genus *Moschus*) are not members of the deer family, however, nor are the so-called mouse deer (genus *Tragulus*).

Cervids, like bovids (cattle, sheep, and goats) are two-toed ungulates (order Artiodactyla) belonging to the suborder Ruminantia, all members of which have four stomachs and chew cud. Cervids and bovids began to diverge in the late Oligocene (20–24 mya) (Figure 9.2). The

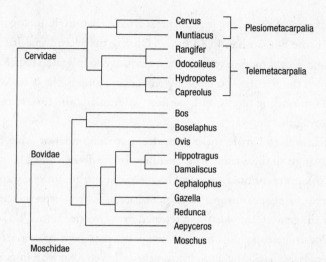

FIGURE 9.2   Artiodactyl phylogeny, showing the bovid-cervid split.
(Adapted from Hassanin and Douzery 2003, 216, fig. 3.)

earliest undisputed deer fossil, *Dicrocerus*, is from about 19 mya; it resembled a muntjac.[9]

What most obviously distinguishes cervids from bovids is the headgear. Recall that bovids, including cattle, sheep, and goats, have horns, which consist of a hollow bony core sheathed with keratin. Cervids, in contrast, have more elaborate branching structures called antlers, which are composed entirely of bony material. Bovid horns are permanent structures; cervid antlers are shed and grown anew each year at a considerable energetic cost. As it happens, reindeer are exceptions to the cervid rule in that both males and females are antlered. Female reindeer antlers are significantly smaller than those of the males, and they are shed later.[10] By the end of December, only the females retain their horns, so Donner and Blitzen, as well as Santa's other reindeer, must be females.

There are two main branches of the deer family: one that evolved largely in the Old World (Plesiometacarpalia) and another that evolved primarily in the New World (Telemetacarpalia). The chief diagnostic character distinguishing New World and Old World deer is the arrangement of two of the bones (called metacarpals) that are homologous to those in the human hand and foot.[11] During the Pleistocene Ice Age, some Old World species, such as the American elk, migrated to the New World via the Bering land bridge. Reindeer are among the New World species—including moose and roe deer—that made the reverse migration during this period.[12]

In addition to female antlers, reindeer have a number of other traits that distinguish them from other members of the deer family. Like other deer, reindeer consume a variety of leafy vegetation, but uniquely among mammals, reindeer consume large quantities of lichens, especially members of the genus *Cladonia*, often referred to as "reindeer moss."[13] Lichen digestion is made possible by the enzyme lichenase, which converts the lichen polysaccharide lichenin into glucose. This enzyme is an important factor in the reindeer's ability to exploit Arctic and sub-Arctic environments. Arctic peoples sometimes take advantage of this unique reindeer digestive adaptation—in perhaps the ultimate manifestation of "nothing is wasted"—by eating partially digested lichens from the stomachs of reindeer killed for meat.[14] Only in this way can humans derive any energy from the otherwise indigestible lichen.

Another notable adaptation of reindeer is their hooves, which change with the seasons. In the summer the hooves are spongy, affording good traction on the soft, wet tundra; in the winter they shrink and harden, particularly the outer rim, which cuts through ice, preventing slippage.[15] Reindeer are also well equipped, furwise, for the Arctic chill. There are two layers: an outer layer of hollow, air-filled hairs—as in polar bears— that function to reduce the conduction of heat from body to atmosphere; and an inner layer of dense hair of a more traditional mammalian type.[16] So well do these two layers insulate that reindeer overheat at temperatures we would consider chilly enough to don jackets.

Cold is not the only characteristic of higher latitudes. There is also a dearth of sunlight for much of the year. Reindeer manage quite well in the dark because they can see in the ultraviolet range of short-wavelength light, which is invisible to us humans and most other mammals.[17] When the sun is low on the horizon, only the shortest wavelengths of light—blue-to-ultraviolet range—reach the surface of the earth, 90 percent of which are reflected by the snow and hence available to the reindeer retina. Urine and feces are particularly conspicuous under UV illumination, which could help reindeer keep tabs on predators as well as each other.[18] The lichens on which reindeer dine also glow in the UV range.[19] It would be interesting to know whether reindeer that dwell in the southern part of the reindeer geographic range are as sensitive to UV as those in the far north.

The geographic range of reindeer fluctuated dramatically during the Pleistocene as glaciers advanced and receded. Reindeer were among the first to occupy, at least seasonally, areas newly vacated by receding glaciers. Then, as now, the colder months were spent in forests (taiga) to the south, from which reindeer seasonally migrated to the tundra to the north.[20] During much of the late Ice Age, reindeer were among the most important meat sources for Neanderthals and, later, humans.[21] At the end of the last ice age (17,000–12,000 BP), Europeans had become so dependent on reindeer that the period is known as the "reindeer age." Reindeer consumption during the reindeer age was greatly abetted by a relatively new technological innovation called the spear thrower, or atlatl, which extended the range over which a spear could be accurately thrown.[22]

The reindeer age corresponds to the Magdalenian cultural period of Europe, during which there was a shift toward an increased use of bones, antlers, and ivory in the manufacture of tools and weapons.[23] These materials were used not only as tools in and of themselves, but also to finely work small stone implements called microliths. Some of the most famous cave paintings were created during this period, including those of Lascaux and Altamira. For all of their importance, reindeer are underrepresented in these paintings, compared to aurochs, horses, and bison, perhaps reflecting a relatively utilitarian view of reindeer.

There are some reindeer paintings in smaller caves of the Dordogne region of southwestern France, but the most famous reindeer representation, and one of the most wonderful pieces of Paleolithic art, is a carved piece of mammoth ivory (from a spear thrower of about 13,000 years ago), now at the British Museum. It depicts a female reindeer—as evidenced by her teats—followed closely by a male, swimming across a river, which is what reindeer often do during a migration. The female is particularly detailed, including her eyelashes, and the composition as a whole is skillfully accommodated to the curve and taper of the tip of the mammoth tusk on which it was carved.

Judging by the male's large antlers, the swimming reindeer are depicted as they would appear during the fall migration. Human hunters would stake out strategic positions where the migrating reindeer tunneled through river valleys, from which they set upon the herd and slaughtered as many members as they could, indiscriminately.[24] Sometimes the reindeer were funneled into killing corrals or into lakes where they were speared from a boat or drowned. Pitfalls were also used, as well as atlatls. These were scenes of great carnage and frenzied activity because the corpses had to be skinned and carved before they spoiled and/or attracted wolves and bears. Perhaps some of the wolves were camp followers, in the process of being domesticated. Before the arrival of humans, wolves were the reindeer's primary predator.

REINDEER DOMESTICATION

As the glaciers steadily retreated northward after the last glacial maximum (around 20,000 BP), the reindeer soon followed. By 12,000 BP reindeer had vanished from France[25] and much of the rest of western Europe. A similar temporal pattern occurred in Siberia and North America. During this northward retreat, some populations remained in relatively southerly latitudes (e.g., Idaho, the Great Lakes, New England, and southern Siberia) in forested environments year-round. These woodland populations tended to be less migratory than tundra reindeer, and they were less gregarious, living in smaller social groups.[26]

Not surprisingly for a species with such an extensive range, a number of subspecies of reindeer are recognized, though their taxonomy is in flux. Recent genetic studies (mitochondrial DNA) indicate that a thorough revision is in order.[27] Moreover, these studies show that tundra and woodland subspecies do not cluster as had been expected. The ancestors of domestic reindeer belonged to the subspecies *Rangifer tarandus tarandus*, which is still common in the Arctic tundra of Scandinavia and Finland (collectively known as Fennoscandia), as well as Siberia.

Reindeer are still in the early stages of domestication,[28] and uniquely among domesticated ungulates, wild and domestic herds often coexist in close proximity. For these reasons, the ongoing process of reindeer domestication can provide us insight into the initial phase of horse, cattle, sheep, and goat domestication, when wild populations and proto-domesticates were genetically quite porous.

Reindeer domestication seems to have commenced when some hunter-gatherers who had particularly specialized as reindeer hunters sought to exert a modicum of control over wild herds. Some of the techniques used in reindeer hunting, such as funneled fences, could also be used for more active herd management, but in addition, new innovations such as holding corrals were required. There is evidence for this type of reindeer corral dating back to at least 3000 BP.[29] But reindeer movements, including their migrations, were still dictated by the reindeer themselves. Therefore, the reindeer specialists had to become more nomadic as well, following the herds.

Gradually, over millennia, the more manageable members of the herd were increasingly separated from the "wilder" ones, but to this day the process is not complete. Throughout reindeer domestication, wilder stock were deliberately introduced into more domesticated herds, perhaps for increased vigor in a challenging environment.[30] Conversely, reindeer of varying degrees of domestication became feral and no doubt interbred with their wild counterparts. This genetic fluidity was probably true of most livestock in the early stages of their domestication, until they were introduced into areas without wild stock and/or the wild populations were exterminated.

Not only were the tamer reindeer a more reliable supply of meat, antlers, and hides, but they could be used as decoys in hunting their wilder counterparts. In fact, most of the reindeer meat of the early reindeer herders probably came from wild reindeer. Domesticated reindeer were too hard-won to kill for that purpose,[31] except under extreme circumstances. The hesitation to kill reindeer remains true of those contemporary reindeer peoples who have best managed to avoid the market economics imposed by modern states.

It is not known at what point reindeer began to be used as beasts of burden, as pack or sled animals. Either activity implies a greater degree of domestication than that required for mere herd management. Males in particular are a threat while antlered, so it was probably at this point that antler trimming became commonplace.[32] Long after sled transport became routine, some reindeer people, notably the Tungusic speakers of eastern Siberia, such as the Evenki, became reindeer riders.[33] To this end they created ingenious stirrupless saddles adapted from the horse saddles with which they had become familiar through contact with Mongol and Turkic horse riders.[34] Special staffs are used, in lieu of stirrups, to vault onto the reindeer and for balance. For the riders' sake, it is especially essential that the rear-projecting tines of the ridden reindeer's antlers be trimmed. Santa doesn't have this problem.

Reindeer milk is very fatty and thick, and reindeer herders have long availed themselves of this resource to varying degrees.[35] Reindeer milking is a much more arduous process than cow milking, and far more time-consuming. Moreover, the yields are paltry. Nevertheless, any reindeer milk consumption is an indication of a fairly high degree of reindeer domestication.

Throughout history, "reindeer people" have been tribal nomads, and they remain so to this day, but they comprise a wide variety of ethnicities and cultural practices. In the Western Hemisphere, human-reindeer interactions were confined to the hunting of wild reindeer, until domestic reindeer were introduced into Alaska a century ago.[36] But in the Eastern Hemisphere, where reindeer were domesticated, reindeer peoples exhibit interesting differences in their particular reindeer-related practices.

The Sami (Finno-Ugric language family, formerly called Lapps) maintain the largest herds, sometimes in the thousands, and use their reindeer primarily for transport (by sled) and beasts of burden.[37] Sami herds are loose—that is, widespread—which makes them more vulnerable to predators than those of many other reindeer peoples. The Sami have traditionally used many parts of the reindeer, including meat, hide, and sinew, but generally not milk.

The Nenets traditionally kept small family herds that were more tightly guarded. But the Nenets (and their reindeer) were collectivized during the Soviet period, and they now maintain large, commonly held herds.[38] Traditionally, the Nenets hunted wild reindeer while maintaining domestic herds. To help manage domestic reindeer, Nenets deployed Samoyed dogs, a breed that they and other Samoyedic peoples created, to herd and perhaps guard against predators.[39] The Nenets use reindeer primarily for meat and transport. Like the Sami, they follow their herds over large distances on the open tundra, seeking fertile pastures.

Unlike the Sami and Nenets, the Evenki are a taiga people, and their reindeer are adapted to warmer temperatures and forested conditions. The Evenki keep small herds (60 or fewer) and eat reindeer meat only when no other food is available.[40] The Evenki are also reindeer riders, having perfected the soft, stirrupless saddles, and they utilize reindeer milk to a greater extent than do the Sami or Nenets.

The Eveny are closely related to the Evenki but live much farther north in the tundra. They, too, are reindeer riders and traditionally maintained small family herds. Unlike the Evenki, the Eveny used dogs to transport goods by sled.[41] The language of both the Eveny and the Evenki is in the Tungusic family, and their reindeer practices are often referred to as Tungusic, which seem to involve the highest degree of domestication. Of all the diverse reindeer peoples, the Tungusic peoples are the most adept reindeer riders.[42]

Do these diverse cultural practices reflect independent reindeer domestication, or the vagaries of cultural diffusion from a single domestication event? One hypothesis is that reindeer domestication occurred only once, in the deer stone region of the taiga of southern Siberia, perhaps in the Altai Mountain region.[43] From here it was thought that the use of domestic reindeer diffused northward, to the tundra peoples of Fennoscandia and northern Siberia. This is known as the "monocentric hypothesis."

Dissenters from the monocentric consensus argued that reindeer domestication occurred independently in several parts of Eurasia.[44] This is known as the "polycentric hypothesis." Recent genetic research supports a modified polycentric hypothesis, with at least two independent reindeer domestications: one in Fennoscandia and one in Russia.[45] In Fennoscandia the Sami were the domesticators, and their reindeer-related practices developed independently from those in Russia.[46] In Russia, though, there are diverse practices and ethnicities among reindeer peoples, from Nenets to Evenki. It is noteworthy, in this regard, that there is evidence—not conclusive at this point—for independent domestications in western and eastern Siberia,[47] which might explain, in part, the differences in the reindeer practices of the Nenets and other Samoyedic tribes, on the one hand, and those of the Evenki, Eveny, and other Tungusic tribes on the other.

DOMESTIC VERSUS WILD REINDEER

Reindeer are at an early stage of domestication, and most domestic populations experience some gene flow from wild populations, mitigating the effect of natural selection for domesticated traits and, of course, artificial selection as well. How representative are reindeer of the early domestication of pigs, cattle, sheep, and goats, all of which, during the early stages of domestication, were in the proximity of wild counterparts? There are reasons to believe that reindeer domestication has a special feature, most notably the harsh environment in which they dwell, that may limit certain domestication-related alterations.

Yet reindeer domestication does share features of the early domestication of all wild ungulates, when they were beginning to experience human control. As we have seen, this period of early management of wild populations may have been protracted. During this phase there is much genetic introgression between wild and semidomesticated populations, until the latter are moved beyond the natural boundaries of the species. Yet, though reindeer in the process of being domesticated continue to remain in close proximity to their wild counterparts, some divergence has occurred. How much this divergence merely reflects phenotypic plasticity—especially related to dietary changes—or genetic differentiation remains to be established for most traits.

It is only recently that reindeer have experienced significant artificial selection. Moreover, artificial selection regimes vary, not only regionally but often within tribes. Traditionally, for example, many Sami populations exerted little control over mating.[48] Nevertheless, domestic reindeer do evidence certain features of the domesticated phenotype to varying degrees, including reduced body size, shorter snouts, and perhaps shorter legs.[49] And reindeer herders do not seem to have a problem recognizing the difference between wild and domesticated individuals. Traditionally, when wild individuals entered domestic herds, it was with the herders' blessing.

Reindeer coloration is probably the trait most affected by the domestication process. Wild-reindeer coloration varies among populations from near black to near white. The underside, rump, and often neck are lighter, while the legs are typically darker. (I have seen only New World reindeer [caribou] in the wild, but the difference in coloration between Alaskan tundra reindeer [quite light coloration] and those of the mountainous forests of Newfoundland and especially the Gaspé Peninsula, Quebec [quite dark coloration] were striking.) But within a particular reindeer population, there is very limited color variation.

In domestic reindeer the coloration is much more variable within herds—fur of many shades and the occasional pinto, a pattern never seen in wild reindeer.[50] Such high variability, as we have seen, is an indicator of relaxed natural selection. Generally, when it comes to mammalian coat coloration, relaxed selection means reduced selection for camouflage, or crypsis. For domestic reindeer, relaxed selection may reflect reduced predation.

One of the advantages of being a domestic reindeer is that human protection makes you less vulnerable to predation.[51] Hence, color mutations that make you more conspicuous are not at as much of a disadvantage in domestic reindeer as in wild reindeer. But the degree of human protection varies widely. The Sami herds are wide-ranging and harder to protect than the tightly monitored Evenki herds, and Sami reindeer experience significant predation. Nonetheless, Sami reindeer exhibit the same high degree of color variation, in contrast to their wild counterparts. So in this case, and perhaps others, there must be other selective forces, the elimination of which would explain the color variation.

One important source of selection on reindeer coloration is insect parasites, including blackflies, mosquitoes, and, in Fennoscandia, especially warble flies.[52] In midsummer, mosquitoes often torment reindeer to the extent that they cannot eat as they seek shelter on a few remaining snowy patches. These nasty buggers have even been known to cause caribou stampedes. But warble flies (genus *Hypoderma*) are worse. They lay their eggs on the reindeer's forelegs. When grooming, the reindeer licks the eggs and ingests them. The larvae then migrate to the skin, burrowing through muscle and other tissue on the way. The swellings on the skin sometimes become infected and pus filled, and when the fly exits the skin it leaves a hole in its wake. Because warble flies damage meat and hide, and stunt growth as well, domestic herds are treated with broad-spectrum antiparasitic drugs, such as ivermectin.[53]

For whatever reason, warble flies preferentially attack light-colored reindeer, which may explain why wild reindeer in Fennoscandia tend to be dark.[54] But treated domestic herds are protected from the warble fly scourge, so lighter-colored reindeer are no longer at a selective disadvantage. Relaxed warble fly selection may explain in part why Sami reindeer have high levels of color variation despite heavy predation.[55]

Relaxed selection, whether from reduced predation or parasitism, is only part of the explanation for the color variation in domestic reindeer. Some color mutants are deemed more attractive than the wild type in certain cultures. Among the Sami, for instance, white fur is valued.[56] Genetic bottleneck effects, common in small domestic populations, can also magnify color variation through chance increases of one or more color mutations.

DOMESTICATION AND SEXUAL SELECTION IN REINDEER

Reduced sexual selection, and hence a reduction in sex differences, is also part and parcel of the domestication process. Antlers and body size are the most obvious sex differences in the deer family. Among deer species, male antler size is correlated with the degree of male-male competition for mates. In species like roe deer and muntjacs, in which even the most successful males mate with relatively few females, the horns are small. In wapiti and red deer, male-male competition is much stiffer; a relative few males monopolize most matings. In these species, for which much more is at stake matingwise, males have larger antlers. I have a picture of my then 18-year-old son standing next to a wapiti (American elk) horn that we found in the Lamar Valley of Yellowstone. My son was over 6 feet tall, but the antler, its base stuck in the ground, was nearly as tall, despite its curve. And it was quite heavy. Male caribou antlers are even larger and heavier. Male caribou antlers are, in fact, the largest relative to body size of any deer species, from which we can infer that male caribou experience intense male-male competition for mates.[57]

It would seem, then, that we should be able to discern a noticeable decrease in the antlers of domesticated reindeer. I looked long and hard for any evidence of such but could find none. There seemed to be an utter lack of research on the subject, so I looked at as many photos as I could find. Of the hundreds of photos I examined, no really obvious differences stuck out, except maybe a few more oddly angled and shaped tines in the domesticates. At this point in their domestic evolutionary history, it appears that sexual selection has not yet been much reduced in reindeer. Perhaps this is to be expected at this early stage of domestication. Reduced sexual selection usually begins with the culling of young males, and it becomes most pronounced when humans take over the mate choice process; the latter happens only when the domestication process is fairly advanced. Reindeer domestication, it seems, has generally not progressed to that point.[58]

There is a broader evolutionary point to be made regarding sexual selection and sexual dimorphism, for which reindeer are particularly instructive. Sexual dimorphism arises when selection drives male and female phenotypes in different directions. Dimorphism can result from sexually divergent standard natural selection or from the special category of natural selection called "sexual selection." Here I will consider only the latter.

To create sexual dimorphism, it is not enough that only one sex benefits from a particular trait such as antlers. The trait that benefits one sex must, in addition, negatively affect the fitness of the other. In other words, what's good for the goose must be bad for the gander, or vice versa. This is called "sexually antagonistic selection." Sexually antagonistic selection is rare when you consider all traits in all organisms; it's a special case. More often, when one sex experiences selection for a trait, the phenotypes of both sexes change because they belong to the same species; hence, any phenotypic changes in one sex will cause the other sex to phenotypically hitchhike along for the ride because of genetic and developmental correlations.[59] Unfortunately, this important point is missed in many popular accounts of evolution, especially, as I will discuss in Chapter 13, human evolution.

Here I will discuss situations in which sexual selection does involve at least some sexually antagonistic selection. The varying degrees of sexual antagonism are evident when we compare bovid (cattle family) horns and cervid (deer family) antlers, both of which function critically in male-male competition for mates (intrasexual selection). Recall that bovid horns are permanent structures that require little in the way of energy to grow. Antlers, however, are shed yearly and must be grown anew each year at considerable energetic cost. Since horns cost less, there is less evolutionary sexual antagonism in bovid headgear. Not coincidentally, many female bovids—cattle, sheep, goats, and antelopes—have horns, albeit smaller ones. There is evidence that female horns in bovids may be adaptive in some cases,[60] but not much consideration has been given to the role of phenotypic hitchhiking in the high incidence of female horns in bovids and low incidence of female antlers in cervids.

There is much more sexual antagonism with respect to cervid antlers because they are so costly. Hence we should expect less phenotypic hitchhiking. The reindeer is the only member of the family in which females are antlered. The presence of antlers on female reindeer is intriguing. Those with an adaptationist bent will look for evidence that antlers are advantageous to female reindeer in some way. For example, it has been suggested that their horns function in establishing female dominance hierarchies.[61] But other social deer have dominance hierarchies and are antlerless; and in some reindeer populations, most females lack antlers.[62] These females seem to be able to sort out dominance relationships without horns. So perhaps it is worth considering whether female antlers in reindeer are a mere by-product, by way of genetic correlation, of the strong selection for male antlers and insufficient selection against female antlers. Antler size varies markedly among reindeer populations worldwide, so the phenotypic hitchhiking hypothesis should be testable. If true, populations in which males have the largest antlers should have females with the largest antlers; and in those populations in which females lack antlers, we would expect the smallest male antlers.

DOMESTICATION AND REINDEER BEHAVIOR

Recall that, according to Belyaev, a behavioral shift toward tameness, not changes in any particular physical trait, is the vanguard of domestication. So even though reindeer are in such an early stage of domestication, we should expect to find some evidence that domestic reindeer are more tame than wild reindeer, and not simply because the former are raised in the presence of and hence habituated to humans. That is, we need evidence of a genetic alteration—not just phenotypic plasticity—from the wild-type condition. Unfortunately, reindeer are not suitable for the requisite controlled experiments to demonstrate such—at least not directly. Fortunately, however, there are indirect ways to assess the impact of domestication on tameness.

Throughout Europe there have been attempts to reestablish reindeer populations in parts of their former range from which they have been extirpated or severely reduced in number. The most expedient way to accomplish a reintroduction is to transplant domestic reindeer. The result is populations with varying degrees of wild and domestic ancestry, which can be determined because the domesticated individuals are genetically distinct from the remaining wild individuals in the populations. It is noteworthy, therefore, that in Norwegian reindeer populations, flight distance—the measure of how close a human can approach a reindeer before it flees—is highly correlated with percentage of domestic ancestry. Individuals in completely wild populations have a much greater flight distance than those with a high percentage of domestic ancestry.[63]

Hence, the now feral reindeer remain tamer than truly wild reindeer, retaining the behavioral signature of domestication. This despite the fact that these herds are now extensively hunted, some for nearly a century. So even at this early stage of domestication, the domestication process, like any evolutionary process, has a lot of inertia and is not easily reversed.

AN UNDERAPPRECIATED ANIMAL

For those of us whose experience of reindeer does not extend much beyond Rudolph, it is difficult to comprehend their central place in the lives of many northern peoples. Many simply could not exist in the harsh environments they inhabit without these spectacularly cold-adapted creatures, wild and domestic. In Fennoscandia and Siberia, reindeer remain to this day the only means of land transport in many areas.

During much of the late Paleolithic, reindeer were a primary food source for Eurasians, but judging by the relative scarcity of their representations in cave paintings, they were not as highly respected as aurochs, horses, and bison. They don't seem to have been deemed sacred. By the time domestication commenced, that attitude had changed, as evidenced by the Bronze Age megaliths depicting flying reindeer—a motif that still figures prominently in the religion of contemporary Siberian tribes such as the Evenki and Eveny.[64] Some believe Santa's flying reindeer ultimately derive from these myths. I don't, but I have been called Scrooge more than once.

Reindeer are latecomers to domestication, the latest of any artiodactyl. Because they are at such an early stage in the domestication process, domestic reindeer provide unique insights into how cattle, sheep, goats, and horses were initially brought under human suzerainty. Domestic reindeer also provide valuable information on the timing of the appearance of some features of the domesticated phenotype. First, they provide additional confirmation of Belyaev's hypothesis that behavioral change, specifically tameness, comes first. In feral reindeer, tameness is also the last feature of the domesticated phenotype to go. The first physical changes in the domesticated phenotype of reindeer are a modest reduction in size and a substantial increase in color variation. The reduction in body size accords with the archaeological record for other domesticated species, from dogs to cattle. The increased color variation is also consistent with what occurred in other species and in the fox experiment. Other features of the domesticated phenotype, such as reduced sexual dimorphism, are at best inchoate, but that should soon change, as reindeer husbandry inevitably becomes industrialized and reindeer breeding decisions increasingly removed from the reindeer themselves. The first artificially inseminated reindeer may be a significant signpost on the road toward the end of a form of human cultural life, one that should be cherished in and of itself.

CHAPTER 10

CAMELS

CAMELS, BOTH DROMEDARY (ONE HUMPED) AND BAC-trian (two humped) are among the more improbable candidates for domestication. Dromedaries (also called Arabian camels) are renowned for their irascibility, which they often express with an impressively tooth-some bite, or, if merely annoyed, a well-aimed stream of foul phlegm. They are also really big, second only to the aurochs among our domesti-cated fauna. The legs of a dromedary are longer than those of any horse, and they support a massive body of up to 1,300 pounds. Bactrian camels are even more massive, though shorter of limb. In retrospect, the domes-tication of these imposing creatures seems a remarkable achievement.

My firsthand experience with camels was, until recently, limited to zoos, where, I noticed, they were usually displayed with wild mammals, not placed in the "petting zoo" area where their South American cousins, the llama and alpaca, typically reside along with sheep, goats, and other domesticated fauna. That seems appropriate, given the camels' physical formidability and indecorous temperament. There is still something wild about domesticated camels. For this reason I never aspired to ride one. Yet one day in April 2011 I found myself seated on a dromedary camel.

As I anticipated, it was not a pleasant experience. A scuba safari with camel transport had seemed a good idea when I signed up for it—a chance to visit more remote, relatively undived parts of the Red Sea from the shores of Egypt's Sinai Peninsula, which were otherwise inaccessible. (Or so I thought, until a four-wheel-drive vehicle pulled up to our campsite on the second day.) Within minutes I regretted my decision. Trepidations first emerged when I boarded the camel. Though she was in the standard kneeling position, I was already farther from the ground than I preferred. And my legs were splayed awkwardly and uncomfortably astride the camel's huge chest. My groin muscles in particular seemed in imminent peril. The saddle—wooden, with a veneer of rough cloth—was not at all comfortable. (See Figure 10.1.)

I have never particularly enjoyed riding horses, mules, or donkeys, but this was shaping up to be far worse. When the beast took to its feet, rear legs first, I pitched forward and almost smacked my face on its neck. I was saved from this indignity by the high front pommel wedged securely but far from comfortably in my crotch. The dive master, a local Bedouin, found it amusing, which was annoying, but he did advise that in the future I keep both hands on the pommel when the camel rose to its feet. And thereafter, I did. The dive master also advised me to relax and enjoy. I did not.

When my camel was fully afoot, I seemed two stories off the ground. This effect was greatly enhanced whenever the camel climbed the numerous escarpments seemingly directly above the Red Sea. The escarpment climbing came as a particularly unpleasant surprise, as I had always assumed that camels were built for flat, sandy surfaces, not hard, slippery rock. It turns out I was basically right about that, judging by the camel's own behavior.

Whenever we came to a rocky escarpment, she and all her sisters would stop, not at all eager for the challenge. But we had camel drivers with sticks accompanying us; they employed the latter to urge the camels up the rocky escarpments, and the camels reluctantly complied. If the camels were reluctant to climb up the rocky escarpments, they were loath to go down them. In fact, I came to surmise, the reason camels were reluctant to go up the rocky escarpments was *because* they were loath to go down them. Figuring that the only reason a camel would be

FIGURE 10.1   Camels with unconventional (scuba) loads near Dahab,
Sinai Peninsula. (Photos by author.)

loath to go down a rocky escarpment would be fear of slipping and fall-
ing, I, too, was loath to go down the rocky escarpments.

At one point I seriously considered a dismount, however ignoble in
the Bedouins' eyes, but that was impractical: camels won't kneel on hard
rock, and there is no safe way to dismount from an un-kneeled camel.
So, when the camel drivers had sufficiently thwacked with their sticks

the flanks of their charges, we (I identified with the camel at this point) descended awkwardly with the predictable slipping and sliding. I gritted my teeth and looked for a place to fall, ideally in the opposite direction of the camel. The plan was for the camel to fall into the ocean far below, and for me to fall toward the mountainside.

The flat, sandy stretches were much better, though still far from comfortable. The dive master had likened the sensation of a camel ride to a rocking chair, but that analogy was not apt. If the camel was a rocking chair, it was a defective rocking chair, one with a lateral motion—what mariners call "roll"—as well as forward and backward motion. Only while researching this book did I discover that the source of this extra dimension of camel motion is its idiosyncratic gait, called pacing (see Appendix 10 on page 335).

The vast majority of ungulates (hoofed mammals) use a gait called the "trot" at speeds between a walk and a gallop. In the trot the legs are moved in diagonal pairs: right rear and left front, then left rear and right front, and so on. Because it takes two such movements to complete the cycle, the trot is called a two-beat gait. The pace is also a two-beat gait, but one in which the legs on one side move in unison: left rear and left front, then right rear and right front. Camels, uniquely, pace at every speed, from a slow walk to what for most ungulates would be a gallop. Pacers inevitably sway, side to side, and the faster the pacing speed, the more perilous the gait is for riders, as I discovered every time my mount got goosed by the camel behind, which was frequently. The pacing gate and consequent swaying motion lend new meaning to the camel's sobriquet, "ship of the desert": motion sickness.

One thought that recurred on my rides was the old saw that camels are horses designed by committee. But that is a parochially Western attitude. Camels are not botched horses; they are, in many respects, superior to horses. They can carry much heavier loads, and they can carry those heavy loads to places that no horse could ever go—some of the most inhospitable places on earth. So adept are camels in this regard that for centuries, Bedouins actually abandoned wheeled transport as an inferior mode of transportation in the sandy deserts where they resided.[1] That changed only after the introduction of four-wheel-drive vehicles during World War II.

With respect to physical appearance, camels may suffer by comparison with horses. Their loose-lipped, mushy-mouthed faces look kind of goofy; and the hump can seem goiterish to the Western eye, especially when it leans to one side. Those knobby "knees" certainly don't help matters aesthetically either, nor does the extensive storkish neck. Overall, there is something Dr. Seussian in the camel physique. But I have come to see something regal in camels. Their general bearing is self-assured—the way they hold their heads, slightly uplifted. And the way they walk when unburdened is quite stately. That pacing gait is, to my eye, more elegant than a trot—when viewed from the ground, that is.

The most amazing attribute of the camel, though, given its size and strength, is—despite the occasional spit or bite—its docility. How could humans ever come to exert sufficient control over their presumably much less tractable wild ancestors to make of them beasts of burden and, more impressive still, a source of milk? But domesticated they were—to the point, in fact, that there are now no truly wild dromedaries.

CAMELID BIOLOGY

The dromedary camel (*Camelus dromedarius*) is one of only a few surviving members of the family Camelidae. The camelid family was formerly much more prosperous, especially in North America, the land of its origin. From the Miocene to the Pleistocene (20–2 mya), camelids were some of the most common and varied herbivores on this continent. The current absence of camelids in North America is evolutionarily a quite recent state of affairs, as is their presence in South America, Asia, and Africa.

The dromedary camel, the one with one hump, is the camel that generally comes to mind when the word "camel" is mentioned. This species inhabits the hot, arid regions of North Africa, Arabia, and western Asia

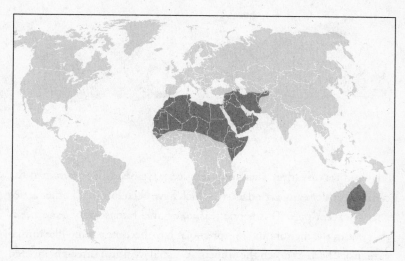

FIGURE 10.2   Current geographic distribution of dromedary camels.

(Figure 10.2). The two-humped Bactrian camel, *Camelus bactrianus*, is found in the cold, arid Gobi desert of Mongolia, northern China, and adjacent regions. These two Old World species—dromedary and Bactrian—are often referred to as "true camels." Their ancestors migrated out of North America less than 3 mya, via the Bering land bridge, to Northeast Asia and subsequently south and west.

The other branch of the camel family, the llama-like camelids, exited North America to the south, via the Panamanian isthmus, where they came to thrive in cold, arid regions of South America, including the Andes and Patagonia. There are two wild species of llama-like camels (the guanaco, *Lama guanicoe*; and the vicuña, *Vicugna vicugna*) and two kinds of domesticated llama-like camels (the alpaca and, the most llama-like of all llama-like camels, the llama). These are all that remain of the camel family.

All camelids, wild or domestic, have the typical two-toed feet and double-pulley ankles of artiodactyls. In many other respects, however, camels are atypical artiodactyls—not as atypical as pigs, perhaps, but atypical nonetheless. Their teeth, for example, differ markedly from those of the ruminants described in earlier chapters: bovids (cattle, sheep, goats,

FIGURE 10.3   Resting postures of cows and camels.

etc.) and cervids (deer, including reindeer). Camelids have retained the canine teeth of early artiodactyls, which have been lost in all other artiodactyls except pigs. This is one reason a camel bite is nasty. Another is that one of the incisors and one premolar have become canine-like miniature tusks. Farther down the alimentary canal we find a three-chambered stomach rather than the four-chambered affair of ruminants. Camelids evolved rumination (cud chewing) independently of ruminants.

Also atypical is the way the leg attaches to the hip in camelids, which is responsible for the distinctive way they lie down by resting on their knees, front and back. Other artiodactyls rest on their front knees but back hips (Figure 10.3). Camelids are unique among ungulates in their lack of hooves. Instead, the two toes terminate in soft leathery pads, where they can splay much wider than hoofed toes can, even those of caribou. This is a secondary condition, evolutionarily. The first camelids actually had hooves.[2] The replacement of hooves by wider pads is alleged to help stabilize the pacing gait.[3] Judging by my experience, I would have to say that the camel footpad is an imperfect adaptation in this regard.

CAMELID HISTORY

Camelids first appeared on the evolutionary scene in the Eocene, about 45 mya.[4] Like cattle (Bovidae), they began to diversify in earnest in the early Miocene, about 20 mya, as open savannahs replaced woodlands in North America. Widely varied camelids continued to thrive throughout the Miocene, including such notables as the huge semi-saber-toothed *Titanotylopus* and the even huger and humped *Megatylopus*, which weighed nearly 2 tons, and largest of all, *Megacamelus*. On the other end of the size spectrum, *Stenomylus* was about as big as a small gazelle. Among the odder camelids was *Floridatragulus*, which had an extremely long snout. There were also several very long-necked species commonly referred to as "giraffe camels."

The two remaining branches of the camelid family emerged in the mid-Miocene (about 11 mya).[5] The ancestor of all llama-like mammals, *Hemiauchenia*, migrated to South America about 3 mya (Pliocene-Pleistocene boundary); the ancestors of all Old World camels migrated out of North America to Asia at roughly the same time, via the Bering land bridge. The split between Bactrian camel and dromedary may have occurred in North America prior to their migration to Asia.[6]

DOMESTICATION OF THE DROMEDARY

The physical traces of dromedary domestication are paltry, which is not surprising, given the inhospitable regions they frequent. There seems a consensus that camels were domesticated on the Arabian Peninsula, but there is disagreement about the dates, ranging from as early as 5000 BP to as late as 3000 BP.[7] From there, they gradually spread to North Africa and West Asia, and eventually to northern India and Pakistan.

It seems likely that camels were originally domesticated for their meat, though it has also been claimed that their original function was as pack animals.[8] Most parts of the camel were utilized; the hides, in particular, were an important secondary product, used for clothing, blankets, and shelter. Camel dung was also an important resource (as fuel), as well as camel milk. But it was when their transport capacities were exploited that domestic camels really came into their own and their usage spread. A dromedary can carry up to 600 pounds over long distances. That's a huge load. It takes a tame camel to accept any load, because it requires at a minimum the camel's grudging cooperation—not least, continuing to kneel during the loading process. Those pioneer transport camels were probably also the least likely to spit and bite. Once loaded, there is also the issue of getting the beast to stand up. To remain kneeling is a very effective form of camel protest.

It was these transport camels that opened up the trading routes—for transporting first copper ore and smelted copper, then incense from the southern Arabian Peninsula to Egypt and the Levant, and later salt and slaves across the Sahara.[9] Perhaps most historically important was the route from Arabia across West Asia to Persia (modern-day Iran), where dromedaries traded loads with their Bactrian cousins, which had arrived from the eastern portion of what came to be known as the Silk Road, a network of trading routes that extended from China to Arabia and, eventually, Rome.

The Silk Road remained unsuitable for wheeled transport throughout its existence (about 150 BCE–1450 CE). It was, rather, a network of trails—camel trails. And camel power was a crucial factor in the unprecedented globalization that resulted from this trade network.

At some point a brave soul decided to mount a camel. Or perhaps it was an expendable slave doing so under duress. Mounted camels eventually served an important military function as cavalry. Imagine the first confrontation between a camel cavalry and a horse cavalry. The all-important psychological advantage was surely on the side of the camel cavalry, and not just for the riders. The horses could not have been eager to confront their much larger foes. The Persian king Cyrus the Great was among the first to exploit this psychological advantage in his battles with Croesus of Lydia (547 BCE). During one campaign, somewhat desperate because he was vastly outnumbered, Cyrus converted camels from the supply train to cavalry, which rode in front of the infantry. As he had anticipated, the Lydian horses panicked and fled.[10]

The success of Arab camel troops against European horse cavalry during the Muslim conquests and the Crusades was not lost on the Europeans. Napoléon deployed camel cavalry in his conquest of Egypt,

and later, French camel cavalry proved invaluable during the "pacifica-
tion" of Algeria. The British also managed to train camel cavalry, which
proved crucial in their North African conquests, notably the Battle of
Omdurman during the Mahdi revolt.

The attempt to establish a camel corps in the United States in the
middle of the nineteenth century was a failure, however.[11] At the onset
of the Civil War, these camels were released to fend for themselves in
a feral state. Somewhat surprisingly, given the amazing adaptations of
dromedaries to arid conditions, the feral camels eventually died out in
the American Southwest. In Australia, however, feral camels continue
to thrive in the outback, to the point, in fact, that they have begun to
overgraze native plants.[12] Today, the wildest dromedary camels reside in
the Australian outback.

It was the Bedouins who domesticated the dromedary, and camels
retain a central role in Bedouin culture, though it is now largely symbolic. In earlier times, though, camels played an essential role in Bedouin ascendancy, beginning with their takeover of the incense trade
from settled merchant communities. The Bedouin period culminated
in the construction of the marvelous desert cities of Petra and Palmyra.
Tribal pastoralists, the Bedouins never established anything like a kingdom or state, however. Instead, they warred with each other, eventually
losing their grip on the incense trade.

Having abandoned the wheel soon after domesticating the dromedary, the Bedouins remained steadfast in their resistance to wheeled
transport until the end of World War II, when four-wheel drive became
widely available. Dromedaries then became increasingly superfluous
throughout much of their former range, including the land of the Bedouins. But even as their utility waned, camels retained their central
symbolic cultural role. Bedouins continue to refer to themselves as "the
people of the camel."

Camel fairs have a long history, but recently they have acquired the
trappings of dog shows, in which judges carefully examine the minu-
tiae of camel conformation. But much more money is at stake at a camel

fair than at any dog show. Even more money is at stake in camel races, which are a recent development, not at all a traditional Bedouin activity[13] (though it now serves as an important symbolic source of Bedouin identity). If horse racing is the sport of kings, camel racing is the sport of sheikhs. Some of these sheikhs are among the most prominent racehorse owners as well. The current ruler of Dubai, Sheikh Mohammed bin Rashid Al Maktoum, heads Godolphin Racing (named after one of the three stallions that founded the Thoroughbred breed), which is based in Newmarket, England. The sheikh is no less involved in camel racing, having established a state-of-the-art breeding facility, the Dubai Camel Reproduction Centre, where in vitro fertilization is routine and camel cloning has been achieved.[14]

Camels can reach speeds of up to 45 miles per hour, which is comparable to that of a horse. But riding a camel at that speed, given its pacing gate, is much more difficult than riding a horse at that speed. Moreover, camels far exceed horses in endurance. A camel can run at speeds of 25 miles per hour for up to an hour—a feat that would leave a horse quite dead. So, camel races are also longer than horse races. Camel races are therefore much more dangerous for the riders than are horse races. Camel jockeys have traditionally been "recruited" from the poor populations of Pakistan, Bangladesh, Sudan, and Mauritania. As in horse racing, smaller jockeys are strongly preferred, which means the recruits tend to be young, very young—children, in fact. In the late 1990s, the not infrequent demise of child jockeys became a huge public relations problem for the United Arab Emirates, which responded by introducing robot jockeys.[15]

FROM WILD DROMEDARY TO DOMESTICATED DROMEDARY

As was true of reindeer, nomadic hunter-gatherers cum pastoralists—not sedentary agriculturalists—domesticated dromedaries.[16] Moreover, like reindeer, camels inhabit harsh environments, in which they must continue to survive without much in the way of human provisioning. To this day, most domestic dromedary camels forage much as they did before domestication. Once unloaded at our campsites, the camels in our little caravan dispersed widely to find whatever scanty vegetation they could, completely unhindered by human control. At one point I could locate only one of the dozen with my 10X binoculars, despite unobstructed views, and it was quite distant. This unchecked wandering made me a tad nervous at first, because however aversive I found riding them, the thought of a return by foot was even less appealing. The camel guy assured me that they would be there when we needed them, but given their self-sufficiency, I couldn't imagine why. And in terms of the bigger picture, it was hard to fathom how such seemingly self-sufficient creatures would ever have come under the human yoke. It had to be somewhat voluntary—camel initiated.

But what could have attracted camels to humans sufficiently to overcome their presumably long-standing motivation to avoid human hunters? For, just as reindeer domesticators were reindeer hunters, camel domesticators were also camel hunters. Perhaps human-controlled water supplies were enticing, or food provisions during lean periods, or perhaps it was salt or some other essential mineral. Whatever the bait, dromedaries eventually became thoroughly hooked, to the point that the wild dromedary is extinct. Any camel you find in Arabia, North Africa, or western Asia, no matter how unfettered and free-ranging, is a domesticated animal.

We can infer that domestic dromedaries probably closely resemble their wild ancestors, physically, because they still live much like their wild ancestors; they are still subjected to much the same natural selection regime. There is good reason to believe that the weight of artificial selection on the camel physique remains light. The exceptions are racing camels, which have recently been subjected to intense artificial selection.[17]

Even the color of domestic dromedaries, which ranges from light buff to dark brown, seems to have remained close to the wild type. White camels, though highly prized in some areas, remain rare.[18] One discovered in Rajasthan, India, was the first in living memory for that area. On the Arabian Peninsula, where white camels are actively bred for, they are more common but still rare.[19] Also in Arabia, black (melanistic) camels are prized but remain relatively rare.[20] In the United States, paint camels—white and brown, like pinto horses—have been produced, but paint camels are virtually absent in Arabia and North Africa.

The wild ancestor of the dromedary was called Thomas's camel (*Camelus thomasi*),[21] which seems to have been significantly larger than its domesticated descendants.[22] The degree of diminution varies among dromedary landraces, in relation to habitat (natural selection) and, more recently, artificial selection.[23] Breed development in Africa and Asia is still in the early stages, but some impressive phenotypic differentiation has emerged in a few areas. In Arabia, for example, camels from the Red Sea coast tend to be considerably smaller than those from the interior. (I am thankful that my camel was not from the interior.) The Asail, a racing breed, possesses a very thin neck and legs and an underdeveloped udder.[24] The latter trait has become problematic in developing the racing breeds.

Early attempts to classify dromedary breeds were based on the ecological conditions in which they lived (flatlands, hills, coast, etc.).[25] Subsequent attempts to classify dromedary proto-breed types have focused on functional (for humans) attributes such as milk, meat, transport, and racing.[26] In one recent classification, four breed groups were recognized: meat, milk, dual-purpose, and racing.[27] Some breeds, such as the Magaheem of Arabia, are in the process of further differentiation into subtypes and, eventually, distinct breeds.[28] Camel breed development offers a window into past breed development in other ungulates.

THE BACTRIAN CAMEL

Camel caravans that set out westward from China along the Silk Road generally did so in winter, braving the subzero temperatures of the high steppes, Gobi desert, and high mountains of inner Asia.[29] No drome-dary could withstand such conditions. For this stretch, and indeed most of the Silk Road, it was the Bactrian camels (*Camelus bactrianus*) that carried those loads.[30]

Both dromedary and Bactrian camels thrive in extremely harsh con-ditions, where few mammals, certainly very few large mammals, could long endure. To this end they evolved some special anatomical and phys-iological attributes. The hump, of course, is iconic in this regard. It is a fat-filled energy reserve on which camels can subsist for weeks without food when conditions are especially bad. Less externally obvious are some special physiological features. Unlike most mammals, camels can vary their body temperature—from 34°C to 41°C—during the course of the day, thereby reducing the heating and cooling energy bill con-siderably.[31] Camels can tolerate much more salt in their food and water than can any other artiodactyl—eight times that of sheep or cattle.[32] Moreover, their blood glucose levels are twice those of other ungulates. This combination of high salt intake and high blood glucose levels would send most mammals, including us, into a state of extreme hypertension and diabetes. Camels, though, seem immune to these disorders.

These are adaptations shared by dromedary and Bactrian, but the latter has, in addition, other adaptations for the cold. For though Bac-trian camels, like dromedaries, must endure extreme heat and drought, they must also endure long periods of subfreezing temperatures, fully exposed to the winds and snow. Hence, their dense, shaggy winter coats are quite beneficial, as is their ability to rapidly shed them in large swaths when conditions warrant. There is no water during the winter in the land of the Bactrians. It is so cold and dry that the snow never melts; it just turns directly to vapor—a process called sublimation. Bactrian cam-els must therefore derive all of their water from eating snow in quantities that would cause hypothermia in noncamels.

The term "Bactrian" is a misnomer. It was bequeathed by Aristotle[33] under the mistaken assumption that two-humped camels originated in the northern region of present-day Afghanistan—bounded by the

Pamir mountains to the north, the Hindu Kush to the south, and the Amu Darya (Oxus) river to the west—that the Greeks called "Bactria." Bactrian camels could certainly be found in Bactria in Aristotle's time (fourth century BCE), but they were not native to that area. Domestication probably commenced in an area farther to the east: southern Mongolia and northern China to eastern Kazakhstan.[34]

DOMESTICATION OF THE BACTRIAN CAMEL

Little in the way of specifics regarding the time and place of Bactrian domestication can be asserted, given the present state of knowledge. But by about 6,000 years ago (4000 BCE), Bactrians began to appear outside of their native range, notably in Turkmenistan.[35] Their westward (and southward) movement continued, gradually; domestic Bactrians reached Afghanistan sometime in the third millennium BCE, and Pakistan by about 2000 BCE.[36] It was only later that they arrived in Iran, which has been claimed by some as their place of origin.[37] By about 1000 BCE, Bactrian camels were present in Assyria; they probably arrived from Iran as tribute to a succession of powerful Assyrian rulers, including Shalmaneser III.

According to Daniel Potts, in both Iran and Assyria Bactrian camels were used primarily to hybridize with dromedary camels, which had preceded them to these areas.[38] Generally, Bactrian males were mated to dromedary females. The result was a one-humped supercamel, larger and more powerful than either parent species. These hybrids were prized for their strength and consequent carrying capacity (over 1,000 pounds). The hybrid camels may also have had intermediate physiological traits that rendered them more successful in environments outside of the normal ranges of both dromedaries and Bactrians. Eventually, the practice of hybridization extended from Syria and Anatolia in the west to Afghanistan.

East of this hybrid zone, virtually all camels are Bactrians, the vast

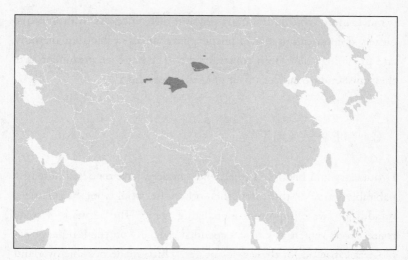

Fɪɢᴜʀᴇ 10.4   Current geographic distribution of wild Bactrian camels.
(Courtesy of Oona Räisänen and IUCN Red List 2010.)

majority of them domesticated. Wild Bactrian camels are among the most endangered large mammals. There are fewer Bactrian camels than giant pandas, less than 1,000, confined to a few of the remotest places on earth[39] (Figure 10.4).

The most striking difference between wild and domestic Bactrians is stature. Wild Bactrians are quite lithe and long legged, much like dromedaries; domesticated Bactrians are shorter of limb and considerably heavier than their wild counterparts.[40] Shortened limbs, as we have seen, are characteristic of domesticates. Generally, there is an overall size reduction as well, but the opposite seems to have occurred during the domestication of Bactrian camels. This may reflect selection for greater load-bearing capacity in Bactrian camels. Something happened to the humps during domestication as well. Those of wild Bactrians are quite conical, wide at the base, tapering toward the top. The humps of domesticated Bactrians are much more cylindrical.[41]

Behaviorally, domestic Bactrians are, of course, much tamer than wild Bactrians, which are extremely wary and human averse, reflecting many centuries of intensive predation by humans. This behavioral difference is much more striking than any physical difference. As in dromedary camels, domestication-related phenotypic changes in Bactrian anatomy and physiology have been constrained by the harsh environment that they must endure much as their wild ancestors did.

CAMELS' LOT

Dromedary and Bactrian camels, like reindeer, have rendered humanly habitable parts of the earth that otherwise would not be. Camels, in addition, were important globalizing agents. The network of trade routes from Yemen to China depended crucially on the unique physiological capacities of these two species, which made possible overland travel across vast deserts, frozen steppes, and forbidding mountains.

Both of the true camel species seem to have been domesticated at about the same time, which is quite recent compared to other livestock. Only reindeer, which according to some are only semidomesticated, are a later addition to the livestock ark. And, as for most of the other species that have preceded them there, only the domesticated dromedary and Bactrian camels will perpetuate the species. The wild dromedary camel is long extinct; the wild Bactrian camel is hovering on the brink of that sad fate. Their domesticated kin, though, continue to thrive. As of 2010, domesticated dromedary camels numbered about 15 million, and there were about 2 million domesticated Bactrian camels.[42] (Compared to fewer than 1,000 wild Bactrians.) In a man-made world it pays to be domesticated; wildness is a luxury that fewer and fewer large mammals will be able to afford.

CHAPTER 11

HORSES

T HE HORSE IS OUR MOST CHARISMATIC DOMESTIC animal, and the most esteemed. Dogs, of course, have fervid advocates, as do cats. But cats, as we have seen, have vocal detractors, and though there are many fewer who can't abide dogs, the vast majority of humankind is indifferent to domesticated canines. Horses evoke, at the very least, admiration from those who interact with them, however casually, but usually it's more than that—often something like veneration.

Though I certainly admire horses for their grace and beauty, I find myself on the low end of the horse appreciation scale, probably because of an aversive experience at a formative stage in my life when I was trying to impress my first girlfriend.[1] Hence I am positioned to view with uncommon detachment and no little skepticism some of the more extravagant claims about the significance of horses, to the effect that they are responsible for civilization as we know it. Such hyperbole aside, there is no doubting the considerable influence of horses on human history.

Without horses, nomads would not have long posed such a threat to urbanized cultures from China to Cambodia and from Austria to Rome, culminating in the destruction wrought by Genghis Khan and his descendants. The initial spread of Islam was as much horse-borne as camel-borne. Moreover, this Islamic conquest and the concomitant spread of the Arabic language eventually came face-to-face with the original, even more massive horse-borne cultural waves from the north, and with them, that family of languages known as Indo-European, which includes virtually all of the European languages, as well as many of those spoken from Iran to India. The geographic spread of this language family, as well as many other cultural elements—such as the wheel—from its center of origin in the central Asian steppes, owes much to horses.[2]

Then there was the European conquest of the New World. Native Americans, who had no experience of horses, were initially awed and intimidated by the beasts. This psychological element was an important factor in the success of the conquistadors against the Aztecs and the Incas, despite being vastly outnumbered. But more important in the long term was the practical advantage that horses provided for warfare. It didn't take long for the American natives to adopt horses into their cultures, with profound consequences. Newly horse-riding tribes moved into the previously uninhabited American steppes, known as the Great Plains, to become the iconic buffalo-hunting, teepee-dwelling nomads of the eighteenth and nineteenth centuries. They also became some of the most celebrated equestrian warriors in history, a relative few of which were able to heroically fend off for years the European onslaught, despite the ravages of smallpox and other exotic European diseases.

The influence of horses on human history can be traced to the unprecedented mobility they afforded. Distances that formerly had taken weeks to travel could be traversed in days, and distances that formerly had taken days to travel could be traversed in hours. Horse capacities set the limits on human land transport for thousands of years, until the advent of train travel in the nineteenth century. And it was only after automobiles became commonplace during the twentieth century that horses were effectively replaced in that role. Today, this horse-powered legacy persists in the layout of many highways, which follow old horse paths; and the term "horsepower" itself has been retained as the measure for the working capacity of automobile engines.

With increased mobility came expanded networks of cultural exchange, transport of goods, and economic interdependencies—in a word, "globalization." As always, with globalization came unprecedented opportunities for a powerful few to control the relatively powerless, often through force—a force that itself was conveyed on horseback. For horses revolutionized the practice of war.

The important role of the horse in transport and warfare would have come as a surprise to the original horse domesticators. In those early days, horses were viewed in much the same way as cattle and other domesticated ungulates. Horses—like pigs, cattle, sheep, and goats—were originally domesticated for their meat.[3] Horse domesticators were horse hunters.

Horses had been hunted since the time of the Neanderthals, at least 60,000 years ago, but horse hunting became more intense once early modern humans replaced the Neanderthals 40,000 years ago.[4] Only reindeer remains exceed those of wild horses—among large animals—at Cro-Magnon encampments in southern Europe. Wild horses were memorialized in cave paintings from Pech Merle (31,000 BP) during the Aurignacian period to Altamira (18,000–14,500 BP) during the Solutrean and Magdalenian periods.[5] (See Figure 11.1.) But horse con-

FIGURE 11.1   Cave painting of a horse at Altamira.
(© iStock.com/siloto.)

sumption declined in western Europe thereafter, as horse populations decreased. The last of the horse hunters dwelled in the Eurasian steppes to the east.

Prior to their domestication, wild horses had long been among the most important sources of protein in the Eurasian steppes.[6] Domestication made this food supply more reliable. But quite early in the domestication process, attitudes changed, and unlike the other barnyard animals, domestic horses fairly rapidly transitioned from a meat source to other, more prestigious functions. Today, most Europeans and their derivatives would be about as aghast at the thought of eating horse meat as they would dog meat.[7] How did horses so quickly escape the larder?

Part of the explanation is timing. Horses are relative latecomers to the domestication process. It was only thousands of years after pigs, cattle, sheep, and goats had entered the barnyard that the first tentative steps were made toward taming the wild horses—also called tarpans[8] (*Equus ferus*)—of the Asian steppes. Thus the meat role had been largely filled in most of the areas that domestic horses found themselves. But another factor, certainly, was the fact that horses proved more useful for transport, both through their attachment to wheeled vehicles and through more direct attachment to human riders. This was the beginning of the horse's rise to a prestige status that far exceeded not only the other barnyard animals, but dogs as well.

The next step was their incorporation into warfare, a transformative event, on a par with the invention of ballistic weapons. From Bronze Age chariot battles to the jousts of medieval knights, even to the much more massive nineteenth-century conflicts such as the Crimean War, the quality of horse loomed large in the outcome. And those without horses—the infantry—were at a decided disadvantage against any horse-borne foe. Inevitably, the cavalry-infantry distinction was also one of status. The higher your status, the more likely you were to find yourself on horseback. The nobility went to battle with horsepower; the peasants, on foot. When the nobility no longer actively participated in battle, their commanders and a relative few select cavalrymen assumed the privilege. And long after cavalry ceased to play any significant role in battle, high-ranking commanders were memorialized in bronze on horseback, for symbolic heft.

Equestrian statues have fallen out of favor, but the value of horses as status symbols remains, however anachronistically. Though fox hunting has become more than a bit unsavory for all but the most unreconstructed British aristocrats, dressage and especially thoroughbred horse racing remain for the celebration of elite status. The very term "thoroughbred" bespeaks the aristocratic ideology that undergirds the "sport of kings." As we shall see, this ideology runs counter to evolutionary wisdom. For Thoroughbred breeders have managed to produce a singular exception in the annals of artificial and natural selection. In stark contrast to, for example, the recent success of cattle breeders selecting for milk production and dog breeders selecting for size, hair quality, skull shape, or whatever they choose, Thoroughbred horse breeders have not been able to produce faster horses for decades. Thoroughbreds seem to have arrived at an evolutionary dead end.

But it has been a glorious ride. The evolutionary saga of horses is one of the best documented of any animal species; and it is a long one. It begins at a point in time when dinosaurs still roamed.

HORSE EVOLUTION

Horses represent one of the few remaining members not only of their family (Equidae), but of an entire order of mammals, called Perissodactyla (which means "uneven [number of] toes"), that were once among the dominant land animals. The first fossil evidence for perissodactyls is from about 55 mya (early Eocene),[9] but by this time there were already several distinct perissodactyl families (including that of the horses and asses), indicating that the first perissodactyls must have evolved millions of years earlier. Molecular evidence indicates that the first members of this mammalian order appeared prior to the mass extinction of dino-

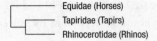

FIGURE 11.2   Perissodactyl phylogeny, showing extant families.

saurs 65 mya.[10] But it was only after the dinosaurs vanished that perissodactyls, like many other mammal groups, came into their own and experienced an explosive proliferation of species.

Perissodactyls partially filled the large herbivore void in the wake of the dinosaur extinction. Some groups, such as the titanotheres (brontotheres), chalicotheres, and indricotheres, were among the largest land mammals that ever lived. But early perissodactyls came in all sizes and exploited many habitats; they were the dominant herbivores for millions of years. Things began to change, though, about 25 mya. From then to the present, there has been a steady decline of perissodactyl species to the relative few that survive today. These include members of the horse and ass family (Equidae), four species of tapirs (Tapiridae), and five species of rhinoceroses (Rhinocerotidae), for most of which humans have greased the skids toward oblivion. In fact, with the exception of domestic horses and one zebra species, every surviving member of this once large branch of mammals is about to fall off the tree of life entirely (Figure 11.2).

Perissodactyls and artiodactyls are collectively referred to as "ungulates," which roughly means "hoofed." The two orders of ungulates are distinguished by one primary feature: the axis of symmetry (where most of the weight is carried) through their legs. In artiodactyls, as we have seen, the axis of symmetry runs between the third and fourth toes, while that of the perissodactyls runs through the third toe. That such a seemingly simple and inconsequential difference is a defining feature of these two large groups of herbivores again testifies to the conservative side of evolution. After the primary weight-bearing axis was established, ungulate evolution took two different paths, which diverged through time because of the cumulative nature of evolution—its building on what already exists—to the point that these two lineages came to constitute two distinct orders (groups of families) of mammals.

Perissodactyls lack two key advanced features of the artiodactyls: the double-pulley system for moving the toes, and the complex stomach found in all but the pig family. The horse family (Equidae) is the only group of perissodactyls with the mobility and speed of artiodactyls, which may be one reason they thrived long after other perissodactyl groups withered as grasslands increased. Zebras are about as mobile as wildebeests, but only because of their relatively longer legs. This lengthening of the legs is one theme in the evolution of modern horses.

Zebras, which graze the same grasses as wildebeests, digest these grasses in a completely different way. Wildebeests, like all bovid artiodactyls, have complex stomachs, where most digestion occurs. Zebras, like all perissodactyls, have simple stomachs. Their primary adaptation to a vegetarian diet is in the hindgut, where a particularly dense fauna of protozoans and bacteria breaks down cellulose.[11] This adaptation is not as efficient as the ruminant stomach, though, as is obvious from an examination of the feces. Wildebeest (and cattle) feces are almost liquid when fresh, with only the merest hint of the food source; zebra (or horse or ass) feces consist of a bolus bound together with undigested grass. But by way of compensation, zebras and other equids consume larger volumes of forage aided by efficient chewing and more rapid digestion. In fact, equids can subsist quite nicely on nutritionally poor grasses that would cause cattle and other ruminants to starve.

Horse evolution, much like human evolution, was long viewed as progressive in a more or less linear way, from the fox-sized omnivores that frequented the tropical forests of the Eocene, to the large grazers of the Pleistocene plains.[12] It took a long time for even evolutionary biologists to adopt Darwin's treelike conception of evolution for horse evolution, to view it as an ever-branching process, to view horses and asses as but terminal leaves of one branch of mammals that was formerly much more bushy and better foliated (see Appendix 11A on page 337).

The beginning of the Pleistocene (about 2 mya) was the time of the final flourishing of the true horses. Some estimate that there were as many as 50 species in North America alone; others condense things radically to a few species.[13] However many species existed during the early Pleistocene, horses were a remarkably diverse and abundant part of the vertebrate fauna in North America. By the end of the Pleistocene (12,000 BP), though, things had changed markedly.[14] The last North American horse died about 10,000 years ago, along with a host of other large mammals, such as the mammoths, camels, and ground sloths.[15]

So, for the last 10,000 years the only wild horses—indeed, the only wild equids—have dwelled in the Eastern Hemisphere. This is painfully ironic because, as for camels, the center of horse evolution, the center of equid diversity and abundance, had always been North America (for about 55 million years).[16] In relatively recent times—within the last 5 million years, some species crossed over to the Old World via the Bering land bridge when sea levels were low. One of those migrations eventuated in modern zebras, as well as Asiatic and African wild asses.[17] A later migration of the wild ancestors of the domestic horse, *Equus ferus*, greatly expanded the range of this species to much of the Northern Hemisphere, which bought the species another 10,000 years.[18] For it was only in the Old World portion of their range that horses survived long enough to be domesticated. Horses returned to the New World courtesy of Columbus and subsequent Spanish explorers. Some escaped, and their descendants, called mustangs, are often referred to as "wild horses." Mustangs are not wild horses, however; they are feral horses.

There is disagreement as to what to include in the species *Equus ferus*. Some experts include not only the direct ancestors of domestic horses (tarpans), but also the Mongolian wild horse, often referred to as "Przewalski's horse." The latter is deemed a subspecies, *Equus ferus przewalskii*, and the Eurasian wild horse is classified as *Equus ferus ferus*. Others give the Mongolian wild horse species status: *Equus przewalski*, in which case there are two surviving members of the genus *Equus*.[19] Whatever the case, all domestic horses derive from the Eurasian wild horse (tarpan), not the Mongolian wild horse. The latter, though, provides useful opportunities for comparisons by which to judge the effects of domestication.

HORSE DOMESTICATION

It was probably through hunting horses that humans learned the nuances of horse behavior that was a prerequisite for horse domestication. Horse hunting, though, was not nearly as widespread as, say, cattle hunting or pig hunting at the time, because wild horses preferred habitats that most humans found unsuitable for much of the year: the cold temperate grasslands of Eurasia, called steppes. Moreover, these grasslands, and hence horses, had been in steady decline since the end of the last major glacial advance about 20,000 years ago. Throughout much of the Northern Hemisphere, the steppes were increasingly replaced by dense forest unsuitable for the grass-loving tarpan.

Though some wild horse populations persisted outside of the steppes, they were largely confined to fairly small, often isolated pockets of pasture—such as alpine meadows and marshes—in Europe, Turkey, and the Caucasus.[20] By about 8000 BP, the only remaining large horse populations were in the still extensive Eurasian steppes. These steppe horses had long been hunted by diverse steppe peoples.[21] For some steppe populations, horse meat made up much of the animal protein in their diet.[22]

Around 7800 BP, domestic cattle and sheep were introduced into the steppes from the west (lower Danube valley), and they spread rapidly eastward to the Volga-Ural steppe region.[23] This caused some significant cultural changes for the western steppe peoples, as domestic animals became connected with wealth and power. If the first horse domesticators already had cattle and sheep, why did they feel the need to add horses to the stable? According to David W. Anthony, the reason was that horses were better adapted to these grasslands, and could better withstand the winters without needing food supplementation.[24] Horses were therefore a more reliable supply of winter meat than sheep and cattle were. Who were the first horse eaters to successfully manage wild horses to this end? Opinions vary,[25] but some recent findings point to the Botai of northern Kazakhstan,[26] though there may have been independent domestications elsewhere in the vast steppe region.[27]

The earliest phase of horse domestication was simply some degree of management of wild horses. Such management may have been practiced by a number of steppe populations over thousands of years. The Botai were the first steppe culture to perfect and exploit horseback riding, around 5500 BP.[28] Just as early reindeer peoples used domesticated reindeer to hunt wild ones, the Botai used their horse-riding skills primarily to hunt wild horses.[29] The Botai were beyond the range of domestic sheep and cattle and relied heavily on horse meat. Their domesticated horses were more valuable in procuring horse meat than in being horse meat. The Botai also valued their domestic horses for the milk they provided, especially in the fermented form known as koumiss.[30]

As horseback riding spread westward, it was put to different uses. In the Volga-Ural steppes, horses were used primarily to herd sheep and cattle, the efficiency of which was greatly enhanced thereby.[31] Far from being the preferred mode of transport, much less sport of kings, horseback riding was eschewed by the elites for over 2,000 years. Instead, they preferred the chariot as an expression of their status.[32] Chariot travel was made possible by the invention of the spoked wheel, a marked improvement on the solid wheels that had formerly graced ungainly battle wagons, until about 4000 BP.[33] The lighter and speedier chariots proved especially useful in warfare, not only in the steppes, but also to the south for the urbanizing Near Eastern cultures.[34] Horse-drawn chariots became the foundation for elite chariot corps, which played a dominant role in warfare in the Near East for centuries. Once the horse-drawn chariots had become associated with status and wealth, chariots, with and without horses attached, became increasingly popular burial objects in the graves of high-status individuals throughout the Bronze Age, from Scotland to China.[35]

It wasn't until about 3000 BP that organized horseback warfare, as opposed to spontaneous raiding, came to the fore.[36] A key development in this transition was the invention of short bows that made it possible to shoot accurately from a moving horse.[37] Previous bow technology had produced much larger bows that were better suited for chariot archers. With the advent of cavalry, warfare took a fateful turn with incalculable effects on the course of world history. Horses experienced a huge bump in prestige, and horseback riding was well on its way to becoming the sport of kings.

SOME PECULIARITIES OF HORSE DOMESTICATION

The process of horse domestication has some distinctive features. As for reindeer, wild stock were deliberately introduced into domestic herds until they were extirpated, which happened relatively recently.[38] It seems that from the early days of horse domestication, horsekeepers sought to enhance certain qualities in their domesticates—such as speed, strength, and intelligence—by keeping the wild blood flowing.[39]

The obvious drawback to this practice is reduced tameness. Ancient horse breeders managed to ameliorate this problem somewhat by introducing only wild females into their domestic herds.[40] In many wild progenitors of domestic animals, males are much more aggressive and less tractable than females. For this reason, relatively few wild males are used in constructing domestic herds of cattle, pigs, sheep, and goats. But the exclusion of wild males was extreme during horse domestication—to the point, in fact, that the genetic contribution of wild male horses was confined to perhaps only a few individuals at the outset.[41]

One effect of the more than occasional introduction of wild females into domestic horse herds was that the early domesticates may have diverged relatively little from their wild ancestors. This complicates things for archaeological investigators, but they are an ingenious lot and have managed, through nonskeletal clues—such as burials, dung heaps, and tooth wear—to determine the timing of key events in the domestication process. As in sheep and goats, one of the key indicators of horse domestication is the appearance of horse remains outside of their normal range, or in areas where they were never hunted.[42]

About 5,000 years ago, horse bones began to appear—in association with human settlements—in the lower Danube valley,[43] as well as in regions around the Caucasus Mountains, such as present-day Georgia, Armenia, and Azerbaijan.[44] It is significant that this range expansion occurred at the time of the Botai culture and the advent of horse riding. Soon thereafter, horse bones began to appear throughout much of Europe, Turkey (Anatolia), Iran, India, and Mesopotamia.[45] Domestic horses reached Egypt before 3550 BP (1675 BCE, Second Intermediate Period).[46] They then spread south to Nubia during the New Kingdom, and west throughout much of northern Africa.

Meanwhile, domestic horses were also expanding eastward across the steppes toward northwestern China.[47] By the time domestic horses arrived in China—about 4,000 years ago—they were probably hitched to chariots.[48] During the eastward (and northward) expansion in particular, there was much interbreeding with wild mares.[49]

SIGNATURES OF DOMESTICATION

Like the Mongolian wild horse, the Eurasian wild horse was stocky and relatively short legged compared to its domesticated descendants.[50] They also had larger heads and thicker necks. Tarpan coloration probably varied somewhat, as suggested by cave paintings,[51] general biogeographic considerations, and genetic reconstructions.[52] Throughout much of the steppe portion of their range, Eurasian wild horses probably resembled the Mongolian wild horse: dun or brown with grayish notes, called grulla coloration. A dark eel stripe ran along the backbone to the base of the tail. Wild horses also exhibited some dark striping along the shoulders and darker legs. Of particular note, all wild horses had a dark

FIGURE 11.3   Mongolian wild horse.

brushy mane consisting of short, stiff hairs—not the long flowing manes of domestic horses.[53]

We can infer the social behavior of the ancestors of domestic horses from Mongolian wild horses and feral horses. They lived in small herds consisting of one dominant male, several females, and their offspring. The life of a dominant male was full of incident. In addition to avoiding predators, from wolves to humans, they were constantly under pressure from bachelor males seeking a takeover. Fending off these rivals required considerable time and energy—and physical risk from teeth and hooves in brutal and bloody battles. Sometimes stallions had to fend off challenges from maturing males within the herd. Usually, though, such males were driven out once they became too frisky with the females or insufficiently deferential to the stallion. Stallions also had to keep a constant eye on their females, who were not immune to the wiles of charismatic bachelors seeking a tryst. A successful stallion was diligent and belligerent. No wonder horse domesticators wanted to limit the access of wild stallions to their herds.

Female tarpans were much more sociable and compliant than stallions, and much easier to control. The females that were most predisposed toward tameness, the least aggressive, and the most tolerant of human proximity were, of course, most fit for the human environment, favored by both natural and artificial selection. At some point, selection for tameness resulted in horses that were increasingly sensitive to human behavioral cues, culminating in Clever Hans, the horse that seemed capable of basic arithmetic, but only in the presence of his owner, from whom he picked up subtle—and unconscious—cues.[54] Horse sensitivity to human cues has never reached the level of dogs, but it exceeds that of cats, pigs, cattle, sheep, and even goats.[55]

It is likely that long after the process of domestication commenced, the only way to distinguish domestic from wild horses was behaviorally: the tame ones were the domesticates. By the time horses were first ridden, the palette of their colors had broadened considerably and often mixed in a single individual. The uniform grulla coloration gave way to brown, chestnut, black, white, and various combinations thereof, such as roan, paint, and tobiano.[56] The usual factors were at work here: relaxed natural selection for the wild-type coloration; and genetic drift, which increased the frequency of rare mutations.[57] But artificial selection also figured in the new palette. Even in these relatively early days, horse breeders sought to increase the frequency of formerly rare coat colors and color patterns. In part, this was simply the very general human predilection for novelty. In part, no doubt, aesthetics also played a role. And there were practical pecuniary reasons as well: horses of a different color were probably worth more than horses of a typical color.

As with dogs, this artificial selection must have been intense enough to override natural selection against some of the new color patterns that were linked to deleterious traits such as deafness, night blindness, and colon defects.[58] Such linkages, called negative pleiotropy, reflect the fact that the genes involved are not specifically coat color determinants in a developmental sense but, rather, more general physiological actors that influence the development of a number of traits. The package deal again.[59]

One of the more prominent alterations in domestic horses was the mane. Wild horses—like all wild equids—had a stiff, brushy mane, but the mane hairs lengthened during the domestication process to the soft flowing garnish we see today. It has been speculated that the stiff mane of wild equids protects them to some degree from the deadly neck bites of large predators. If so, the long manes of domestic horses would be the result of relaxed natural selection. It seems likely, though, that human aesthetic preferences figured more prominently in mane elaboration, since domestic asses have largely retained the wild-type mane.[60]

LANDRACES AND BREEDS

Domesticated horses differentiated into landraces as their range expanded, and these landraces were the foundation for breed construction beginning in the early nineteenth century. A number of so-called primitive breeds actually remain landraces to this day, including the Exmoor pony, Icelandic (the pacer), Norwegian Fjord, Welsh Mountain, and Camargue. Their names reflect the geographic regions where they evolved. In the New World the Mustang, Chincoteague Pony, and various Criollo types in Central and South America should also be considered landraces, as is the Australian Brumby. Most of these landraces became feral at varying points in their history and remain feral or semi-feral to this day.

Some feral landraces, such as the Exmoor pony, have the coloration and conformation of their wild ancestors to varying degrees. Two particularly wild-type landraces are the Hucul pony from the Carpathian region (Romania, Poland) and the Sorraia horse of Spain. It has been suggested that the latter may be a direct descendant of the tarpan and never domesticated.[61] Its mane, though, belies that claim. For the Sorraia, like all of the "primitive" landraces, has a long mane, a legacy of its previous domestication.[62] (See Figure 11.4.)

The Eurasian wild horse was probably effectively extinct by the eighteenth century, through a combination of overhunting and interbreed-

FIGURE 11.4   Sorraia horse.

ing with domestic horses. Yet there are claims that it survived until much later. It is commonly stated that the last one died in a Moscow zoo in the late nineteenth century.[63] But this individual had a long mane. The last putative wild horses (aside from the Sorraia) to survive in the wild—the Bialowieża forest of Poland—were captured and dispensed to local farmers in 1806. But though they exhibited the dun coloration and dark stripe of true wild horses, they had probably interbred with local domestic horses by the time of their dispersal; and after their dispersal they were deliberately interbred. (See Figure 11.5.)

The Bialowieża horses caught the eye of the biologist Tadeusz Vetulani, who endeavored to genetically reconstruct the wild horses from them. The result of his effort is the Konik horse, which, though resembling the wild horse in coloration and size, is no wild horse.[64] Again the mane is a giveaway. Later the Heck brothers—whose efforts at auroch resurrection were documented in Chapter 7—applied their dubious genetic methods to wild horse reconstruction, using the Konik among other "primitive breeds."[65] The Heck horse was no more a successful reconstruction of the wild horse than were the Heck cattle a re-creation of the auroch. The

FIGURE 11.5   The "last" tarpan. Note the long mane.

horse is yet another case in which domestication creates its own momentum and is not easily reversed. (See Figures 11.6 and 11.7.)

Aside from coloration, the trait on which breeders have had the most obvious impact is size. Eurasian wild horses probably showed considerable size variation over their wide range, but on average they would fall in the lower end of the height scale of contemporary horses.[66] Initially, domestication seems to have resulted in even smaller sizes, such that, had the early horsemen been the size of an average Dutch male, they would have been dragging their heels.[67] But things have changed dramatically since then. Some modern draft breeds, such as the Belgian, Clydesdale, and Shire, are giants, weighing over a ton and having a height to match. The Percheron, another large draft breed, are said to have descended from medieval war horses and jousting horses.[68] At the other end of the scale are the Shetland pony, Dartmoor pony, and Haflinger, among other breeds that are considerably smaller than wild horses. Recently, true dwarf and miniature horses, the equivalent of toy dogs, have been developed to serve as companion animals or for biomedical research.[69] (See Figure 11.8.)

FIGURE 11.6   Konik horse. (© iStock.com/RuudMorijn.)

FIGURE 11.7   Heck horse.

The sizes of particular breeds in large part reflect the functions for which they were developed. Horses used for draft purposes tend to be among the largest. Midsized breeds—for example, the Lusitano (used in bullfighting), Lipizzaner, Arabian, Quarter Horse, Standardbred, Morgan, and Appaloosa—were developed for riding or propelling small

FIGURE 11.8   Toy horse.

carriages. Smaller (pony) breeds were often deployed as pit horses in mines—a horrible existence. Émile Zola, in his masterpiece *Germinal*, powerfully depicts the plight of one such pit horse that never saw the light of day once he first entered the mine.

In the English-speaking world, breeds are frequently distinguished by temperament, which is often quite erroneously attributed to geographic origin. Draft breeds are called "cold-blooded" because they are docile. The riding horses just mentioned are called "warm-blooded" because they are somewhat more tightly wound. The third category in this scheme is "hot-blooded," which includes breeds such as the Arabian, the Barb (for Barbary Coast of North Africa), and the celebrated Akhal-Teke of Turkmenistan. Hotbloods, which, as the name implies, are high-strung and difficult, tend to come from warmer climates to the south.

It is from these slender and leggy hot-blooded breeds, especially the Arabian, that Thoroughbreds derive many of their celebrated traits. It has been suggested that the reactive temperament of hot-blooded breeds, along with their long legs and small heads, is a typical juvenile trait that in these breeds is retained in adulthood.[70] On this view, Thoroughbreds owe much to the process of paedomorphosis. More generally, it will be worthwhile to investigate the role of heterochrony in all stages of horse domestication and breed differentiation.

THE STRANGE CASE OF THE THOROUGHBRED

For many, the Thoroughbred represents the pinnacle of horse evolution. It certainly reflects the apex in human contributions to horse evolution; whether or not you consider it a pinnacle is a matter of perspective. For the Thoroughbred, perhaps more than any noncanine domestic breed, reflects a problematic, archaic ideology: genetic purity. This ideology is problematic from a biological, and hence evolutionary, perspective because genetic variation is the sine qua non of a healthy population. It is archaic because it reflects a view of inheritance that was abandoned by all but the aristocracy—that of a blending of parental traits. This view of inheritance as a blending process implies that breeding outside of the pure line can only dilute the good traits upon which the breed—or royal family, as the case may be—was built. But in the modern, particulate view of inheritance, outbreeding is generally a very good thing, resulting in unprecedented genetic combinations, some of which could hit the jackpot in the genetic lottery.

Another upshot of the modern particulate view of inheritance is a statistical phenomenon called "regression to the mean," according to which exceptional parents tend to produce offspring that are closer to the average. The reason for this tendency is that a seemingly simple trait, such as speed, is actually a complex of exercise physiological traits, each influenced by many genes, often in a nonadditive way.[71] Therefore, champion racers represent a fortuitous and improbable combination of many genes, each of which is inherited independently. And breeding for complex traits, as opposed to simple traits like coloration, is somewhat like a lottery. You can increase your odds in the lottery by restricting the matings to parents with desirable traits. This approach worked especially well in the early stages of Thoroughbred breeding, but eventually you arrive at a point of diminishing returns. Once this point is reached, the best way to improve the breed is to expand the lottery by introducing "new blood." Thus, random breeding with non-Thoroughbred horses— perhaps Arabians, Barbs, and Akhal-Tekes, to name a few—would be the best way to improve performance in the long run.

There is abundant evidence that the Thoroughbred has reached this point. Racing times have not improved in 50 years, despite improved training methods and all manner of pharmaceutical assistance.[72] And even in this restricted lottery, the regression to the mean continues. The most successful horses, including Triple Crown winners, often prove much less successful studs;[73] an inordinate number, in fact, don't even contribute to the lottery at all, because they exhibit various forms of infertility.[74] They are ruefully referred to as "stud duds."[75] Moreover, up to 20 percent of horse pregnancies fail to result in birth, many pregnancies that do go to term end in stillbirth, and an inordinate number of the live born don't make it through the first year.[76] Stud duds, miscarriages, and stillbirths are a sign of excessive inbreeding in which the genetic lottery has become too constricted to maintain the health of the breed— another reason to open up the breed.

Yet the lack of improvement in race performance reflects more than bad genetics. Recent studies have found a fair amount of variation in genes associated with racing performance.[77] We must also look to the level of the phenotype, and specifically the phenomenon called "phenotypic integration."[78] Organisms are not just collections of parts; they are systems of integrated parts. As such, a change in one part can affect a number of other parts. Thoroughbreds have evolved inordinately large hearts and lungs, increasing their aerobic capacity. But these oversized organs require a huge chest cavity. The huge chest cavity, though, is too big for the stomach and intestines, which can shift around in hazardous ways. Moreover, the bodies of Thoroughbreds are too large for their slender legs and relatively small feet, which have remained like those of a typical Arabian. Hence they are extremely top-heavy, which goes a long way toward explaining the high frequency of leg injuries, often catastrophic.[79] This configurational defect may set the limit on performance, especially as concern for animal welfare increases. We are now at a stage where a complete rethinking of future Thoroughbred evolution is in order.

HORSE GENOMICS

In 2009 the first more or less complete horse genome was obtained from a Thoroughbred, a mare named Twilight.[80] As is the case for the genomes of all domesticated animals, this landmark was hailed for its potential to illuminate human diseases, but the immediate upshot was a race to identify the bits of the "equinome"—more aptly "thoroughbredome"—relevant to racing performance, so as to more efficiently breed winners.[81] Basically, the idea was to do genomic screening. But given the numerous genes potentially involved in horse exercise physiology and their complex interaction,[82] combined with our ignorance of said genes, the excitement generated was a tad premature. The first of potentially thousands of candidate genes was inevitably labeled "the speed gene" nonetheless.[83] A more accurate description is "a gene, one variant of which enhances speed over short distances but is far from beneficial over longer stretches, given the genetic background (allele configurations of other relevant genes) of the majority of Thoroughbreds." This gene encodes myostatin, a significant actor in muscle development.[84]

There are two variant alleles of the myostatin gene in Thoroughbreds: one allele, which I will call the wild type, promotes endurance; the second allele, a mutation, promotes speed over shorter distances. I will call the allele associated with endurance the "e allele" and the mutant allele associated with sprint-like speed the "s allele." Horses that get the e allele from each parent (ee) tend to be more successful in longer races; those that get the s allele from each parent (ss) tend to be better in shorter races; and those that end up with one of each (*es*)—called heterozygous—tend to do best in intermediate-length races.[85] (Similar results have been obtained for a different gene involved in muscle development in human racers).[86]

I call the *e* allele the wild type because it is the allele found in other equid species, and until recently the vast majority of Thoroughbreds were *ee*.[87] This was a winning combination when, as in the nineteenth century, horse races were much longer—more like camel races—often up to 10 miles long. Races were progressively shortened to the point that today the longest distance any American Thoroughbred could ever run (Belmont Stakes in New York) is 1.5 miles; most races are closer to a mile. Obviously, if you were a Thoroughbred in the nineteenth century you wanted to be *ee*. But one of the founding mares for the Thoroughbred breed had the *s* mutation, which remained in the population at low levels until the 1950s, when the stallion Nearctic inherited it from both parents. He mated with a mare with the mutation to produce Northern Dancer (born 1961), who became one of the most influential sires of modern times.[88] There ensued what is known as a "selective sweep" during which the frequency of the *s* allele rapidly increased as a result of the altered racing environment.

Since 2009 a number of other horse breeds have been partially sequenced, providing a foundation for exploring the genetic basis for functional breed differences. Thousands of point mutations (SNPs) have been identified, as well as numerous insertions and deletions of bases (indels). Of particular note are a number of SNPs associated with coloration, and others with gait (trotting versus pacing, for example) and body size.[89]

As in a number of other domesticated animals explored in this book, copy number variations are also evident.[90] CNVs are particularly common in genes related to sensory perception and metabolism. It has been speculated that the CNVs associated with sensory perception could be related to differences in temperament among breeds, but that connection isn't apparent to me.

In 2012 the genome of the Quarter Horse was published.[91] The Quarter Horse is the quintessential working horse of American ranches, and the standard rodeo horse. About 3 million are registered in the United States, by far the most of any breed. The Quarter Horse was selected for speed over short distances, with genetic contributions from Thoroughbreds; but more important, Quarter Horses were selected for working cattle and other ranch activities, which require a calm demeanor and a strong disposition to cooperate with their riders in an intuitive and often improvisational way.

Because of their divergent selection, a comparison of Thoroughbred and Quarter Horse genomes could prove instructive. As of this writing, only some preliminary results are available, which indicate, unsurprisingly, that Quarter Horses are much less inbred than Thoroughbreds. In the near future, though, comparative horse genomics should prove fruitful in determining how artificial selection, founder effects, genetic drift, and hybridization influenced the development of not just Thoroughbreds and Quarter Horses, but a wide array of other horse breeds. The new genomic information also greatly augments previous information from mitochondrial and Y-chromosome DNA in documenting the history of horse domestication (see Appendix 11B on page 340).

BELOVED

As I write this, there is a controversy in New York City regarding the famous horse-drawn carriages of Central Park. At issue is whether the conditions under which these large draft horses labor—hard pavement, air pollution, traffic, and so on—are cruel. The new mayor has promised to ban the Central Park carriages. I wonder whether, if the carriages were donkey drawn, the opposition would be so intense, or whether it would exist at all. But then, if the carriages were donkey powered, the attraction of carriage rides would be greatly diminished. Donkeys just aren't as charismatic as horses.

Part of the charisma of horses is simply their stately and beautiful physical presence and graceful behavior. Uniquely among our domesticated mammals, the domestication process has only enhanced these traits in horses. To most Western eyes the stocky, short-legged, and short-maned Mongolian horses, which closely resemble the wild ancestors of domestic horses, would suffer by way of comparison to the draft horses in Central Park, and much more in a comparison with a modern Thoroughbred. The reverse is true if we compare the auroch to any existing cattle breed, including the Spanish Fighting Bull.

But part of horse charisma comes by way of historical associations related to social status, beginning with Bronze Age charioteers, culminating in aristocratic cavalry, and continuing, anachronistically, in fox hunting, dressage, and the so-called sport of kings. The Victorianishly pimped-out horses—and their carriages—in Central Park fall on the low end of these anachronisms, statuswise.

Ultimately, though, it was their utility as human purveyors that propelled horses to the forefront among domestic creatures in our esteem. And even as this utility rapidly diminished, horses have managed to maintain their high regard in our cultural consciousness. Actually, the appreciation of horses seems to have increased as their utility has waned. For this reason alone, domestic horses will abide long after rhinos, tapirs, asses, and wild horses, the last remnants of that once glorious group of creatures called perissodactyls, have completely vanished.

CHAPTER 12

RODENTS

S OON AFTER PIZARRO'S CONQUEST OF THE INCAS IN
1532, strange rodents began to appear in European households,
first in Spain, then in Belgium (a Spanish possession at the time), and
later in England, France, and elsewhere in western Europe.[1] So strange
were these rodents that they weren't considered rodents. They were certainly not like the prototypical European rodents, such as rats and mice.
Their tails, for example, were decidedly non-rat-like; in fact, they were
barely discernible. And the shape of these creatures was not remotely
murine. But probably as important as these particular differences was
the broader gestalt sense that these creatures were not, well, icky. They
were just too cute and cuddly to be rodents. Which is why these new
creatures from the New World were brought inside European households from which their native cousins were vigorously excluded.

The inability of Europeans at that time to grasp the rodentness of
these imports from the New World led them to cast about for other
familiar creatures by way of analogy and a common name. None of the
familiar European creatures were truly suitable for this role; the best the
Germans, French, and English could come up with was the pig; they
were likened to miniature pigs.

These New World rodents were sort of piggish, in some respects. Their
tails certainly looked more like the short, furry ones of pigs than the long,

naked ones of rats. They also had vaguely piglike proportions—stocky, short of limb, the head a bit oversized. Some may even have thought their grunts sounded a bit piggish. In any case, French, German, and English speakers all settled on the pig model and deployed "pig" in their common name for the New World rodents. The French called them *cochon d'Inde* ("[west] Indian pig"); the Germans, *Meerschweinchen* ("sea piglet"); and the English, "guinea pig." The Spanish, though, demurred; their name for this creature, *conejillo de Indias*, means "little rabbit of the Indies." I'm with the Spanish on this; to my eye, the guinea pig looks a lot more like a rabbit than a pig. The genealogical connection is also much closer.

The great pioneer taxonomist Carolus Linnaeus (Carl von Linné), though, was with the English, Germans and Dutch; he saw something distinctly piggish in guinea pigs, as is reflected in his species name for the creatures, *porcellus*, which means "little pig" in Latin. But to the trained eye of Linnaeus, guinea pigs were unmistakably rodents. He even put them in the same genus, *Mus*, as the house mouse, which was taking things too far in the other direction. Guinea pigs were later assigned to the genus *Cavia*; the species name bequeathed by Linnaeus stuck, however—ergo the full scientific name for guinea pigs: *Cavia porcellus*.

The truly head-scratch-inducing part of the guinea pig's name is the "guinea." A number of "theories" have been proposed, none of them approaching the level of consensus.[2] I will not review them here; it is sufficient for our purposes to note that guinea pigs are not from Guinea, which is in West Africa; they are from the land of the Incas in the western Andes.

Given the multiple confusions that the term "guinea pig" reflects, it is best to revert to the scientific name—not the second part (the species), but rather the first part: *Cavia* (the genus). Following the example of those fanciers who value them enough to have created phenotypes undreamt of in their native land, I will refer to them as "cavies" (singular "cavy").

The cavies that arrived in the New World were used to living indoors in their native land. There, they were often assigned special rooms;

some moved freely about human abodes—until, that is, it was time to be cooked. For cavies in Peru were not pets; they were food. They had already experienced domestication to that end for over 5,000 years prior to their arrival in Europe. Before we explore their domestication, though, let's turn to their predomestication evolutionary history for clues as to why cavies were the first rodents to be domesticated.

RODENTS, A MAMMALIAN SUCCESS STORY

In the 1990s a few rogue taxonomists tried to expel guinea pigs from the rodent order, Rodentia.[3] That effort didn't get very far, however. Guinea pigs remain firmly ensconced under the rodent umbrella.[4] But it's a huge umbrella. Forty percent of all mammals, and about half of all placental mammals, are rodents. In addition to mice and rats (the quintessential rodents), there are voles, lemmings, hamsters, gerbils, gophers, and muskrats. Squirrels are also rodents, as are chipmunks, marmots, prairie dogs, and dormice; so, too, are porcupines, beavers, chinchillas, agoutis, nutrias, capybaras, and, of course, cavies. Those are just a few types of rodents, the names of which are familiar to the general reader. Less familiar, but no less rodential, are zokors, aplodontids (mountain beavers), bamboo rats, crested rats, rock mice, pacas, vizcachas, hutias, cane rats, mole rats, spiny rats, kangaroo rats, jerboas, gundis, degus, and springhares, to name just a few. It takes over 30 families—far more than in any other mammalian order—to encompass all of these species.

Rodents are also the most numerous mammals. The secret to their success is their propensity to live fast, die young, and make a lot of babies. A house mouse Methuselah managed to last over four years in captivity; in the wild most don't make it past the first. But they get a lot done, reproductively, in their short life spans. A one-year-old mouse is probably a great-great-great-grandparent and still churning out new babies. That's a recipe for explosive population growth if left unchecked.

But mice and most other rodents are rarely left unchecked. They are on the low end of the food chain, nourishment for a wide variety of critters, from snakes to hawks to weasels and foxes, all of which are mouse population checkers. In the wild, mice don't die of old age.

Cavies and their kin tend to violate rodent rules, or at least bend them considerably. They live slower, die older, and make fewer babies—more like typical nonrodent mammals. One upshot of these differences is that mouse-related rodents have a much greater evolutionary potential than cavy-related rodents. That is, mouselike rodents have experienced much greater genetic change than cavy-like rodents.[5] How this difference in rates of genetic change cashes out in phenotypic evolution is another matter entirely. There is no evidence that mouselike rodents are more phenotypically diverse than cavy-like rodents—quite the contrary. The different rates of genetic evolution do explain in part why there are so many more species of mouselike rodents than cavy-like rodents.[6]

You have no doubt gathered that mouselike rodents and cavy-like rodents represent two distinct ends of the rodent spectrum—two distinct sides of the rodent tree that have long gone their separate evolutionary ways. Despite this long separation, mice, cavies, and all other rodents, from gophers to beavers, have some key characteristics in common—none more key than their amazing incisors, two on the upper jaw and two on the lower. These teeth grow continuously throughout life, so they require constant wear from gnawing hard stuff, without which the bottom ones would grow right through the roof of the mouth and into the brain. Another distinctive feature of rodent incisors is that they are enameled only in front; the backs of the incisors consist of the much softer dentin. When rodents gnaw, the dentin is abraded, resulting in a chiseled enameled surface on the tip of each incisor. So, not only are rodent incisors continuously growing, they are also continuously being sharpened.

Not surprisingly, rodent taxonomy is more complicated than, say, carnivore or artiodactyl taxonomy, and disputes are correspondingly more numerous. Rodent taxonomists are a contentious lot—not because the field attracts contentious people, but because there is a lot to be contentious about. To this day, matters are far from fixed, though an accord is beginning to emerge with respect to the main branches of rodent genealogy. Here I will adopt the genealogy of Dorothee Huchon and her collaborators.[7]

Fossils and molecular (clock) evidence provide widely conflicting dates for the first rodents. Fossil evidence indicates an origin around the time the dinosaurs disappeared, or soon thereafter (65–55 mya).[8] Early estimates based on DNA divergence suggest a much earlier origin, in which case rodents were underfoot while dinosaurs still roamed the earth.[9] Recent estimates based on more and larger DNA sequences are in better accord with the fossil evidence,[10] which is reassuring. Here, then, are some key (approximate) dates in rodent, and particularly cavy, evolution.

Toward the end of the dinosaurs' reign on earth, about 70–65 mya, a new mammalian clade emerged, consisting of rodents and rabbits (lagomorphs), which are collectively called Glires.[11] The rodent-rabbit split occurred fairly soon thereafter, about 63 mya, and rodents began to diversify (Figure 12.1). By 55 mya, three distinct major rodent clades existed: the squirrel-related clade (which was the first to branch off), the mouse-related clade, and the clade that included cavy-like rodents. The clade to which cavy-like rodents belong also includes (Old World) porcupines and mole rats, which are collectively called hystricognaths (for their particular arrangement of jaw muscles).[12]

Hystricognaths, like all rodents, first evolved in Africa. About 43 mya, the cavy-like hystricognaths parted evolutionary ways with Old World porcupines and mole rats when they migrated—probably by island hopping—from Africa to South America.[13] South America, at this time, was almost as isolated from other continents as Australia, and it had its own unique mammalian fauna; as in Australia, many were marsupials. There were no carnivores (cats, dogs, bears, etc.); artiodactyls (pigs, cattle, deer, or camels); or perissodactyls (horses, rhinos, or tapirs) when the cavy-like rodents arrived. (Primates arrived at about the same time as the cavy-like rodents and by the same route.)[14]

There was much more ecological space for small mammals in this new

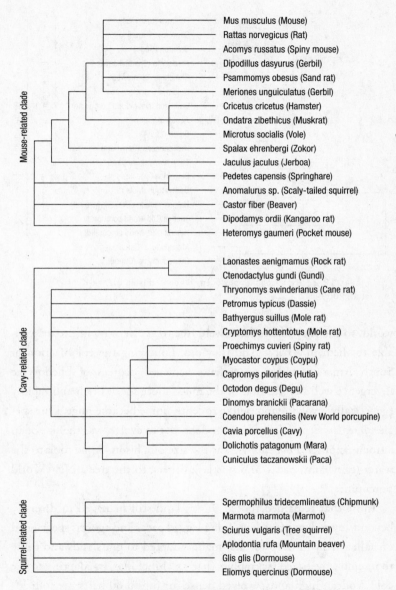

FIGURE 12.1   Rodent phylogeny. (From Blanga-Kanfi et al. 2009, 3.)

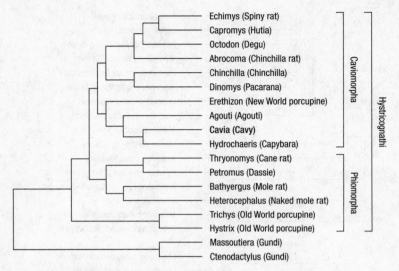

F IGURE 12.2   Hystricognath phylogeny. (Adapted from Blanga-Kanfi et al. 2009.)

world, so the cavy-like rodents were free to evolve ways of life unavailable to them in Africa. And they did. Following the colonization of South America, cavy-type rodents achieved degrees of phenotypic divergence well beyond that of the much more species-rich and rapidly (genetically) evolving mouselike rodents. Some became huge. One species, *Josephoartigasia*, the largest rodent ever to evolve, was the size of an auroch; another, *Phoberomys*, was the size of a bison. Some took to the water (capybaras, pacas, and nutrias); others, to the trees (New World porcupines).[15]

Those that remained rooted to dry land still managed to diversify beyond any rodent precedent. Long-legged pacas and agoutis specialized on fallen fruit in deep forests; maras adapted to grasslands and came to resemble oversized jackrabbits (hares); chinchillas, in adapting to the cold Andean highlands, evolved dense fur (up to 60 hairs per follicle), for which they eventually paid a high price, when Europeans came to covet it soon after Pizarro's conquest.[16] Figure 12.3 lays out the genealogical relationships of these varied cavy-like rodents.

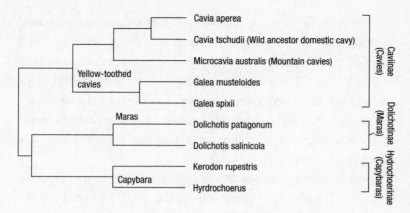

FIGURE 12.3   Caviid phylogeny. (Courtesy of the U.S. National Library of Medicine.)

The cavy family proper (Caviidae) includes maras (genus *Dolichotis*), capybaras (*Hydrochoerus hydrochoerus*)—the largest living rodent—as well as a number of cavy species (genus *Cavia*). Species of *Cavia* are distributed throughout South America—outside of the Amazon basin—in grassy environments. They are primarily grazers, the New World ecological equivalents of cattle[17] (Figure 12.3).

CAVY DOMESTICATION

It has long been a mystery as to which species of *Cavia* is the wild ancestor of domestic cavies. Recent genetic evidence indicates that it was *Cavia tschudii*, the montane cavy, which still inhabits the central-western Andes, from northern Chile to Colombia, at elevations of 3,000–4,000 meters (9,000–12,000 feet).[18] Wild cavies are quite social, typically living in groups of one male and several females and their younger offspring. These groups have well-defined home ranges in which they forage, with extensive tunnel systems in thick grass, to which they retreat for sleep or to escape predators.[19]

It appears that domestication occurred only once, somewhere in southern Peru. The domestication process may have commenced as early as 9000–6000 BP.[20] During this period, wild cavies became a major source of meat for central Andean peoples, and some may have been brought home and kept for future meals. Alternatively, as with dogs, cats, and pigs, the cavies themselves may have initiated the domestication process when they found areas around human habitations particularly hospitable, because of either enhanced food resources (scraps) or protection from predators.

Angel Spotorno proposed a three-stage process of active cavy management.[21] The first and longest stage lasted until the European conquest and resulted in several local landraces, as domestic cavies spread from the initial site of their domestication. These landraces include the Criollo of Colombia, the Nativa of Bolivia, and the Andina of Chile and Peru; they are sometimes referred to as the "creole breeds." As a result of artificial selection for meat production, the creole breeds are considerably larger than their wild ancestors.[22] Compared to wild cavies, the creole breeds have somewhat shorter skulls and greater color variation.[23]

The second stage of cavy domestication occurred when cavies were exported to Europe, where they became pets and, eventually, laboratory animals. In their role as pets, domestic cavies of course experienced quite different selection than they had during their meat phase. The new selection regime focused primarily on increased tameness and the look of the fur. By the time cavy breeder associations—resembling their canine and feline counterparts—were established in the nineteenth century, cavies were distributed worldwide and exhibited a wide range of coat colors and other hair characteristics. From this phenotypic variation, breeders began to systematically select for a wide variety of color patterns and hair lengths.[24]

In what amounts to convergent evolution through artificial selection, the coat characteristics of most modern cavy breeds resemble those of modern breeds of dogs, cats, rabbits, and even sheep and goats. For example, there are now Dalmatian cavies, tortoiseshells, and black-and-tans, as well as a merino breed (curly hair), a rex breed (almost hairless), and

FIGURE 12.4   A cavy.

two fully hairless breeds, not to mention a Himalayan breed the color of a Siamese cat, an Abyssinian breed, and even a ridgeback breed with the same ridge of spinal hair as the Rhodesian Ridgeback dog breed.[25] (See Figure 12.4.)

Such convergent evolution is to be expected, given that the molecular pathways involved in hair development are highly conserved in mammals.[26] This is another example of convergence promoted by homology. In the early twentieth century, some pioneering geneticists, including William Castle and Sewall Wright, took advantage of this conservatism. Wright used cavies as model organisms to work out the complicated genetics of coat coloration in all mammals.[27]

Scientists employed cavies for many other ends as well, to the point that the term "guinea pig" came to have the connotation of unwitting experimental subject. Cavies still remain useful as laboratory animals for certain forms of medical research.[28] But for geneticists, the cavies' role as experimental subjects of choice was soon usurped by their much faster-breeding distant cousins: rats, and especially mice.

Meanwhile, back in their native South America, new, improved cavy breeds were being created to quite different ends. These new South American breeds were improved with respect to meat quantity and hence larger size.[29] (This is Spotorno's third stage of domestication.) Improved meat breeds, such as the Tamborada of Peru and the Auqui of Ecuador, can be well over twice the size of creole and European breeds. Efforts to increase meat production in this way are accelerating rapidly.

EFFECTS OF DOMESTICATION ON CAVY BRAINS AND BEHAVIOR

As we have seen, domesticated mammals typically have proportionally smaller brains than their wild ancestors. This seems to hold true for cavies as well.[30] In one study it was discovered that, despite having smaller brains, domestic cavies performed just as well as wild cavies on certain standardized tests of rodent learning.[31] This should not be surprising, because the brain size reduction in domestic animals need not involve cognitive regions. In fact, it more likely involves areas of the brain related to sensory capacities and motor skills.

Domestication, as we have seen, results not only in tameness in relation to humans, but also increased social tolerance and sociality generally. Increased social tolerance was especially noticeable in formerly solitary creatures, such as cats and polecats, but even already social creatures such as dogs become more sociable than their wild ancestors. Like dogs, cavies were already social prior to domestication, but their sociality may have increased further during the domestication process. Domestic cavies are less aggressive than wild cavies, and their stress response is lower during social interactions.[32] Moreover, although adult male wild cavies are intolerant of other males, to the point that they will kill their own sons when they mature, multimale groups are no problem in domestic cavies.[33]

Let's turn now to studies of some mouselike rodents, which took much different paths to domestication.

HOUSE MOUSE

The house mouse, *Mus musculus*, epitomizes the live fast, die young, and make a lot of babies approach to evolution. The capacity of house mice for explosive population growth is one key element of their success, but so, too, is their remarkable behavioral phenotypic plasticity.[34] This combination of traits makes house mice quintessential evolutionary opportunists, the mammalian equivalent of weeds. And like weeds, mice have spread far beyond their already extensive natural range in Asia; they have come to occupy every continent except Antarctica, and most islands around the world. Also as with weeds, the spread of mice was unwittingly human aided. For house mice were the first mammals to fully exploit the new kind of niche created by human settlements.

The newly developed commensal habits of house mice were an expression of their phenotypic plasticity. These former dwellers of rocks and caves found human habitations and granaries more congenial and never looked back to their old haunts. Their destiny was to be tied to that of humans and, so far at least, it has worked out quite well for them, despite the outright antagonism of their unwilling human partners. To this day we consider the vast majority of house mice to be pests, and we have put much effort into extirpating them. We have failed utterly.

As we have seen, many of our domesticated mammals have gone through a commensal stage—dogs, cats, and pigs, for example. Only a minority of commensals, though, become domesticated. We have many commensals, from cockroaches to house sparrows to raccoons, that remain essentially wild. Nevertheless, commensal creatures are vulnerable to domestication, should the opportunity arise, simply by virtue of their proximity and accessibility. For the house mouse, that opportunity arose quite recently.

FROM MOUSE-TYPE RODENTS
TO THE HOUSE MOUSE

The house mouse, *Mus musculus*, and all other "true mice" belong to the family Muridae, which includes, in addition, rats, gerbils, and a whole lot more. In fact, Muridae is not only the largest rodent family; it is the largest family of mammals—at over 1,100 species.[35] The murid family split off from other mouselike rodents sometime in the Miocene (20–25 mya); mice and rats parted ways 10–15 mya; and the genus *Mus* evolved about 6 mya[36] (Figure 12.5).

Mus evolved on the Indian subcontinent, which remains the center of distribution for many members of this genus, including the house mouse.[37] By the onset of the Neolithic, about 12,000–10,000 BP, there were four distinct subspecies of house mice: *Mus musculus domesticus* in the steppes of western Pakistan; *Mus musculus musculus* in northern India; *Mus musculus castaneus* in northeastern India (present-day Bangladesh); and *Mus musculus bactrianus* in central and southern India. All four of these sub-

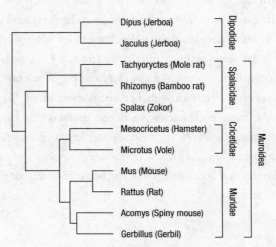

FIGURE 12.5   Murid phylogeny. (Adapted from Blanga-Kanfi et al. 2009.)

species became human commensals, but in different parts of the world. *M. m. domesticus* expanded into the Fertile Crescent as agriculture developed there; *M. m. musculus* invaded central Asia and China, the second cradle of civilization; *M. m. castaneus* expanded into Southeast Asia; *M. m. bactrianus* remained largely within the subcontinent.[38]

By 4000 BP, both the *domesticus* and *musculus* subspecies had arrived in Europe, but by separate routes:[39] *domesticus* via the spread of farming from the Near East, *musculus* by way of horse-drawn carts across the central and western Asian steppes. The eastern German border roughly marks the boundary between the largely western European *domesticus* and the central European *musculus* house mice. In the last 500 years, *domesticus*, but not *musculus*, reached the Americas, Australia, New Zealand, sub-Saharan Africa, and islands of the Indo-Pacific aboard western European ships during the exploration and conquest of these areas.

The association between humans and house mice has served mainly the interests of the mice. Given their long-standing status as pests, it is remarkable that anyone saw in mice potential pets. But some—probably aristocrats who had limited experience of mice as pests and hence could get past the ickiness factor—did, and they proceeded to initiate the process of mouse domestication. It began in China, where, as early as 3000 BP, there is reference to spotted mice owned by a princess.[40] By about 2100 BP (Han dynasty), Chinese mouse breeders had produced not only a yellow mouse, but also the famous "waltzing mouse," so called because of its strange dancing movements, the result of a defective inner ear.[41]

Mice bred as pets are known as "fancy mice" because, unlike their precursors, they are fancied by mice fanciers. By the eighteenth century the Japanese had taken the lead in the development of fancy mice, some of which made their way to Europe; they became especially popular in Victorian England. In Japan, the East Asian *musculus* was interbred to some extent with the Southeast Asian *castaneus* subspecies in producing the varied fancy mice strains there. It was these *musculus-castaneus* hybrids that were imported to England. The native *domesticus* subspecies contributed little, if any, to English fancy mouse strains. But once fancy mice were repurposed for scientific use, the *domesticus* subspecies was added to the mix. In fact, the majority of the laboratory mouse genome comes from *domesticus*, a substantial minority from *musculus*, with a dollop of *castaneus*.[42] The mix does vary somewhat between lab strains.

Though lab mice were inaugurated with a genetically diverse hybrid genome, their genetic variation was soon partitioned into highly inbred strains, as is characteristic of most domestic animals. But in mice this process was carried much further—through frequent brother-sister matings—to the point that all genetic variation was eliminated in many strains. These highly inbred strains have proved quite valuable in biomedical research, especially in isolating genetic substrates for diseases.

Though there is little genetic variation within strains, there is ample between-strain genetic variation. There is lots of phenotypic variation between strains as well. Tameness, for instance, varies markedly among laboratory strains.[43] Yet even the least tame laboratory strain is much tamer than wild *domesticus*. Fancy mice fall on the extremely tame end of the spectrum.

The most noticeable physical difference between wild and laboratory (and fancy) mice is body size. Even the smaller laboratory strains are about twice as heavy as wild mice. Lab mice are also couch potatoes. If they have access to an exercise wheel, wild mice run much longer than lab mice do—and faster. The athletic advantage of wild mice is reflected in their larger heart ventricles.[44] Another indicator of athleticism is muscle strength, as measured by how long a mouse can hang from a dangling cord. Lab mice can dangle for 30–40 seconds. Wild mice don't deign to dangle; they just pull themselves up, climb to the top, and have done with it.[45] Wild mice also have spectacular jumping ability, which Steven Austad likened to popping popcorn.[46] This capacity has been bred out of lab mice. To stay with the popcorn metaphor, lab mice are the unpopped kernels (UPKs) at the bottom of the microwave popcorn bag. Scientists and mouse fanciers alike prefer their mice that way. In any case, the non-UPKs took themselves out of the domesticated breeding pool by popping to freedom.

Diminished athleticism is typical of domesticated mammals, but only in mice has it been quantified to this extent. Another feature of the domesticated phenotype found in lab mice is brain size reduction.[47] Lab mice also have considerably smaller eyes than their wild ancestors, in keeping with the notion that much brain reduction in domesticated animals is in sensory, not cognitive, regions.

Reproductively, lab mice are also typical of domesticated mammals in their early sexual maturation and large litter size relative to their wild ancestors.[48] They also age more quickly.[49] So lab mice have taken the live fast, die young, and leave lots of babies gambit to an even greater extreme than wild mice, which are already very far out on that limb.

PHENOTYPIC VARIATION WITHOUT GENETIC VARIATION

The most highly inbred mouse strains, in which all individuals are virtually clones, are called "isogenic." You would expect isogenic mouse strains to be phenotypically uniform, especially since they are raised in highly standardized ways to eliminate any environmental variation. But surprisingly, these genetically and environmentally uniform lab mice still exhibit large amounts of phenotypic variation.[50] Take body weight, for example. Isogenic mice reared under highly standardized conditions still manage to vary markedly in this trait. This variation is a great inconvenience for those seeking to isolate the effects of genes or environment on phenotypic traits.

There must be a third factor, in addition to genes and environment, to account for this variation. This third factor has traditionally been referred to as developmental noise because of its random nature. But recently, the mechanisms behind this noise have been uncovered. One important clue is that this third factor usually kicks in before birth. What kind of developmental factor is congenital and exerts lifelong effects but is not genetic? (For the answer, see Appendix 12 on page 344).

RATS

In terms of ickiness, rats far exceed mice. Not only are rats much larger, but they have long been associated with disease and pestilence, including the Black Death plagues. For the species under primary consideration here—*Rattus norvegicus*, commonly called the brown rat, Norway rat, or, in New York City, the sewer rat—the disease association is undeserved. In particular, the rats bearing the fleas that carried the bacteria that caused the Black Death plagues were not brown rats; they were black rats, *Rattus rattus*. Brown rats provided an underappreciated service in this regard, by outcompeting the somewhat smaller black rats—which had preceded them as human commensals—over much of the black rat's former range in Europe and the United States.

Compared to house mice, rats are newcomers to human habitations, and unlike mice, brown rats continue to thrive in a noncommensal state in their native habitats: the plains of Mongolia and northern China.[51] Black rats come from forested areas to the south, in southern China and Southeast Asia.[52] Brown rats are burrowers; black rats are arboreal and good climbers. If you find a rat in your basement, it is probably a brown rat; if you find one in the attic, it is probably a black rat. (In California, we referred to the latter as roof rats.) If you find an extremely small rat in either place, it is probably a mouse.

Though some rats remained behind in a "state of nature," others, like the house mice before them, found life with humans more congenial. Those who chose the commensal life soon spread along paths of human migration and trade—first by land, then by sea. Black rats were the first to arrive in Europe, perhaps as early as 2,400 years ago.[53] They reached England during the Roman period (about 1,800 years ago: 200–300 CE). Brown rats did not invade Europe until centuries later, during the medieval period. They then commenced gradually displacing their predecessors in Europe and, later, much of North America.

PEST TO PET AND BEYOND

The scientific species name for the brown rat, *norvegicus*, reflects the mistaken belief—current at the time—of the English naturalist who bequeathed it, that brown rats originated in Norway.[54] The brown rat is still commonly referred to as the Norway rat in the English-speaking world, but I will avoid doing so here.

As you might expect of a member of the mouse family (Muridae), rats—brown and black—are also adherents to the live fast, die young, and leave lots of offspring approach to life, which makes them formidable pests. From the time rats first appeared, most Europeans relied on cats and dogs to keep them in check, but without much success. Aristocrats could afford more effective means of control, and many kept rat catchers in their employ. Not the most scrupulous lot, rat catchers could raise extra cash by selling rats captured live for a blood sport called rat baiting. These rats were placed into a pit with a single dog, usually a terrier, and bets were made as to how long it would take for the dog to kill every last one of them.

Queen Victoria's official rat catcher was a man named Jack Black, who often sold his live-caught rats to the great rat-baiting entrepreneur Jimmy Shaw. But the ever-resourceful Black developed yet another revenue stream by taming some of the more atypically colored rats and selling them as pets, complete with ribbons and sometimes leashes, to aristocratic Victorian women.[55] For a time, beribboned lap rats became a status symbol. Soon, a black-hooded white rat was developed among other color strains, and mouse fanciers allowed fancy rats to be shown with their smaller cousins. The English National Fancy Rat Society was eventually established in the 1970s with an exclusive focus on rat breeding. By this time rats had also become popular lab animals, particularly for (behaviorist) psychology experiments on learning.

In less than 200 years the descendants of Jack Black's terrier baiters have diverged considerably from their pestilent ancestors. Coat coloration, for example, is much more variable in domestic rats. Most note-worthy is the increased frequency and amount of white (depigmented) areas, which are often associated with domestication. White pigmentation has been especially encouraged in lab rats.

As in domesticated mice, the heart and other organs are smaller in domesticated rats.[56] Like domesticated mice and domesticated cavies, domesticated rats reach sexual maturity at an earlier age than did their wild ancestors.[57] Moreover, domestic rats are no longer tied to the seasonal patterns of their wild counterparts; they breed year-round.

Behavioral differences are even more striking. Domesticated rats are notably tamer than wild commensal rats, as we have come to expect, and they are considerably more tolerant of each other's proximity.[58] Domesticated rats are also less anxious or fearful, both in social contexts and in novel nonsocial environments.[59] They tend to be more exploratory and perform better than their wild counterparts in standardized learning tasks.[60]

THE EXPERIMENTAL DOMESTICATION OF THE RAT

Though the farm fox experiments are the most famous studies to come out of Belyaev's Novosibirsk laboratories, equally important was their experimental domestication of the brown rat.[61] Because of their much shorter generation time and hence greater evolutionary potential, rats have a critical advantage over foxes in this regard. And though the rat experiments commenced decades later, artificially selected rat generations now far exceed artificially selected fox generations.

Again, the goal was to select for tameness.[62]

Two selection lines were established: one in which selection was for reduced defensive aggression, which I will henceforth refer to as the "domesticated line"; and one in which there was selection for increased defensive aggression, which I will refer to as the "wild line." The domesticated and wild lines soon diverged dramatically in their response toward humans. Domesticated rats quickly came to display reduced aggression toward humans. By generations 71–72, these rats could be handled with ease.[63] Male rats from this generation were also assessed for aggression toward laboratory rats of the popular Wistar line. Rats from the domesticated line were also less aggressive toward Wistar rats, though the differences between wild and tame rats were less stark in this context.[64]

One of the most noteworthy themes from the farm fox experiments was the correlation of behavioral and physiological alterations caused by artificial selection for tameness. Similar correlated changes were observed during experimental rat domestication. Some of the correlated behavioral changes in artificially selected rats also closely replicate what occurred during the domestication of lab rats. For instance, rats artificially selected for tameness were also less anxious in a novel environment than were rats from the wild line; the tame rats were also more exploratory and exhibited enhanced spatial learning compared to their wild counterparts.[65]

Rats from the domesticated line also exhibit the typical changes in reproductive development that result from domestication. They reach sexual maturity at an earlier age than do rats from the wild line; and they are sexually active year-round, in contrast to wild rats, which are seasonal breeders.[66]

The most noteworthy correlated responses in both foxes and rats are hormonal, because it is the hormonal changes, particularly those involved in the stress axis, that, according to Belyaev, are behind many of the other correlated responses to selection for tameness. And indeed, levels of these stress-related hormones are reduced in rats selected for tameness, as they are in foxes.[67] Every level of the hypothalamic-pituitary-adrenal axis is affected, as well as the hormonal feedback mechanisms that stabilize it. Moreover, the adrenal glands, from which the stress hormone cortisol is released, are actually smaller in rats from the domesticated line than in rats from the wild line.[68]

A number of correlated behavioral alterations are clearly linked to these hormonal changes caused by selection for reduced aggression. The increase in exploratory behavior, for example, probably reflects a diminished stress response. Stressed-out rats tend to freeze or flee in an unfamiliar setting. It takes a certain amount of emotional relaxation to explore novel environments. The enhanced learning capacity may also simply reflect this reduced fearfulness and hence increased focus on the learned task. It has long been known that anxious or "high-strung" rats and mice perform worse than relaxed rats in a variety of learned tasks.[69]

THEMES AND VARIATIONS

Domesticated rodents exhibit many features of the domesticated phenotype found in other mammals, from dogs to horses, but also some interesting differences. Let's start with the differences. Most domesticated mammals become smaller than their wild ancestors, at least initially, with the notable exception of the Bactrian camel. (Through artificial selection in later phases of domestication, breeds that are larger than their wild ancestors have been constructed in dogs, horses, and pigs.) All three rodent species, though, became larger during initial domestication. This difference between rodents on the one hand and nonrodents on the other indicates that body size is evolutionarily a quite malleable trait. In domesticated animals body size may reflect primarily human needs. In the past, humans generally wanted their large domesticated animals smaller and hence more manageable, but their smaller domesticates, like cavies, larger and hence more productive. The explanation for increased size in domestic mice and rats, though, is not obvious.

While cavies, mice, and rats all show signs of accelerated sexual development (progenesis)—a trait that is characteristic of domesticated mammals—there is no reported evidence of neoteny. (Perhaps a closer examination of cavies will reveal such.) In their lack of neoteny, rodents diverge from other domesticated mammals.

The similarities among domesticated rodents and other domesticated mammals are much more numerous than the differences. With respect to coat coloration, wild rodents, including mice and rats, exhibit quite a wide range.[70] But domesticated rodents have many more color variants, some never seen in nature. Moreover, white, the color of domestication, is prevalent in domesticated rodents, as it is in all other domesticated mammals.

The behavioral convergence of rodents under domestication and other mammals is equally striking. The behavioral sine qua non of domestication is tameness; like all the other creatures discussed here, cavies, mice, and rats are tamer than their wild ancestors in the presence of humans. All three rodent species also show reduced aggression toward others of their species. These behavioral alterations are associated with a reduced stress response in cavies and rats, as is the case for domesticated foxes.

The convergent phenotype resulting from domestication of mammals as distantly related as rats, foxes, and camels is truly remarkable. In part, this reflects similar selection for tameness. But selection for tameness alone cannot account for the prevalence of white or the increased sociability of domestic animals. Such convergences reflect homologies common to all mammals. The stress response, for example, is highly conserved, as are the neural substrates for emotions like fear and aggression, as well as the need for social contact.[71] Humans, too, share these neural substrates for emotions and, as we shall see in the next chapter, humans share some emotional traits with domesticated animals, which has encouraged some to claim that humans, too, have been domesticated—self-domesticated.

CHAPTER 13

HUMANS—PART I: EVOLUTION

KANZI IS ONE REMARKABLE APE—BONOBO, TO BE precise. He was adopted—kidnapped actually—by the dominant female, named Matata, of a captive colony at the Yerkes Primate Center in Georgia, who, as it happened, was being taught to communicate with humans through lexigraphs, graphic symbols of single words. Matata was not a great student, a bit too old for that sort of thing, but Kanzi proved extraordinarily apt. He had been just sort of hanging around during Matata's sessions, not paying much attention, seemingly. It soon became obvious, though, that he had observed and learned much, out of the corner of his eye, as it were. When his training became more explicit, he soon learned the meaning of well over 300 symbols, by means of which he could express his desires to his human companions. He also understood spoken English—thousands of words, in fact—which he could translate into lexigraphic symbols by pressing the images on a screen. Even more impressive, Kanzi could communicate verbally, in his normal Chimpanese, with his late stepsister, Panbanisha, over the phone. They had been known to gossip.[1]

There are actually two species of chimpanzees. The one usually displayed in zoos—*Pan troglodytes*—is by far the most common and the one for which the term "chimpanzee" is usually reserved. Kanzi belongs to the other, far less common species (*Pan paniscus*), called bonobos, which have a restricted range in the Congo forests. Bonobos were formerly called "pygmy chimpanzees," though they are no shorter than common chimpanzees. (The name may derive from the fact that they share their forest habitat with small-statured humans formerly known as "pygmies.") Bonobos are significantly leaner than chimpanzees and have longer legs; their necks are thinner, their shoulders more narrow, and their chests less deep. Bonobos are altogether less robust than chimpanzees, less physically imposing. The bonobo head is also significantly smaller than that of the chimpanzee, with a less protruding snout and smaller brow ridges. On top of the head is a distinctive mop of long hair that tends to form a natural part down the middle.

But it is the behavioral differences between chimpanzees and bonobos that have received the most attention of late. These behavioral differences are much more pronounced than the physical differences and have engendered much discussion for their implications regarding human evolution.

Before bonobos took center stage, chimpanzees were thought to be the best models for extrapolating human behavioral evolution, especially once it was discovered that chimpanzees routinely hunt other primates in an organized and premeditated manner.[2] This finding fit in nicely with a man-the-hunter narrative and certain sociobiological elaborations of its implications, especially regarding putative sex differences in behavior.[3] Also conforming nicely to this narrative is the fact that chimpanzee males are violent, not just individually but often in groups, raiding the territories of adjacent groups. These group conflicts can be quite grizzly and have been touted as the wellspring of human warfare.[4]

Enter the bonobo, the hippy ape. Actually, bonobos embrace the hippy ideal far more than any human hippies ever did, for bonobos are not only pacifist; they also engage in more sex than the most nympho-maniac members of our own species. Rarely do a couple of hours pass without at least some serious foreplay. This rampant sex is also quite promiscuous—partners are ever changing—and indiscriminate with respect to gender. Bonobo lesbian sex is especially rife, and many believe it is the behavior around which bonobo society is organized.[5] For bono-bos, in stark contrast to the macho patriarchal chimpanzees, live in a female-dominated society. Male bonobos, like Kanzi, are larger and stronger than female bonobos, but not as much larger and stronger as are male chimpanzees in relation to female chimpanzees. Put another way, bonobos are less sexually dimorphic than chimpanzees. Moreover, whatever size advantage a male bonobo enjoys is outweighed by genitally cemented female bonobo solidarity.[6]

Inspired by the farm fox experiments, Brian Hare and coworkers proposed an overarching explanation for the physical and behavioral differences of chimpanzees and bonobos: that bonobos are, in essence, self-domesticated.[7] They hypothesize that bonobos have experienced nat-ural selection for tameness, from a more chimp-like starting point. Hare equates tameness with lower levels of aggression, but, as we saw with the fox domestication project, a reduction in fear is at least as important an element in tameness as lowered aggression is. Indeed, lowered aggression may be largely a by-product of the reduced fright response, since most animal aggression toward humans is defensive in nature.

Nonetheless, the self-domestication hypothesis is intriguing and worth exploring here. Hare's thesis begins with the observation that infant and young juvenile bonobos and chimpanzees are equally socia-ble and nonaggressive. As they age, however, chimps become increas-ingly intolerant and aggressive, while bonobos largely retain juvenile levels of sociability.[8] Hare then infers that, as in the farm foxes and dogs, bonobos were selected for juvenile levels of aggression, and this selection caused correlated paedomorphic alterations in morphology.[9]

The notion that bonobos are paedomorphic chimpanzees actually long precedes the self-domestication hypothesis, and bonobos may be paedomorphic without being self-domesticated.[10] Nonetheless, Hare's self-domestication hypothesis predicts some degree of paedomorphosis.

Brian Shea found evidence for neoteny in the arms, legs, trunk, and skull of bonobos.[11] It is the latter that has received the most subsequent attention; the evidence for skull paedomorphosis to date is equivocal at best.[12] In fact, it is safest to say that there is only moderate evidence for heterochronic differences in the skulls of chimps and bonobos, less evidence that bonobo skulls are paedomorphic, and very little evidence that bonobo skulls are neotenic.[13]

The best candidate paedomorphic features of the bonobo skull are a relatively small head, flatter face, and reduced brow ridges.[14] Beyond the head lies a better potential neotenic feature: many bonobos retain a white tuft of hair near the base of the tail, which in chimpanzees is confined to juveniles.[15] This sort of depigmentation, as we have seen, is characteristic of the domesticated phenotype and often a by-product of selection for tameness. The reduced sexual dimorphism in bonobos may also be construed as evidence for the self-domestication hypothesis.[16] Bonobos exhibit reduced sex differences with respect to not only body size, but also the size of their canine teeth.[17] But the best evidence for self-domestication by way of paedomorphosis is behavioral.

Bonobos seem more reliant than chimpanzees on their mothers throughout childhood and adolescence; they are less independent than chimps. In this respect bonobos are also more humanlike than chimpanzees.[18] This dependency may be related to the fact that their cognitive development is retarded relative to chimps for both social and nonsocial tasks.[19] With respect to the former, it is especially interesting that bonobos are slower to attend to social status cues. Young chimps quickly learn that some group members are more tolerant and likely to share food than others. They learn to avoid begging from the intolerant individuals, from whom begging elicits aggression. Bonobos are much slower to learn such discrimination, to self-inhibit in the presence of potential aggressors.[20] This delayed learning may reflect retarded development of the fear response. Bonobos are also developmentally delayed with respect to spatial tasks,

Paedomorphosis (trait of juvenile ancestor)	Peramorphosis (new trait not present in ancestor)
Progenesis (earlier offset)	Hypermorphosis (delayed offset)
Neoteny (reduced rate)	Acceleration (increased rate)
Postdisplacement (delayed onset)	Predisplacement (earlier offset)

FIGURE 13.1   Categories of heterochronic alterations in evolution.

tool use, and mastering causal relationships—all crucial in understanding the physical world, especially as it relates to acquiring food.[21]

Hare argues that this delayed cognitive development is linked to delayed emotional development (possibly fear) and that it is the retention of juvenile emotions that renders bonobos so sociable and cooperative.[22] Adult bonobos certainly seem to retain juvenile levels of play[23] (high) and aggression (low).

To test this hypothesis, we would have to know much more than we do about the common ancestor of these two species. Bonobos and chimpanzees began to go their own evolutionary ways 1–2 mya, and given the paucity of fossil evidence, it is not obvious which of the two more resembles the common ancestor.[24] So we need to consider whether chimpanzees exhibit peramorphosis (predisplacement and/or acceleration and/or delayed offset [hypermorphosis]) relative to the common ancestor, not just the hypothesis that bonobos are paedomorphic (postdisplacement and/or neoteny [deceleration] and/or early offset [progenesis]) (see Figure 13.1).

HUMAN SELF-DOMESTICATION?

Hare and others have also proposed that a very similar form of self-domestication occurred during human evolution and explains many of our singular features—not least, our unrivaled (among primates) capacity to cooperate with each other, and our hypersociality. In essence,

according to the self-domestication hypothesis, dog domestication reca-
pitulates many important features of both bonobo and human evolution. Since we know much more about human ancestry than we do about
chimpanzee/bonobo ancestry, human evolution provides the best test
for the convergent evolution of apes and dogs. In this chapter I will focus
on the evidence that humans are paedomorphic apes; in the next chapter
I will consider the role of self-domestication in human social behavior.

The German anthropologist Albert Naef was the first to suggest a
role for paedomorphosis (neoteny) in human evolution.[25] He noted
that infant common chimpanzees much more resemble human infants
than adult chimps resemble adult humans. Stephen Jay Gould explored
this idea in much more detail.[26] Gould argued that such features as the
upright posture, sparse hair, small teeth, large eyes, and large head so
characteristic of humans are the result of a slowing or retardation of
development relative to other apes.[27] Subsequent research paints a more
complicated picture. Before I treat evidence for human paedomorphosis
in detail, though, let's situate the human species in its more broadly primate context, genealogically.

THE PRIMATE BRANCH OF THE MAMMALIAN TREE

With bonobos and *Homo sapiens* we enter a new order of mammals:
the primates. As Figure 13.2 shows, primates are more closely related
to rodents and rabbits than to any of the other domesticated species we
have considered. The first incontrovertible primate fossils, such as *Plesiadapis*, date from 58–55 mya (Paleocene).[28] Molecular evidence points
to a much earlier date of origin: 90–80 mya, before the end of the age of
dinosaurs.[29]

We primates are distinctive among mammals in a number of ways. Our
olfactory sense is poor compared to that of most mammals, but our vision
is state-of-the-art, especially our capacity to perceive depth and color.

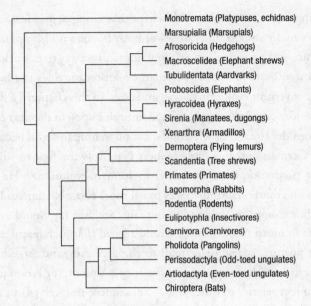

Monotremata (Platypuses, echidnas)
Marsupialia (Marsupials)
Afrosoricida (Hedgehogs)
Macroscelidea (Elephant shrews)
Tubulidentata (Aardvarks)
Proboscidea (Elephants)
Hyracoidea (Hyraxes)
Sirenia (Manatees, dugongs)
Xenarthra (Armadillos)
Dermoptera (Flying lemurs)
Scandentia (Tree shrews)
Primates (Primates)
Lagomorpha (Rabbits)
Rodentia (Rodents)
Eulipotyphla (Insectivores)
Carnivora (Carnivores)
Pholidota (Pangolins)
Perissodactyla (Odd-toed ungulates)
Artiodactyla (Even-toed ungulates)
Chiroptera (Bats)

FIGURE 13.2   Mammal phylogeny.

One morphological corollary is a relatively flat face with forward-facing eyes. Like rodents, primates have retained the primitive condition of five digits per limb. But unlike rodents, primates have an opposable thumb and big toe, useful for grasping branches. For primates initially evolved as arborealists, and most still are. Primates are also distinguished from other mammals by our fingernails—which replaced claws—and the soft, tactilely sensitive pads beneath those nails. Humans are not distinctive in leaving individually identifiable fingerprints; the vast majority of primates have those dermal ridges with snowflake-like variation.

We primates are most celebrated for our brains, which are larger, relative to body size, than those of other mammals.[30] One part of the brain in particular, the cerebrum, which supports much of what we consider intelligent behavior, is better developed in primates than in other mammals.[31] Perhaps related to this evolutionary brain development, primates

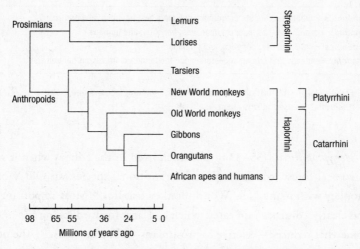

FIGURE 13.3   Primate phylogeny, showing major branching events.
(Adapted from *LIFE* 8e, Fig. 33.27, W. H. Freeeman & Co. 2007.)

generally grow slower and live longer than other mammals of comparable size,[32] affording ample opportunity for learning. Some believe primate cortical development is also related to the fact the primates are the most social mammalian order.[33] Much primate learning is social learning.

The primate branch of mammals can be divided into two sub-branches or suborders (Figure 13.3). One branch includes the so-called prosimians—lemurs, bush babies, lorises, and pottos—which are the most primitive primates. Some lack one or more of the typical primate features already described; in others they are less well developed. The second branch contains everything we know as monkeys and apes, including ourselves, collectively referred to as anthropoids. The split between prosimians and anthropoids occurred in the Paleocene epoch, between 60 and 55 mya.[34] About 35 mya (Oligocene), another split occurred, between those anthropoids that were to inhabit the New World, called platyrrhines ("broad nose"), and those that remained in the Old World, called catarrhines ("downward-facing nose").[35] Only the latter will concern us further.

Hominoids = gibbons, orangutans, gorillas, chimpanzees, bonobos, and humans
Hominids = orangutans, gorillas, chimpanzees, bonobos, and humans
Hominines = chimpanzees, bonobos, and humans
Hominins = humans and human ancestors since the human-chimpanzee/bonobo split.

FIGURE 13.4   Definitions of "hominoid," "hominid," "hominine," and "hominin" as used in this chapter.

Aegyptopithecus (35–33 mya), which lived in the Sahara when it was a tropical forest, was an important transitional species, an Old World monkey with many New World monkey features.[36] Most important, it was clearly a diurnal fruit eater, which had large forward-facing eyes and a somewhat reduced—relative to prosimians—olfactory sense.[37] The next major branching event occurred about 22 mya, separating Old World monkeys, such as baboons, guenons, colobuses, and mangabeys, from the hominoids: gibbons, orangutans, chimpanzees, gorillas, and humans. *Proconsul* was a transitional species that combined some features of Old World monkeys and hominoids. Most notably in the latter regard, it had lost all but a vestige of its tail, a distinctively hominoid trait.[38]

The meanings of the terms "hominoid," "hominid," and "hominin" have never been stabilized. Here, "hominoid" refers to a superfamily of primates, living members of which include gibbons, orangutans, gorillas, chimpanzees, bonobos, and humans. (The term "hominoid" is therefore roughly equivalent to the colloquial term "ape.") On the hominoid branch, gibbons were the first to split off from the rest to go their own evolutionary way, 19–16 mya. All of the remaining species— orangutans, gorillas, chimpanzees, bonobos, and humans—belong to a single family, Hominidae, and are collectively referred to as "hominids." Orangutans split off from other hominids 15–13 mya, then the gorillas (9–7 mya), and finally the chimpanzees and bonobos (7–5 mya). Humans, bonobos, and chimpanzees comprise a subfamily called Homininae and will be called "hominines," while humans and other extinct members of our direct lineage will be referred to as "hominins" (Figure 13.4).

Like most other primates, hominoids evolved in dense African tropical forests, but by about 16–14 mya, some began to move into more open woodland as the climate dried somewhat. It was also around this time that hominoids first moved into Asia.[39]

HOMINIDS TO HOMININS

In 1950, Ernst Mayr, one of the chief architects of the modern synthesis, gave an important talk at Cold Spring Harbor Laboratory that was to impede progress in the study of human origins for decades.[40] Fully innocent of any firsthand knowledge of human fossils, he managed to squeeze all that had been discovered to that point into three unwieldy species in a linear series, much as in the old-fashioned view of horse evolution (discussed in Chapter 11). Because of his stature, anthropologists were too intimidated even to give names to subsequent discoveries, much less assign new scientific names, giving the impression that human evolution was a ladder-like progression, not the branching affair we find in other species' genealogies.

Fortunately, though it took far too long, anthropologists finally did free themselves from the shackles that Mayr had bestowed on them. The result is a quite different view of human evolution—one of multiple branching events and numerous lineages, to the point of bushiness. If we take our species to be a point of reference, some branches have a vertical ("progressive") component, but most go sideways. It is important to note as well that several species of hominins coexisted at any given time throughout the last 7 million years until the extinction of the Neanderthals.

Unfortunately, the fossil record for the period of time when human and chimpanzee/bonobo lineages are thought to have split is quite poor. Two rather different species existed in East Africa during this period: *Sahelanthropus tchadensis* and *Orrorin tugenensis*. *Sahelanthropus* (7–6 mya) is known only from a skull, which was neither chimpanzee-like nor particularly hominin-like. It had a chimp-sized brain.[41] *Orrorin*, for which we have more of the skeleton, may have been partly bipedal,[42] but both *Sahelanthropus* and *Orrorin* probably spent more time in the trees than on the ground.

The third contestant for the earliest hominin and the one for which we have the most fossil evidence is *Ardipithecus*, commonly referred to as "Ardi." There were actually two species of Ardi—*A. ramidus* and *A. kadabba*—which lived in East Africa from about 5.8 to 4.4 mya.[43] This was after the human and chimpanzee lineages split. Ardis were about the size of a chimpanzee, with a chimpanzee-sized brain. They had an interesting combination of traits—their upper bodies quite ape-like, their lower bodies more hominin.[44] They seem to have been about as equally comfortable on the ground as in the trees.

As of this writing there is no consensus about whether *Sahelan-thropus*, *Orrorin*, or *Ardipithecus* qualify as hominins. The majority of anthropologists seem to believe that Ardi is the best candidate, but the rather contentious debate continues. A definitive evaluation requires more fossil discoveries from the period of roughly 7–4 mya.

From 4 mya to the present, the fossil record improves with the first appearance of the genus *Australopithecus*, of which there were at least eight species. The oldest of these (4.2 mya) belong to the species *A. ana-mensis*.[45] With respect to its skull, *A. anamensis* was quite apelike, but it was fully bipedal, with a humanlike upright gait. *A. anamensis* dwelled in a forest environment, so it was walking between trees, to which it no doubt fled at any sign of one of the numerous large predators for which it was on the menu. The teeth of *A. anamensis* indicate a diet of mostly grains, nuts, tubers, and fruits.

The most famous australopithecine fossil was a female belonging to the species *A. afarensis* and nicknamed "Lucy."[46] Lucy and her kin lived in East Africa from about 3.8 to 3.0 mya in open woodland and probably foraged both on the ground and in the trees for fruits, leaves, and other relatively soft vegetable matter. But it was recently discovered that Lucy's diet also included meat. Moreover, members of her species used stone tools to scrape the meat from bone and hide and to crack large bones for the nutritious marrow inside.[47] Thus, *A. afarensis* was perhaps the first hominin to both eat meat and use tools, some of which they may have carried with them on their travels. Prior to this discovery, it was the almost universal consensus that stone tools were first made about 1.5 mya and only by members of our genus, *Homo*.

What may have been the last *Australopithecus* to walk the earth is the most recently discovered. *A. sebida* is the name assigned to this species, and it lived in South Africa as recently as 1.9 mya, far later than australopithecines were thought to have lasted, and at about the same time that the first members of the genus *Homo* evolved.[48] Like all other australopithecines, *A. sebida* embodied a mosaic of humanlike and apelike traits, but more of the pieces are humanlike than in other members of the genus. Of particular note is the increased curvature of the lumbar spine, indicating that *A. sebida* was even more bipedal than previous members of the genus. Also significant were modifications of the hand that suggest increased manual dexterity and perhaps more intensive tool use.[49] The *A. sebida* brain, however, while on the large side for the genus *Australopithecus*, was within the australopithecine range.

THE FIRST MEMBERS OF THE GENUS *HOMO*

The genus *Homo*, to which we belong, evolved near the end of the australopithecine reign. Two species of *Homo* were near contemporaries of *Australopithecus sebida* during their early evolution: *Homo ergaster* (1.9–1.4 mya) and *Homo erectus* (1.8 mya to 40,000 years ago).[50] A third, even older species often assigned to the genus *Homo*—as *Homo habilis*, which roughly means "handy man"—was long thought to be the first hominin to make stone tools.[51] While the "handy man" no longer holds that distinction, he did advance the stone technology, to what is known as the Oldowan, which consisted of primarily two tool types. Using suitably sized quartz and quartzite rocks, *H. habilis* would expertly chip off sharp flakes to be used in scraping meat from hides. After a number of flakes were removed, the remaining core rock, called a chopper, could be used as a weapon, for hunting or conflict resolution.[52] The brains behind this technology were about 50 percent larger than those of australopithecines.[53]

The face of *H. habilis* was also flatter than those of previous hominins, but it retained many primitive features. *H. habilis* was less than half the size of modern humans, with apelike short legs and long arms. While meat was an important part of the *H. habilis* diet, *H. habilis* was itself an important prey species for numerous predators, including a saber-toothed cat. Meatwise, *H. habilis* was more dined upon than dining.

Homo ergaster ("working man") is for me one of the most intriguing hominins. A particularly outstanding specimen—because it is nearly complete—can be found at the American Museum of Natural History in New York. Called "Turkana Boy" (1.5 mya), after the site of its discovery (Lake Turkana, Kenya), it embodies a number of advanced features, including long legs and a humanlike chest. Turkana Boy and his kin were probably long-distance runners and proficient hunters.[54] The most striking feature of *H. ergaster* was its size, which was comparable to that of modern humans. Turkana Boy is variously estimated to have been between 8 and 12 years old at death, based on his bone and molar development, and he was already 5 foot 3 inches tall; so *H. ergaster* was tall and lean, as befitted the hot climate. Turkana Boy probably didn't have much hair and sweated profusely.[55]

Another advanced feature of *H. ergaster* was the reduced sexual size dimorphism, relative to earlier hominins.[56] The size difference between male and female *H. ergaster* was less than in *Australopithecus* but still greater than in humans.

H. ergaster used an advanced stone technology, called Acheulian, with much longer cutting edges than previous Oldowan tools.[57] Acheulian hand axes were also characteristically symmetrical and were often finished with softer material such as bone or antler, which provided more precision.

OUT OF AFRICA

Homo erectus was the first of several hominins to exit Africa, eventually spreading throughout Asia. The oldest hominid fossil found outside of Africa—about 1.8 million years old—was discovered at a site in the small town of Dmanisi, Georgia.[58] Dmanisi Man had many primitive features for a member of the genus *Homo*, representing something of a transitional state from the australopithecine condition. *H. erectus* differentiated into a number of distinct populations in Eurasia; opinions differ as to their taxonomic status. Java Man (1.6 mya), the first to be discovered, is the prototype for Asian *H. erectus*;[59] Peking Man (700,000–800,000 years ago) is a later representative of the Asian lineage.[60]

The recently discovered diminutive *Homo* fossils from the island of Flores, Indonesia, sometimes colloquially referred to as the "hobbit," may be the last of the *H. erectus* line, but this specimen is so morphologically distinctive that it has been assigned to a new species, *Homo floresiensis*.[61] Aside from its dwarfism, *H. floresiensis* was notable in that it continued to exist until about 14,000 years ago, long after *H. erectus* was thought to have been replaced by *other* members of the genus *Homo*; by this time, in fact, *Homo sapiens* had long occupied adjacent areas (Figure 13.5).

Some of the first European *Homo* fossils (1.1–1.2 million years old) are often assigned the species named *Homo antecessor*.[62] Some see this species as a link between *Homo erectus* and *Homo heidelbergensis* (Heidelberg Man), the latter celebrated for its significant brain enlargement relative to preceding *Homo* species.[63] *H. heidelbergensis* originated in Africa but, like its *H. erectus* predecessors, migrated out of Africa. This species eventually occupied all of Eurasia as well as Africa.

Both Neanderthals (*Homo neanderthalensis*) and humans (*Homo sapiens*), as well as another species referred to as Denisovans, are thought to have evolved from *H. heidelbergensis*. Neanderthals were the first to split from the *H. heidelbergensis* lineage, probably from European populations (600,000–350,000 years ago).[64] About 200,000 years ago,

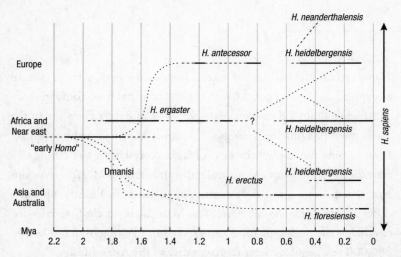

FIGURE 13.5   Phylogenetic tree for genus *Homo,* with approximate timelines. (From "Before the Emergence of Homo Sapiens" by Giorgio Manzi 2011.)

Homo sapiens split from other *H. heidelbergensis* populations in Africa; their (our) epic migration out of Africa commenced between 70,000 and 60,000 years ago. Neanderthals and humans are of about the same height, but Neanderthals were heavier and stronger. The brain size of the two species is about the same at birth, but adult Neanderthal brains were somewhat larger than those of humans.[65]

This concludes our brief survey of human evolution. We can now turn to the self-domestication hypothesis, with particular emphasis on what occurred after the split between the human and chimpanzee lineages about 7 mya.

HETEROCHRONY IN HUMAN EVOLUTION

Let's first consider the role of heterochrony in human evolution. It needs reiterating at the outset that heterochrony is only one broad category of evolutionary developmental alteration,[66] and it would be a huge mistake to shoehorn all of human evolution into it. Nonetheless, there is ample evidence of heterochronic alterations during human evolution, though not nearly as much as Gould averred. Given the self-domestication hypothesis, we will be particularly concerned with evidence of paedomorphosis and, more particularly still, evidence of neoteny, or decelerated development.

Humans are certainly born at a much less developed state than either chimps or bonobos. While bonobo neonates seem almost comically uncoordinated, they are balletic compared to the completely hapless and helpless human neonates, for whom frowning, smiling, vocalized distress, and spastic limb movements are pretty much the extent of neuromuscular development. This difference is especially impressive when you consider that humans have slightly longer gestation periods than bonobos and chimps.[67] Therefore, human fetal neuromuscular development is decelerated (neotenic) relative to our closest relatives. So, too, is much childhood development. For example, bone development (ossification) is markedly delayed,[68] and growth is retarded relative to other apes.[69]

Both humans and chimpanzees experience an adolescent growth spurt, but it commences at an earlier age in chimpanzees.[70] It is noteworthy that the onset of the growth spurt in Turkana Boy (*Homo ergaster*) and *Homo erectus* was closer to that of chimpanzees than that of humans (judging by bone development).[71] Adolescence is also protracted in humans relative to our closest living relatives, so we reach sexual maturity at a later age. In this respect we certainly differ from domesticated animals, for which not only neoteny, but also accelerated sexual development (progenesis), contributes to paedomorphosis. Human paedomorphosis lacks the element of progenesis.

So there is an important sense in which overall somatic development is neotenic, but the picture becomes more complex when we get down to details. Let's now consider neoteny with respect to some distinctively human features of functional anatomy, beginning with bipedalism. Bipedalism occurs with more frequency in immature great apes than in adults. This is why Naef and Gould included bipedal locomotion as a neotenic trait. But human bipedalism involves a suite of adaptations that are distinctly non-apelike.[72] As we have seen, *Ardipithecus* already exhibited some of these modifications more than 4 mya.[73] These traits, including pelvic modifications, increased curvature of the spine, increased leg length, and changes in foot anatomy, are further developed in australopithecines, and by the time *Homo erectus* arrived on the scene, bipedal locomotion was at a near humanlike state.[74]

Some of the most marked anatomical alterations related to bipedalism are in the pelvis. Do these pelvic alterations indicate neoteny or any other component of paedomorphosis? The short answer is no. In fact, recent research suggests quite the opposite. The pelvis of the human infant closely resembles that of adult australopithecines (which in turn resembles that of chimps and bonobos).[75] As such, the human pelvis is peramorphic relative to the last common ancestor of humans and our closest ape relatives. In fact, there is evidence for all three elements of peramorphosis in human pelvic evolution: hypermorphosis (delayed growth termination); acceleration (increased growth rate); and predisplacement (earlier onset of growth).[76]

Other anatomical alterations related to bipedalism concern the positioning of the skull on the top of the spine. Humans have distinctive modifications to a structure called the occipital condyle that allow us to face forward while walking upright, which do not at all resemble those of a juvenile ape.[77] Our long legs and large feet are also peramorphic, not paedomorphic, features.

WHAT ABOUT OUR BRAINS?

The distinctively human feature of which we are justifiably most proud is our outsized brain. Brain growth in humans is extremely fast—accelerated—relative to other apes, during fetal development and the first few years of life. But as a percentage of adult brain size, this rapid growth is actually slower than that of chimps and bonobos.[78] So, is our brain development in fact paedomorphic? No, not at all. The reason our brain growth is slower than that of apes relative to adult size is that our adult brains are so much larger. Our adult brains are larger not only because they grow rapidly, but also because they sustain this rapid growth for a longer period, which is called hypermorphosis (delayed offset), a form of peramorphosis.

The evolutionary alterations of development in the human lineage after the split from bonobos/chimps were not gradual and constant, but rather characterized by periods of relatively rapid change followed by extended periods of stasis. Australopithecines had chimp-sized brains. So, for the first 5 million years after the split, there was little alteration in brain size in the human lineage. The first members of the genus *Homo* (*habilis, ergaster,* and *erectus*) had significantly larger brains than the australopithecines (via acceleration, hypermorphosis, or both), but there ensued another long period of stasis until the advent of *Homo heidelbergensis*. The next marked change occurred when the Neanderthals evolved about 500,000 years ago.

Human brain evolution also nicely illustrates the conservative side of evolution and the tinkering nature of natural selection. The contingent fact that bipedalism long precedes brain enlargement had important consequences for human evolution, including the "obstetric dilemma":[79] a narrow pelvis is required for efficient bipedal locomotion; large brain size, though, means large neonate heads, which need to exit through the pelvis during birth. Since the human pelvis evolved before large brains, it set an upper limit on neonate brain size. Even so, a head the size of a human neonate's must make a precarious quarter turn to exit the pelvis at its widest point.[80]

Human childbirth is much more difficult and dangerous than that of other living apes, in which the pelvic girth is ample for the neonate head. Australopithecines also had ample pelvic girth. Early members of the genus *Homo* (e.g., *ergaster* and *erectus*) would have found themselves in an intermediate state birthwise: they had a narrow pelvis and brains larger than *Australopithecus* brains but smaller than the brains of later mem-

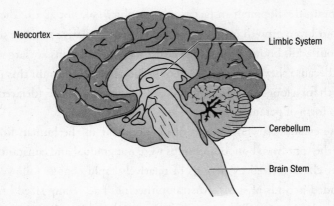

FIGURE 13.6   Schematic of the human brain.

bers of the genus. The large-brained Neanderthals, however, seem to have experienced every bit as much difficulty during childbirth as we do.[81]

There is, of course, more to human brain evolution than an increase in brain size. Various forms of neural reorganization have no doubt occurred since we diverged from the chimpanzee lineage, but such alterations leave no trace in the fossil record. Recent research has, however, uncovered some interesting differences in the development of our brains when compared to our closest living relatives. Of particular note is the fact that some of these differences point toward a neotenic pattern in the human brain. I will focus on the neocortex and especially the prefrontal cortex, which mediates planning, executive functions and many of our most sophisticated cognitive capacities generally (Figure 13.6).

Figure 13.7 depicts a typical cortical neuron. The single long projection is called an axon; it conducts electrical signals from the nerve cell to adjacent or sometimes distant neurons. The speed of conduction depends crucially on a substance called myelin, which enwraps the axon and acts as an electrical insulator, preventing electrical leakage. The timing of neuronal myelination is a key developmental event, because prior to myelination, conduction is slow and inefficient, while after myelination, electrical conduction is much more rapid. However, prior to

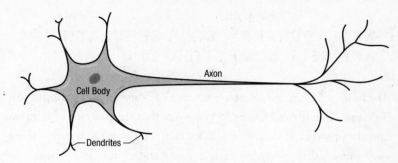

FIGURE 13.7   Cortical neuron, illustrating dendrites, cell body, and axon.

myelination the axon is much freer to "explore" a wide variety of connections with other neurons, which facilitates learning. After myelination this plasticity is lost, and with it, the childlike capacity to absorb new information. It is noteworthy, therefore, that the axons of the neurons of the prefrontal cortex become myelinated much more slowly in humans than in chimpanzees during childhood.[82] This, then, is a case of neotenic development in a particular part of the brain involved in cognition.

The neotenic development of this part of the brain is not confined to myelination. Gene expression refers to the degree to which a particular gene is actively engaged in the synthesis of the protein that it encodes (see Appendix 12 on page 344). The expression of a particular gene in a particular type of cell—such as a cortical neuron—changes as that cell lineage differentiates from a stem cell. These changes enable scientists to look for heterochrony in gene expression patterns.

A particularly striking case of such heterochrony occurs in the prefrontal cortex. A number of genes involved in neural development exhibit a heterochronic shift in humans relative to chimpanzees, for all of which adult humans resemble juvenile chimpanzees in their expression profiles.[83] In the prefrontal cortex, at least, development is slowed (neotenic) relative to chimp development, even while brain growth is accelerated. This neotenic dynamic is not brain-wide, however. In another part of the brain, called the caudate nucleus, there is no evidence of neotenic gene expression.

HETEROCHRONY, GENE REGULATION, AND SELF-DOMESTICATION

The human lineage bifurcated from that of bonobos and chimpanzees 7–5 mya. The discovery, over 35 years ago, that the proteins of humans and chimps are virtually identical, led to the proposal that much of the evolutionary divergence between humans and chimps must be due to alterations in gene expression—that is, mutations in noncoding DNA sequences, not mutations in protein-coding DNA sequences.[84] This idea was quite controversial at the time but has steadily gathered adherents, approaching something like a consensus view. Recent research in genomics, whereby DNA of chimps and humans can be directly compared, certainly seems to reinforce this consensus view, though there are vociferous dissenters who argue that the pendulum has swung too far toward an emphasis on noncoding DNA.[85] I have some sympathy for the dissenting view, partly because many coding DNA sequences are templates for proteins, called transcription, that affect the expression of other genes.[86] Therefore, gene expression is regulated by coding as well as noncoding sequences.

What matters more than this particular debate, in my view, is the shift since the 1970s toward an emphasis on changes in gene expression in explaining evolutionary divergence, which is in keeping with the developmental turn in evolutionary biology. For evolution in complex species is largely a matter of tinkering with developmental processes. One important category of such tinkering involves changes in the timing and rates of development for various anatomical and physiological systems, including the brain and behavior—in a word, heterochrony. We are now in a position to identify some of the genes whose expression has been tinkered with since humans diverged from other apes to produce these heterochronic shifts, not least in the brain.

The picture that emerges is certainly not one of monolithic neoteny —or even paedomorphosis—of the sort Gould proposed. Human evolution evidences as much peramorphosis—including acceleration—as paedomorphosis. In fact, the traits that most distinguish us from chimps and other apes—such as bipedalism and large brains—are peramorphic. Nonetheless, there is plenty of evidence for paedomorphic heterochronic tinkering in human evolution as well, including retarded physiological development in parts of the brain that are perhaps related to our unprecedented ability to learn.

There is also the intriguing reduction in human sex differences relative to apes, which appears to have commenced by 2 mya.[87] As we have seen among domestic animals, reduced sexual selection often results from particularly pronounced neotenic changes in males, which render them more like females. In this regard, the reduction in overall size differences, as well as canine length, are particularly noteworthy.[88] Human males are a lot less brawny than males of most other apes, including bonobos. With respect to sex differences, humans more closely resemble gibbons—in which sex differences are minimal—than any other apes. It is noteworthy, therefore, that gibbons are dyed-in-the-wool monogamists, and hence experience less intense sexual selection than the polygynous gorillas and orangutans, or promiscuous chimpanzees and bonobos.

Overall, there is much less evidence of neoteny in human evolution than in most domesticated species—certainly nothing approaching the degree of neoteny we find in dogs. This fact isn't devastating for the self-domestication hypothesis, but it does suggest that the amount of convergent evolution of dog and human may have been overstated. The most important convergence predicted by the self-domestication process, however, is behavioral and emotional. Does the evolution of human emotions, since our last common ancestor with chimps and bonobos, recapitulate the evolution of dogs from wolves, driven primarily as it was by selection for tameness? Are humans domesticated in this sense? If so, does this thing we call human nature more resemble bonobo nature than chimpanzee nature? These questions are the subject of the next chapter.

CHAPTER 14

HUMANS— PART II: SOCIALITY

ONE DAY IN 1986, CHRISTOPHER KNIGHT LEFT HIS home in the small town of Rome, Maine, and headed for the adjacent woods. He was not seen again—by friends, relatives, or any other acquaintances—until March 2013. His family, evidently not close-knit, never reported his absence, which was fortunate for Christopher because he didn't want to be found. He just wanted to be alone, free of any human contact. And in this he succeeded remarkably well. Twenty-seven years without a social interaction, save one brief exchange with a solitary hiker sometime in the 1990s. So intent was Christopher on avoiding human contact that during the winters he never left the area immediately surrounding his encampment, for fear he could be traced through footprints left in the snow. He never lit a fire during those 27 years lest the smoke give away his location.

Christopher would have continued in this way for the rest of his life, had he been self-sufficient in more than social ways. But he neither fished nor hunted, so he subsisted entirely on goods pillaged during the night from cabins and camps in the area. Finally, after over a thousand illicit foraging expeditions, he was caught in the act.

That he could survive all those Maine winters without fire is remarkable. Much more remarkable was his ability to survive 27 years without any direct human contact, without human interaction. For we humans are extremely social creatures. Most of us require a large dose of human interactions daily for our psychic well-being, which is why solitary confinement is tantamount to torture, so aversive are its effects on the typical human mind. Christopher Knight, though, would consider solitary confinement a state of grace.

While human sociality is the most intense among primates, there is a considerable amount of variation. At one extreme are those who require human interaction every waking minute; Christopher, along with some other true hermits—religious or otherwise—is at the extreme opposite end of the spectrum. The hypersocial, though, are much closer to the norm than are the hermits. Christopher Knight is an extreme outlier, even given the broad range of human sociability.

The personality dimension—introversion versus extroversion—can be used as a rough indicator of sociability. Extroverts require more social interactions, for their mental health, than do introverts. In fact, introverts require a certain amount of solitude, or freedom from human interaction, to feel psychologically at ease.[1] Being an introvert myself, I probably find Knight's choice somewhat less incomprehensible than most. I often crave solitude, especially after a period of extensive or intensive social interactions. I am most content when I can spend time alone each day. In my youth I fancied I could happily live a solitary existence on a part-time basis; a winter in an Alaskan cabin had its appeal. But a solo backpacking trip in Yellowstone in 1971 when I was 17 disabused me of that notion, leaving my self-image in tatters.

It wasn't supposed to be a solo trip. A friend with whom I had planned the two weeks of hiking backed out at the last minute. I briefly thought about abandoning the adventure but, fortified by a misplaced confidence in my self-sufficiency, I proceeded. Since my friend owned the car we were to use, I needed another form of transportation, which on my budget meant a Greyhound bus—a more than 40-hour bus journey with multiple stops, from Sacramento, California, to Jackson Hole,

Wyoming. There I hopped on a national-park bus bound for my trail-head near Old Faithful, Yellowstone.

Except for the driver, the full-sized Yellowstone bus was empty. I chose to sit in the very back. Having been tormented for much of the previous 40-plus hours by a hyperactive and super-chatty young woman with some serious psychiatric issues, I was ready for some quiet time, and the empty national-park bus felt like a haven. The driver, a woman on the elderly side of middle age, was an extrovert, however. Over the intercom she beckoned me to come forward; I reluctantly complied.

She—I will call her Judy—was a retired school bus driver from Dade County, Florida. Judy soon engaged me in conversation, which was pleasant enough, since she was quite self-sufficient in that regard and required little input on my part. Nevertheless, when Judy asked if I wanted the audio tour, I readily acceded, welcoming the excuse to avoid even the minimal role in the dialogue that had been mine theretofore, and hence completely relax.

It was about a 90-minute trip to my campground, so there was a lot of information, mainly natural history and local lore regarding various sites. The one I have always remembered concerned a negligible lake sitting astride the Continental Divide. It goes something like this: One day in the early nineteenth century, the famous mountain man Jim Bridger sat down for a rest there and contemplated the lake. Having struck up a conversation with the lake, at some point he queried, "Is you a pond or is you a lake?" The lake, evidently indignant, reared up and emphatically replied: "I's a lake." Hence the strange name for this unimpressive body of water: Isa Lake.

I remember thinking, "Bridger must have been pretty garrulous for a mountain man, nattering on like that, needing to hear his own voice." But two days later, after nearly 48 hours of complete solitude, I was conversing nonstop to nothing remotely human but myself. The departure point for my multinight hike wasn't more than 10 miles from Old Faithful, but the backcountry was deserted. It was after Labor Day, which in those days meant the end of the season. There were very few people in the park at all, and only a skeleton staff, except for the Old Faithful Lodge and vicinity.

I had not counted on this. I meant to go solo, but not that solo. I soon became unnerved; after two days I was a psychological mess. My increasingly bizarre conversations with myself functioned partly to keep in abeyance paranoid thoughts about what a grizzly bear might do to me.

I had intended a seven-day one-way trip, ending with a hitchhike back to Old Faithful. After three full days of solitude, I cracked. I began to retrace my steps back to the trailhead with alacrity, the volume of the conversation with myself cranked up. I managed to do the ingoing in one day what it had taken me three days outgoing. Upon finally spying the paved road, my sense of relief was almost as intense as my prior anxiety. It was dark, so I camped by the roadside—something I would never do, normally, or even, to that point in my life, could have conceived of doing.

The next morning, up at the first hint of light, I packed and headed for Old Faithful, the paved road my trail. A few miles down the road I saw a most welcome sight: a park bus. It pulled up; I climbed in. I was greeted by Judy's friendly face. When we arrived at Old Faithful, people were assembling for the geyser's next scheduled eruption. Completely out of character, I rushed to join them, reveling in the human company, however anonymous, eagerly eavesdropping on quotidian conversations. I even struck up a conversation with a stranger—something I rarely do. That night I splurged, spent my last dollar in fact, on the cheapest room at the lodge.

The next day I began my return home with Judy, who kindly waived the nominal charge. We talked for the entire trip, this time with my full participation; I almost teared up when we parted at Jackson Hole. There I found a Western Union office and sheepishly wired my parents for money to pay for the bus ride home; and I was out of there the next day, humiliated but enlightened.

My teenage misadventure in Yellowstone illustrates that even someone at the low end of the sociability scale requires a lot of social interaction for his/her mental health. For social interactions are part of the very core of our humanity. Moreover, our sociability, not our vaunted intelligence, is the single most important factor in our domination of the earth.

Our nearest relatives, chimpanzees and especially bonobos, are highly social creatures, but they are recluses compared to us. At densities in which we humans commonly reside, even the peaceful bonobo would be so stressed that there would be no hint of social order; no amount of penis fencing, female genital-to-genital rubbing, oral sex, or *Kama Sutra*–level variations of heterosexual intercourse could prevent carnage from ensuing. But until recently, our sociality has received short shrift in treatments of our differences from other apes.[2] The emphasis is generally on our larger brains, language, manual dexterity, and bipedalism— important factors all. But even given these traits, if we were merely as social as bonobos, we humans wouldn't have become "masters of the planet," to use Ian Tattersall's phrase.[3] Did our hypersociality come by way of self-domestication, through natural selection for tameness?

"IT AIN'T NECESSARILY SO"[4]

To delimit with confidence the role of natural selection in our social proclivities is extremely difficult if we are to apply anything approaching the normal standards of evidence in evolutionary biology. In part, the difficulty is a quite general one: It requires a lot of ingenuity and effort to isolate the causal role of natural selection in the evolution of any trait in any creature. First, it must be determined whether the existing trait was the product of selection of any kind; second, selection *for* a trait must be distinguished from selection *of* a trait; that is, the target of selection must be distinguished from correlated by-products. This discrimination is easiest to accomplish in a controlled laboratory setting by means of artificial selection, but a number of compelling cases have emerged from careful and laborious field studies as well, sometimes with experimental manipulations.[5] Research on domestication, as we have seen, combines elements of both.

Knowing the ancestral condition and that of the descendant, as is the case in evolution by domestication, is, of course, extremely helpful. For human traits that fossilize, such as teeth, bones, and brain size, the situation is much better than for traits—such as cognition and emotions—that don't. Behavior is somewhat intermediate but closer to cognition than to bones. Behavior doesn't fossilize, but some behavioral artifacts do.

Short of experimental control, whether in the laboratory or in the field, the next-best approach to the problem is to combine genealogical (phylogenetic) and ecological information—the "comparative method":[6] the comparison of related species that vary with respect to a particular aspect of the physical, biological, or social environment. From these genealogy-based ecological comparisons we can make preliminary inferences about selection for a particular trait. Unfortunately, humans often do not lend themselves to the comparative method, because we are the only surviving members of our genus and we diverged from our closest surviving relatives 5–7 mya. Call this the "N of 1" problem, where N stands for the sample size to be used for comparisons.

The "N of 1" problem is not universal for studies of human evolution; it depends entirely on the trait under consideration. As we saw in the previous chapter, the comparative method can be effectively applied to traits like brain size and even neocortex size. In these cases it is entirely legitimate to compare humans and other primates, and indeed all other mammals, because the developmental processes involved are deeply conserved. Where the "N of 1" problem looms particularly large is for uniquely human traits, such as language, and for the numerous human cognitive adaptations posited by evolutionary psychologists.[7] We will consider shortly whether the human self-domestication hypothesis involves traits that are more like language, where the "N of 1" problem is acute, or whether human self-domestication is more like brain size, where the "N of 1" problem is not a problem.

But there is another factor that greatly complicates the evaluation of any proposed human cognitive adaptations: the huge role that culture plays in our cognitive lives. The self-domestication hypothesis purports adaptations that are biological, the product of natural selection. But culture plays a formative role in the development of all human cognitive capacities. Given the prodigious cultural forces that influence human cognitive development, it is no easy task to distinguish a purely biologically adaptive (as opposed to culturally adaptive) component in the development of complex human cognitive and emotional traits. Usually such attempts simply involve a search for "cultural universals," which are supposed to implicate natural selection.[8] But this inference is far from straightforward. Cultural universals can be universal for reasons other than natural selection.[9] Conversely, some of the best evidence for active selection in humans comes from traits, such as lactose intolerance, that are not culturally universal.[10]

For all of these reasons, many questions about human cognitive evolution may remain forever unanswered, by the standards of mainstream evolutionary biology, especially those concerning what past selection was *for*. We need to acknowledge at the outset that the self-domestication hypothesis may fall into this category.

This counsel of humility, given the obstacles just described, goes completely unheeded by many, especially those prone to making bold claims about human nature based on the slightest evidence that a trait is adaptive. From human ethology to sociobiology to evolutionary psychology, there has been an overwhelming tendency to blithely ignore the evidential standards of mainstream evolutionary biology. Evolutionary psychologists are in many ways the worst offenders; the assumptions that distinguish evolutionary psychology from its predecessors provide quite effective insulation from empirical assault (see Appendix 14 on page 347).

In lieu of the biological information required to apply the comparative method, evolutionary psychologists adopt a first-principles approach, in which they make assumptions about the human environment, usually an unspecified period of the Pleistocene, and then infer, through a process called reverse engineering, what natural selection "must have done to the human mind/brain.[11]

REVERSE-ENGINEER THIS

Reverse engineering is the practice of taking apart human artifacts, such as clocks, radios, and even computer programs, to figure out how the parts contribute to the functioning of the whole. In its evolutionary application, the first step of reverse engineering is to determine the problem—some feature of the environment—for which the phenotypic "design" is the solution. Then, optimality criteria are applied to establish the best solution. Finally, the design of the parts (phenotypic traits) is explained in terms of their contribution to this optimal solution.

The extension of reverse-engineering analyses into the realm of biology depends crucially on the degree to which evolved biological structures resemble designed human artifacts. And there are certainly similarities enough to justify at least a limited role for reverse engineering in suggesting putative biological adaptations. But these similarities are often quite superficial,[12] not nearly enough to carry the weight required for the adaptationist claims of evolutionary psychology.

Let's consider just one kind of limitation on reverse engineering in the biological realm that stems directly from the conservative nature of evolution: the fact that evolution doesn't optimize from the ground up, but only tinkers with developmental processes already in place. Because of this deep historical contingency, complex phenotypic traits are cringe-worthy from an engineering perspective—none more so than the human brain.

The German anatomist Ludwig Edinger was a pioneer in the search for homologies in vertebrate brains; from these homologies he sought a historical reconstruction of brain evolution.[13] Edinger conceptualized brain evolution, from fishes to mammals, as a series of layers accreted over the brain stem. In the 1960s the neuroscientist Paul MacLean proposed an overarching hypothesis for the evolution of the human brain beginning with what he conceived as the reptile condition. MacLean divided the brain into three components corresponding to three stages of evolution.[14]

The most primitive part of the forebrain, which MacLean referred to as the "reptile brain," includes structures that, according to him, provided the neural substrates for species-typical ritual or instinctive behavior, and basic homeostatic functions like breathing. Anatomically, you can think of the reptile brain as the brain stem and the associated foundational structures. Built on top of the reptile brain is the second component, which MacLean called the "limbic system"—a number of interconnected neural regions that support emotional responses from sexual to aggressive. According to MacLean, the limbic system evolved with the first mammals, so he referred to it as "paleomammalian." The third component, called the "neocortex," is found only in primates, according to MacLean. The neocortex, which surrounds the limbic system, is where the neural substrates for language, abstraction, planning, and other executive functions reside.

There are a number of problems with this "triune brain" hypothesis. For example, parts of the reptile brain are found in all vertebrates, including fishes;[15] parts of the limbic system occur in reptiles;[16] and the neocortex is not confined to primates.[17] But by far the biggest problem with the triune brain framework is that it begins the story of brain evolution far from the beginning. Recent evo-devo research has revealed that not only our brains, but our vaunted neocortex, share homologies with the first brained creatures, called flatworms, with whom we share a common ancestor that existed prior to the Cambrian explosion over 500 mya.[18] A lot happened in brain evolution between flatworms and fishes, and much more between fishes and reptiles. Our brains, perhaps more than any of our other organs, reflect the conservative tinkering nature of the evolutionary process.

Actually, we can trace brain tinkering to a prebrain stage, when the nervous system consisted solely of a network of a few neurons, such as we find today in jellyfish. Those neurons remained essentially the same, even as neural nets were transformed into complex brains like ours, and not because they are particularly efficient. The neurons we inherited from jellyfish are quite leaky with respect to the electrical charges they conduct.[19] These leaky neurons were fine for jellyfish because jellyfish require only a few neurons to do what jellyfish need to do. For us and countless other more complex creatures, they are quite suboptimal, and the myelination of some axons is a make-do evolutionary response to ameliorate the problem.

If our neurons were more efficient electrical conductors, we wouldn't require such big brains. And if we didn't have such big brains, the neurons in our cortex wouldn't have to make such perilous journeys from their place of origin to their final resting place, leaving us vulnerable to all manner of neurological disorders.[20] The fact that even the neurons on the surface of our brain must migrate from generative zones deep in the brain is also an inefficient homology.[21] Any decent electrical engineer would produce neurons closer to where they ultimately need to be.

Much of our brain wiring is also inefficient, because it was not designed from the ground up; it is, rather, a patchwork of repurposed elements and jury-rigged add-ons, what Gary Marcus has so aptly labeled "kluges."[22] These kluges extend to the bits of the brain we hold most dear, such as the circuits involved in language.[23] One of the main organizing principles of the brain is called "neural reuse," in which neural circuits that evolved to serve one function are recycled, repurposed, and redeployed for completely different ends during the course of evolution.[24] Brains designed by an omnipotent God should be reverse engineerable; brains that are products of evolution, not so much.

IS THE SELF-DOMESTICATION HYPOTHESIS MORE THAN A JUST-SO STORY?

Given the stringent standards of mainstream evolutionary biology and the impossibility of circumventing them through reverse engineering, what are the prospects for testing the self-domestication hypothesis? Is it just another just-so story? I think not. In principle, at least, conserved homologous structures in the brain and endocrine systems can be used to advantage here, to surmount the "N of 1" problem and legitimately apply the comparative method.

The weak version of the self-domestication hypothesis is that humans and bonobos have experienced selection similar to that observed in many domesticated animals—with respect to tameness—relative to a chimpanzee-like state. In essence, chimpanzees are treated like the wild ancestors of our domesticated fauna, the standard by which we measure the effects of self-domestication. Therefore, we should expect humans and bonobos to be more prosocial than chimpanzees.

Hare further claims that tameness was achieved through paedomorphosis.[25] Here he clearly has in mind dogs and the farm fox experiment; indeed, he is particularly concerned with demonstrating the convergent evolution of humans and dogs. We have seen that many other domesticated mammals evidence paedomorphosis as well, but it is not universal. Experimentally domesticated rats became tame but not paedomorphic, at least not obviously so. Given the emphasis on genealogy throughout this book, it is worth noting that humans and other primates are more closely related to rats and other rodents than to dogs, foxes, and other carnivores. Hence the paedomorphic element might be viewed as dispensable for a weak version of the self-domestication hypothesis.

The strong—and I think more interesting—version of the self-domestication hypothesis predicts that convergent evolution for tameness was achieved by parallel neuroendocrine alterations in humans (and bonobos) on the one hand, and dogs, cats, rats, and other domestic creatures on the other. In other words, we should expect convergent evolution of homologous elements in the endocrine system and the brain, not just behavior.

The endocrine system of particular relevance is the hypothalamic-pituitary-adrenal axis that undergirds the stress response, which is highly conserved not only in mammals but in all vertebrates.[26] With respect to the brain, the focus should be on the limbic system, a mammal-wide suite of neural circuits that are the substrates for emotional behavior.[27]

The centrality of the limbic system in our emotional life is demonstrated in those rare individuals who are born with no cerebral cortex. It is noteworthy that these individuals display the full range of human emotions and affective behavior.[28] In general, they seem quite happy and socially well adjusted, perhaps a bit more affectionate than the fully brained. Experimentally decorticated mice also exhibit the full repertoire of mouse affective behavior.[29] If selection for tameness in humans and bonobos has truly caused human behavior to converge on that of dogs and other domesticates, it should leave a mark in the limbic system.

I find even the weak version of the self-domestication hypothesis intriguing because it suggests a novel explanation for the success of our species: our hypersociality and unprecedented capacity for cooperative behavior. Heretofore, the overwhelming emphasis has been on human intelligence in explaining how we came to dominate the earth. Our seemingly outsized neocortex is where the evolutionary action has been on this view. Lately, one particular aspect of our intelligence—our social intelligence—has been receiving the most attention.

SOCIALITY AND SOCIAL INTELLIGENCE

Primates are among the most social mammals; primates are also the most intelligent mammals. We humans are the smartest primates; we are also the most social primates. Is there a connection? Many believe there is. In surveys of primate species, a number of studies have found a correlation between the size of the neocortex and the size of the groups in which they live.[30] The correlation between neocortex size and group size suggests to some that the huge neocortex in humans resulted from selection for the management of complex social interactions.[31]

On this view, larger groups result in more complex social interactions requiring greater political dexterity, which in turn requires more neural resources of the kind the neocortex provides. We humans have the largest neocortex because we have the most complicated social relationships and hence require the most "Machiavellian intelligence" to navigate our social environment and therefore also the most neural resources.[32]

This is certainly not a just-so story; the comparative method has been put to good use here. But from a neurobiological perspective, there are some problems. The first is the assumption that brain size or the sizes of particular brain regions are causally significant with respect to any particular behavior or cognitive skill. Size is a gross measure that does not directly relate to computational capacity.[33] And establishing cause and effect with respect to correlations between brain size measures and cognition is not at all straightforward.[34] That is a particular problem in this case because the amount of variation in neocortex size associated with group size is actually quite small,[35] which is what you would expect of a brain region involved in all learning, about diverse matters— including the physical and ecological environment—not just the social environment.

A recent primate-wide survey found little evidence that change in the size of the neocortex resulted from selection for the ability to manage social relationships.[36] Moreover, the human neocortex, relative to total brain size, is what you would expect of not only any primate, but also any mammal with a brain the size of ours.[37] That is, our neocortex, where all the social calculations would occur, is not particularly large, relative to the rest of our brain, including the parts that have no cognitive function at all, given the highly conserved pattern of primate brain scaling. Nor is the frontal lobe—purportedly the site of our most executive cognition—particularly large by primate standards.[38] Perhaps even more surprising is the fact that only 19 percent of the neurons in the human brain reside in the neocortex, just as in all other primates.[39] Humans have more cortical neurons than any other primate, but not proportionally more.

COMPETITION OR COOPERATION

The social intelligence hypothesis emphasizes competitive social interactions. While it is certainly true that our competitive interactions are the most complex among mammals, it is our cooperative social interactions that most distinguish us from other mammals. There is a computational component to our cooperation, in communicating our goals and coordinating our activities, for example—whether in constructing a house or playing a game. But before there is a way to get things done collectively, there must be a *will* to get things done collectively. That is, the proper motivation must be in place. Undergirding this motivation are emotional predilections. According to the self-domestication thesis, alterations in emotions—primarily fear and aggression—were the prime movers in human cognitive evolution. It is this reduced fear and aggression in the presence of our fellow humans—in a word, tameness—that is the foundation of our cooperative interactions.

It will be useful at this point to compare humans and chimpanzees for their cooperative inclinations. Michael Tomasello and his associates at the Max Planck Institute for Evolutionary Anthropology in Leipzig, Germany, conducted much of the research to which I will refer. Tomasello distinguishes three key features of cooperative interactions.[40] First, the participants must share a joint goal and a commitment to that goal. Second, the participants must be motivated to fluidly adopt complementary roles in achieving that goal. Third is the criterion of mutual aid or support, which means that a participant must be not only prepared to perform his particular role, but also willing to help a fellow cooperator fulfill her role if necessary.

Many primates engage in shared group activities, but few such activities meet these criteria for truly cooperative behavior. The group hunting behavior of chimpanzees is often touted as an example of cooperation in which one individual, called the "driver," chases the prey in a certain direction; others, called "blockers," prevent the prey from changing direction; and finally, "ambushers" silently move in for the kill.[41] But as Henrike Moll and Tomasello convincingly argue, the terms "driver," "blocker," and "ambusher" may well be anthropomorphic overinterpretations of what actually goes on in a chimpanzee hunt.[42]

The group hunt may be accomplished without any shared goal or plan; rather, each individual simply chases the prey from its particular position at that moment. This position is far from prearranged; rather, each chimp simply adopts whatever spatial position is most immediately available. The chimps are certainly responsive to each other's behavior during the hunt, but they lack the "jointness"—or, more technically, "shared intentionality"—that characterizes true cooperation. Laboratory experiments support this leaner interpretation of chimpanzee hunting behavior.

Juvenile human-reared chimps were first tested for their ability to cooperate with humans in a problem-solving game and then in a purely social game. In the problem-solving game the chimps did quite well, but only when the human was constantly present. After any interruption the chimps failed to reengage the human partner to resume the game. In the purely social game, the cooperation never even got off the ground.[43] For humans (18–24 months of age), the results were quite different. They cooperated enthusiastically in both the problem-solving and the social games, and when the adults stopped participating, they sought strenuously to reengage them. So, by the age of 18 months, human infants are eager to jointly commit to a shared goal—the first criterion of cooperative behavior. No such commitment can be found in chimpanzees. Other experiments have shown that human infants but not juvenile chimpanzees also meet the second (reciprocity) and third (mutual aid) criteria as well.[44]

Cooperation requires communication. Human communication via language is, of course, nonpareil. But humans are also superior to chimpanzees with respect to nonlinguistic communication of cooperative intent. Infants who have not yet acquired language skills outperform much older chimpanzees in cooperative communication. For instance, though chimpanzees frequently gesture in a variety of ways, there is not one documented case of a chimpanzee pointing for another to direct its attention.[45] By 12 months of age, human infants point to inform others not only of what the infants want but also of things that the infants believe the adults want.[46] By this young age, human infants have a motivation to inform others for nonutilitarian reasons. Chimps show no interest in sharing information in this way, whether with humans or with other chimps.

So, during the last 5–7 million years, the psychology of humans and chimps has diverged considerably. Was this evolutionary divergence primarily emotional or computational (social intelligence)? This question's premise is actually a false dichotomy; obviously, both emotions and intelligence, as well as emotional intelligence, were altered in humans relative to chimps. But according to the self-domestication hypothesis, the psychological change that got the ball rolling was emotional.[47] At some point in human evolution, some individuals became more tolerant of each other and were less aggressive and fearful in each other's presence. In this respect our evolution is convergent with that of the dog and some of the other domesticated species described in this book.

Bonobos are much more closely related to chimpanzees. Brian Hare and his colleagues conducted an interesting series of experiments on bonobos and chimps in which they tested first their social tolerance and then their ability to cooperate to accomplish a joint goal.[48] Tolerance was tested in the context of joint feeding, which can be emotionally fraught, given the temptation to selfishly hoard or monopolize all of the calories. When presented with fruit in a single dish, chimps did succumb to this temptation. Bonobos, however, had no problem sharing, diffusing the tension with their characteristic sexually charged play. The bonobos were more tolerant of each other and seemingly less stressed than the chimpanzees.

Next, pairs of bonobos and chimps were tested for their ability to cooperate to accomplish a common goal. They were required to jointly pull a rope to attain food dishes. When it was obvious that only one dish contained food, chimps would not cooperate and hence obtained no food. Bonobos, however, cooperated quite effectively to obtain the food, which they then shared. If both dishes had food and hence the food could not be monopolized, bonobos and chimps did equally well on the rope-pulling task. In general, chimps do as well as bonobos in cognitive tasks in competitive situations. Bonobos outperform chimps when cooperation is required. Moreover, bonobos, like humans but unlike chimpanzees, will share food with complete strangers.[49] These experiments demonstrate that the cooperative superiority of bonobos is related to their higher social tolerance. Increased social tolerance is characteristic of domesticated mammals from rats to dogs.

EVOLUTIONARY CONVERGENCE TOWARD TOLERANCE?

Hare and colleagues emphasize that the evolution of tolerance through self-domestication makes no claims about the particular neuroendocrine alterations involved—only that they result from paedomorphic brain development causing juvenilized social behavior.[50] This is the weak form of the self-domestication hypothesis. I call it weak because it simply implies that under a broad range of environmental conditions in which there is restricted mobility, there is selection for reduced aggression by way of paedomorphosis.[51]

Hare discusses two sorts of evidence for the weak form of the self-domestication hypothesis, both of which point to evolutionary convergence of humans, bonobos, dogs, and other domesticated mammals. The first kind of evidence, not surprisingly, consists of behavioral similarities, particularly reductions in aggression.

Dogs are much less aggressive than wolves. Encounters between wolf packs often result in death.[52] Groups of feral dogs, however, rarely engage in physical violence.[53] Within-group aggression is also much higher in wolves than in dogs. In dog parks across the United States, the vast majority of interactions are friendly, though playmates often have little or no familiarity with each other. Wolves placed in a similar situation, even if hand-reared by humans, would make even the most cantankerous pit bulls seem subdued.

We saw similar reductions in aggression during the self-domestication phase of domestication in other carnivores, including cats and ferrets, and incipiently in raccoons. We also saw that cavies are more socially tolerant than their wild ancestors. Humans and, to a lesser extent, bonobos also exhibit reduced aggression if the chimpanzee state is ancestral. This is certainly consistent with the self-domestication hypothesis, but far from compelling evidence.

The second kind of evidence is paedomorphosis. At times Hare seems to take any evidence of paedomorphosis in humans, bonobos, and dogs as support for self-domestication, but the focus should be on paedomorphic emotional development and the relevant neuroendocrine substrates.[54] There is good evidence for both in dogs. In addition, mice that were selected for high and low levels of aggression exhibited the predicted heterochronic alterations in social behavioral development. No such alterations were observed in experimentally domesticated rats, however. Bonobos seem to evidence some paedomorphic behavioral development, but the evidence for such in humans is slim. More generally, while all domestic mammals are tamer than their wild ancestors, degrees of paedomorphosis range from substantial to nil. Paedomorphosis is not a universal aspect of the domesticated phenotype. In my view, paedomorphosis is actually a dispensable element of the self-domestication hypothesis, even for the strong version.

The strong version of the self-domestication hypothesis predicts parallel alterations in the neural and endocrine systems of domesticated species like dogs and cats and putative self-domesticated species such as humans and bonobos. The endocrine component is the most accessible, and the experimental domestications of foxes and rats already provide a useful point of departure for subsequent investigations. Foxes and rats are quite distantly related, having parted evolutionary ways over 60 mya. Yet they exhibit interesting similarities with respect to domestication-induced alterations in their homologous stress physiologies. The strong version of the self-domestication hypothesis predicts similar alterations in humans and bonobos relative to chimpanzees. Some evidence suggests that bonobos and chimps exhibit divergent endocrine responses in social situations. In response to competitive situations, male chimps—but not bonobos—experience elevated testosterone levels.[55] The elevated testosterone levels may render chimps more aggressive than bonobos in these situations.[56]

The limbic system is just beginning to receive attention, with a focus on the amygdala and its connections. Some recent comparative research on hominoids is particularly relevant. Autopsies showed that parts of the amygdala of humans more closely resemble the bonobo amygdala than the amygdala of other apes.[57] Other studies have identified differences in neural connections from the amygdala in bonobos and chimps,[58] and an association, in humans, between both amygdala connectivity[59] and volume,[60] and the size of an individual's social network. The authors of some of these studies have speculated—prematurely, in my estimation—about a connection between these anatomical differences and differences in empathy, prosocial behavior, and even sexual behavior in humans, bonobos, and chimps. Much, much more research of this sort is required for any confidence in such brain-behavior connections in hominoids. With respect to the strong version of the self-domestication hypothesis, it will be important to establish such differences in, for example, wolves and dogs, as well.

DOMESTICATED DOMESTICATORS?

Given the state of our current knowledge, it is much too early to pronounce a verdict on either the weak or strong versions of the self-domestication hypothesis. But as a sort of organizing principle, a stimulus for future research, the self-domestication hypothesis provides a useful counterpoint to traditional views of human evolution in two important respects. First is the shift in focus from our so-called higher cognitive faculties and their neural substrates to parts of the brain we share with all mammals that undergird our emotional behavior. Second is the emphasis on our cooperative, or prosocial, side—rather than, as has been the focus heretofore, our competitive interactions—in driving the evolution of social behavior. For it is our unmatched ability to cooperate for shared ends that most distinguishes us from other primates. And it is through our cooperative endeavors that we have become a force of unprecedented power among any life-forms to have ever evolved. While all living things have the capacity to change their environment, our ability to do so is unmatched and ever accelerating, to the point that we humans have collectively constructed the environment in which most future evolution will occur. The animals we domesticated were in the vanguard of a new phase of life on earth: the Anthropocene.

CHAPTER 15

THE ANTHROPOCENE

THROUGHOUT MOST OF THE HISTORY OF OUR SPECIES, we have been a minor part of the earth's fauna. Had an alien naturalist from some distant planet visited the earth 100,000 ago and done a relatively complete earth-wide survey of all life-forms then, she would have discovered that then, as now, the dominant life-forms were bacteria. In the sea, our alien naturalist—call her Alice—would have been impressed, by first the abundance and biomass of planktonic creatures, and then perhaps by the crustaceans, including krill; among the larger creatures, fishes would have most impressed Alice for their diversity and numbers.

On land, plants and fungi would have figured prominently in the final report. Of those creatures we call animals, Alice would have first been struck by the abundance and diversity of insects, which remain the most successful land animals on earth. Of those creatures we call vertebrate animals, the birds would have been the most obvious to Alice. Mammals, too, would have required a section in the final report, though perhaps not a full chapter. Most of the mammal section would have concerned rodents; bats would have gotten decent coverage; artiodactyls, perissodactyls, and elephants would have warranted a paragraph or two, if only for their large size.

Alice would have noticed the primates certainly, but noted as well their limited numbers and biogeographic distribution. Of these primates, the tree dwellers were obviously the most noteworthy; though a few, Alice would have noted, were firmly committed to life on the ground. Among these well-grounded primates, some had a peculiar bipedal gait. The bipedal primates could be observed throughout much of Eurasia and Africa, but nowhere were they particularly abundant. Had Alice a keen eye, she would have noted several forms of these bipedal primates, perhaps four or more in Eurasia and one in Africa. The African biped would have rated merely a footnote in the final report because of its low population numbers and ecological unimportance. Smart and observant as our alien naturalist was, Alice would never have anticipated that this bipedal African primate would one day dominate the earth.

Had Alice visited the earth 30,000 years later (70,000 BP), humans— the African biped—might not have rated a footnote. For, according to some, a huge volcanic eruption (Mount Toba) had by that time nearly eliminated the human species through its direct and indirect effects.[1] But survive we did, and within 10,000 years of the eruption, humans could be found throughout East, South, and North Africa, though not in numbers to impress any alien naturalist. But the next 50,000 years or so were good for humans. And as the human population increased, so, too, did the human impact on other living things, both direct and indirect. Among the more dramatic direct evolutionary consequences of human thriving was the domestication of a number of plants and animals.

What is it that caused an inconsequential component of the African fauna to become such a powerful evolutionary force worldwide? In particular, how is it that humans became domesticators? Before I address these questions, it will be helpful to continue the story of human evolution from the point where we left things in the previous chapter, about 200,000 years ago when the human species first made its appearance.

THE ASCENT OF MAN

Our species, *Homo sapiens*, originated somewhere in the eastern half of Africa about 200,000 years ago.[2] We entered the African scene with something more like a whisper than a shout. Physically, what perhaps most distinguished us from other contemporary hominids was what anthropologists refer to as "gracility" (as opposed to robusticity), by which is meant a general slightness, thinness of tooth and bone, and lack of powerful muscles. We were also a bit longer of limb than other members of our genus. Behaviorally what distinguished us is, well, not a whole lot—perhaps some proto-linguistic skills.

Here are some highlights of subsequent human migrations based on mitochondrial DNA, which, remember, gives us only the female side of things. There is a fair amount of variation in estimates for the timing of the African exit and subsequent dispersal, based on differing assumptions about mutation rates in particular parts of the mitochondrial genome. What follows is largely based on a review by Phillip Endicott and associates.[3]

There were two possible routes out of Africa: one through the Sinai Peninsula and into the Near East; the other south of the Red Sea, through Saudi Arabia, and into Southwest Asia.[4] From Southwest Asia humans moved rather quickly to India through Southeast Asia and arrived in Australia before 40,000 years ago, probably thousands of years earlier.[5] Another, much shorter route took humans from Southwest Asia through Anatolia and into southeastern Europe by 40,000 years ago. Central Asia seems to have been populated somewhat later— about 30,000 years ago—from the west (Europe) or southwest, as was China through a southeastern route. By 20,000 years ago, humans had reached northeastern Siberia; from there some continued westward across the Bering land bridge into North America. Estimates of the first arrival in North America vary: 14,000 years ago is a conservative estimate; it could have been earlier. As in the migration from Southwest Asia to Australia, humans quickly spread from northwestern North America to southernmost South America, probably along the Pacific coast.

When humans first arrived in Eurasia there were at least two other hominid species there: Neanderthals in Europe and western Asia, and Denisovans to the east. There is also evidence of a third species, perhaps *Homo erectus*. What happened when they met? How did they interact? Archaeological evidence alone is not very helpful in answering those questions, but genomic evidence provides some clues.

We now have fairly complete genomes for both Neanderthals and Denisovans as a result of technological breakthroughs that provide genomic scientists the means to extract DNA from fossilized material.[6] This genomic information indicates that humans, Neanderthals, and Denisovans largely kept to themselves when they came near each other. But not always. Occasionally some trysting occurred. The genomes of Europeans and some Asians have 2–4 percent Neanderthal DNA; Australian aborigines and some Pacific island populations are about 6 percent Denisovan.[7] Most Asians also have a bit (0.2 percent) of Denisovan in them, as do Native Americans. All of this accords with the human migration routes described here.

Some trysting also occurred between Neanderthals and Denisovans after they diverged about 300,000 years ago, and the Denisovan genome has traces of much earlier assignations with a fourth, unknown hominid that could have been *Homo erectus*.[8] It is noteworthy that no African populations evidence any trace of Neanderthal, Denisovan, or the unknown fourth hominid. All of this interbreeding occurred after the migration out of Africa.

FROM HUNTING-GATHERING TO FARMING

By 30,000 years ago, Neanderthals had largely, if not completely, disappeared, as had the Denisovans. There has been much conjecture as to why the Neanderthal in particular fell off the tree of life. Some believe that we humans were the culprits, outcompeting our cousins by virtue of our cultural advantages.[9] Others speculate that our having converted wolves into dogs gave us the crucial competitive advantage.[10] But that would require a more intimate human-dog relationship—far beyond the self-domestication stage—than most experts on dog domestication believe could possibly have existed at that time. Climate change certainly didn't help the Neanderthal cause.[11]

Humans entered Eurasia during a period of tremendous climatic fluctuations, periods of rapid warming and cooling, advancing and receding glaciers, and changing precipitation patterns. About 20,000 years ago, conditions were as cold as they had ever been during the entire preceding 2 million Ice Age years. Ice covered most of Europe and northern Asia. Sea levels were as much as 400 feet lower than they are today. Then it began to warm. Cold-adapted plants and animals retreated northward, to be replaced by flora and fauna from warmer climes formerly to the south.

By 15,000 BP, much of the Near East was in a Garden of Eden–like state.[12] Conditions were ideal for a hunting-gathering existence, which is the only existence humans had ever known. Gathering was particularly good: figs, legumes of various kinds, wild grasses such as wheat, and gourds. The latter made useful containers. Over the centuries, human populations acquired valuable knowledge about the biology of all these plants.[13] This knowledge became the basis for a more settled existence. Surplus gatherings were stored to some degree. By 12,000 BP, the agricultural revolution was under way. Goats, cattle, sheep, and pigs were about to be brought under the human yoke, with fateful consequences for our species and, ultimately, all life on earth.

SO WHAT CHANGED?

How is it that we achieved such a degree of dominion in an evolutionary eyeblink? In addressing this question it will help if we first consider one of the seminal developments in human evolution, which anthropologists refer to as cultural (or behavioral) modernity.[14]

There was a time lag of at least 100,000 years between the first appearance of our species in Africa and archaeological evidence of the sort of cultural sophistication we think of as distinctly human. Among the indicators of such sophistication that archaeologists look for are evidence of burials, travel by boat, the extensive use of nonstone materials such as bone and antlers, and especially symbolic behavior in the form of self-adornments, such as pigmented and decorated objects.

This lag between the appearance of the first human fossils, and of the first artifacts that we think of as distinctively human, led to a distinction between anatomically modern humans and behaviorally or culturally modern humans, and a lot of speculation as to what caused the transition to the latter state.[15] Central to the dispute is when and how quickly the transition occurred. One school of thought is that there was "a great leap forward" or "human revolution" about 50,000 years ago.[16] Many archaeologists dissent from this view, however; evidence for symbolic behavior has now been pushed back to at least 100,000 years ago, and some think much earlier, suggesting a much more gradual transition to the culturally modern condition.[17] How you come down on the timing issue is a good predictor of the sort of explanation you will champion as to the cause of the transition from anatomically modern humans to culturally modern humans. Generally, the explanations fall into two categories: biological evolution and cultural evolution.

There is a range of opinions within each camp; it is largely a matter of emphasis: most "biologizers" recognize some role for culture, while most "culturalists" acknowledge some role for biology. Moreover, a third position has become increasingly popular that explicitly combines elements of biological and cultural evolution. Call this third category "biocultural evolution." The biocultural approach, while acknowledging the paramount role of cultural evolution in human dominion, sees a continuing role for biological evolution and, more important, a complex interaction of biological evolution and cultural evolution.

SEARCHING THE HUMAN GENOME FOR ANSWERS

Biologizers generally are advocates for the "great leap forward" view of the transition from anatomically modern humans to culturally modern humans. One popular view is that this transition resulted from a mutation that significantly altered brain development in ways that made language possible.[18] The discovery of the *FOXP2* gene was a most welcome development for advocates of this view. To be more precise, it was the discovery of differences in the *FOXP2* gene of humans and chimpanzees that garnered attention.[19]

FOXP2 codes for a protein transcription factor called forkhead box protein (FOXP).[20] Since it was the second of the FOXP proteins discovered, it is called FOXP2. The *FOXP2* gene is expressed in several types of tissue in humans, most notably the brain. It is considered especially significant that mutations in this gene are associated with severe speech and language deficits.[21] Moreover, inactivation of the gene in songbirds interferes with song development.[22] On the basis of this evidence, researchers looked for *FOXP2* in the chimpanzee genome and compared it to the human *FOXP2*. They identified two differences in the versions of this gene in chimps and humans—two point mutations that resulted in two amino acid substitutions in the FOXP2 protein.[23] Since chimp *FOXP2* resembles that of other primates, it was inferred that an alteration in the *FOXP2* gene had occurred after humans diverged from chimpanzees and, more speculatively, it was proposed that these alterations were somehow involved in the evolution of our unique language skills.

Most anthropologists are of the opinion that language is a uniquely human attribute, one that is not only lacking in chimps but was also absent in Neanderthals. It came as a surprise, therefore, that Neanderthals had the exact same form of *FOXP2* that humans have, and so did the Denisovans.[24] So, either our *FOXP2* is not all it was hoped to be, language-wise; or language evolution commenced before the human species came into existence; or both. In any case, the amino acid substitutions in human *FOXP2* cannot be responsible for the great leap forward if indeed there was a great leap forward.[25]

The failure of *FOXP2* to account for the rise of humankind does not, of course, invalidate the biologizers' position. With the advent of rapid sequencing technologies has come a multitude of data to mine in a quest for the genomic jewels in the human crown. Originally, as in the case of *FOXP2*, the first step was to compare human and chimpanzee genomes to see what makes us different. Several of these whole-genome comparisons have been conducted with an eye toward determining the relative significance of coding versus noncoding genomic alterations in humans relative to chimps. A recent review concluded that noncoding differences are predominant in DNA sequences related to neural development, while human-chimp divergence in immune functions and olfaction is predominantly via changes in coding sequences.[26]

A popular method for distinguishing candidate genomic jewels is to measure the degree of divergence in particular DNA sequences since the human-chimpanzee split. One such study identified 202 genomic elements that are highly conserved in mammals, including chimps, but evidence human-specific divergence.[27] These so-called human accelerated regions (HARs) could be signs of directional selection and therefore an important clue to what makes us different from chimps.[28] Most HARs are in noncoding DNA sequences. The most accelerated of the HARs is HAR1, which, though noncoding in that it is not a template for a protein, does template two large RNAs that are physiologically active.[29] Particularly intriguing is the fact that these RNAs are especially active in the developing fetus during weeks 7–19, a critical period in cerebral cortex development.[30] Is HAR1 one of the genomic jewels in the human crown?

There is a fundamental problem with all of these human-chimp genomic comparisons, with respect to human uniqueness: the split occurred at least 5 mya, and a lot of evolution has occurred since then. First came *Ardipithecus*, *Orrorin*, and *Sahelanthropus*; then numerous australopithecines; then numerous *Homo* species, including *H. habilis*, *H. ergaster*, and *H. erectus*, followed by *H. antecessor*, *H. georgicus*, and *H. heidelbergensis*; then the Neanderthals, Denisovans, and us. Comparing genomes of humans and chimps will certainly help us understand what makes us different from chimps, but much less about what makes us uniquely human and particularly how we came to be domesticators. For that we need genomic information from much more closely related species.

It is fortunate, therefore, that recently both Neanderthal and Denisovan (nuclear) genomes were sequenced and can now be compared to the human genome. So far, over 80 DNA sequences have been identified that vary slightly in humans relative to Neanderthals, Denisovans, or both.[31] Are these the most precious genomic jewels in the human crown? Is this where the answer lies for the evolution of culturally modern humans, for how we became domesticators? Many biologizers would say yes. Others, though, look elsewhere for the evolution of human cultural modernity.

THE CULTURAL DIMENSION

Consider another fanciful thought experiment—one that involves not space aliens, but rather an alien of a different kind. This alien is a time traveler from New York City circa 1776 who visits New York City today. Our alien, Alex, is a Knickerbocker, a member of the closest thing to an aristocracy the city has ever known. Think of his utter disorientation, confusion, and yes, alienation when he finds himself in midtown Manhattan today. His enculturation in 1776 America would have left him completely unprepared to function in the city as it is now.

Consider, first, the changes in social norms. In 1776, women, African Americans, and nonlandowners of any ethnicity were not allowed to vote, much less hold positions of political power. Women were largely confined to their homes, not much in evidence on the street. Nor would Alex have encountered many non-European males on the street. Though New York was already the most cosmopolitan of American cities, there were few if any East Asians, Latinos, or Africans. New York has always had a ("free") African American population, but in 1776 the specter of slavery was omnipresent and they kept a low profile. Only the wealthier New Yorkers would have encountered a black person, primarily as a household servant.

The sheer density of humanity would surely overwhelm Alex. Then there is the technological element. Alex might be killed or maimed the first time he tried to cross a street. Nothing in his experience could prepare Alex for the pace of motorized travel. Indoor plumbing, electrical lighting, telephones, elevators, and sirens would all be bewildering, and much more so airplanes, cell phones, earphones, and iPads. His brain would actually be wired differently.[32] This difference in brain wiring is one manifestation of cultural evolution.

One hallmark of cultural evolution is its speed relative to biological evolution. In fewer than 20 generations, American culture has changed immensely. Not much in the way of biological evolution occurred during that period. The speed with which cultural evolution occurs relative to biological evolution reflects fundamental differences in their dynamics.

Cultural evolution is often said to be Lamarckian in contrast to biological evolution, which is Darwinian. By "Darwinian," it is meant that heritable variants—mutations—are random with respect to environmental conditions and hence adaptation. In this sense it is undirected. Lamarckian dynamics are quite different in that the environment can induce variants (innovations) that have a much-higher-than-random probability of surviving in that environment, of being adaptive. In this sense, Lamarckian evolution is directed, or guided. The progressive increase in cultural capital (knowledge) in any culture is Lamarckian.

Just as important, however, are the differences in the way adaptive variants are transmitted through a population. In the Darwinian dynamic, transmission is strictly from parent to offspring—that is, vertical. As such, it is limited by generation time, which in humans is quite long. Cultural transmission also has a vertical component, but in addition an oblique component (from nonparents to the next generation) and a horizontal component (between same-age individuals). This multimodal transmission hugely increases the potential rate of cultural change to something approaching that of disease epidemics. Most cultural change of course is not nearly that fast but it is always much faster than biological—gene-based—evolution.[33]

One upshot of the relative rapidity of cultural change is that it greatly enhances the evolutionary potential of human phenotypes, especially behavior. Biological evolution in humans is limited by generation time. Human generation spans are among the largest of any mammal, yet cultural evolution affords us the opportunity to respond to the physical environment even more rapidly than mice do. So pervasive is cultural mediation of our response to the physical environment that many human adaptations are adaptations to our cultural environment.

Culture itself is not an individual property; it is a collective property of a society of individuals. By "society" I mean a group of individuals whose activities are integrated and coordinated. This is another fundamental way in which cultural evolution differs from biological evolution. It can't be emphasized enough how the collective nature of human activity has driven human ascendancy. Had early humans been 10 times smarter than us, but solitary, we would never have become domesticators.[34] Whether or not we came by them by way of self-domestication, the emotional alterations that made possible our hypersociality were even more important than our intelligence in our rise to the domesticator state.

Let's now relate these facts about cultural evolution to the question with which we began this chapter: What caused humans to become culturally modern and ultimately domesticators? We begin with the latter part of the question.

The transition from hunting and gathering to farming was a seminal event in human evolutionary history. It was not the discovery of some lone genius that possessed a unique cognitive mutation. Farming was, rather, the result of collective cultural explorations—some say in response to deteriorating conditions for hunting, while others emphasize enhanced conditions for gathering and plant management.[35] In either case, human societies—first in the Near East, later in China and elsewhere—that had acquired through long association intimate knowledge of local plants began to use this collective knowledge to manage some of these plants, such as wheat (and other grasses such as maize, barley, rice, and rye), legumes (peas and beans of various sorts), gourds, and fruits such as figs, in order to guarantee a more constant and voluminous food supply.

For meat, as we have seen, various forms of managing cattle, sheep, goats, and pigs were initiated for the same end, once farming promoted a more settled existence. From that point on, in a sort of feed-forward dynamic characteristic of cultural evolution, humans achieved increasing control of these plants and animals to the point that we recognize as domestication. Neither the onset of domestication nor the various steps along the way require anything more than cultural evolution as the driving force. Why, then, are biologizers so convinced that something more than cultural evolution is required to explain the evolution of culturally modern humans from anatomically modern humans? Why couldn't cultural evolution do all that work? Before we take this question head on, we need to consider the third major option in explaining cultural modernity: biocultural evolution.

BIOCULTURAL EVOLUTION

The biocultural approach, remember—while acknowledging the paramount role of cultural evolution in the evolution of culturally modern humans and our rise to domesticator status—sees a continuing role for biological evolution as well; more important, the biocultural perspective emphasizes a complex interaction of biological and cultural evolution. Several distinct theoretical frameworks fall within this rubric; here I will consider the one called gene-culture coevolution,[36] which is best illustrated by a sort of paradigmatic example: the evolution of lactose tolerance in dairying populations.

Recall from Chapter 7 that the first dairying peoples in Europe, India, and Africa were lactose intolerant like the rest of humanity. Therefore, they could not make full use of the benefits of milk—calcium, hydration, milk fats and proteins—because of the acute gastric distress caused by indigestible lactose sugars. The cultural innovation of dairying created a new cultural environment in which there was natural selection for the capacity to digest lactose. Those who could not do so, those for whom any milk consumed exited the gastrointestinal tract quickly and violently, were at a selective disadvantage and contributed fewer offspring to the next generation of the dairying population. As a result of this selection, a mutation that conferred lactose tolerance rapidly spread through the populations that most relied on dairying. A related evolutionary phenomenon that also occurred with the advent of farming is the evolution of gluten tolerance.[37]

It is important to note that in these two examples of gene-culture coevolution, cultural evolution came first, and biological evolution followed. That is true of any example of gene-culture coevolution I have been able to uncover. Perhaps a better term would be "culture-gene coevolution," or better yet, simply "culturally driven biological evolution."

Other examples of culturally driven biological evolution are human diseases, many of which derive from our domestic animals. According to Jared Diamond, tuberculosis and measles originated in our domestic cattle.[38] Other diseases hypothesized to have originated in our domestic animals include rhinoviruses, influenza A, mumps, rotavirus, pertussis, and diphtheria.[39]

Some human scourges were an indirect consequence of domestication, via the increased settlement and higher human densities, which provided ideal conditions for pathogenic transmission.[40] Commensal rodents were vectors for smallpox, plague, and typhus, three of the most significant pathogens in human history.[41] Villages and cities brought humans into unprecedented proximity with each other and these rodents. Malaria is another disease that did not become epidemic until the advent of agriculture and village life.[42]

The human genome has responded to these by-products of domestication. Human populations exposed to these diseases have evolved varying degrees of resistance since the advent of farming and cities. The Black Death epidemics, for example, resulted in evolutionary alterations of the human immune system, specifically in genes coding for proteins, called toll-like receptors, that recognize and block certain pathogens.[43] The genomes of Eurasian populations with a history of plague epidemics exhibit selective changes in these proteins, which conferred resistance. Other adaptive genomic alterations were caused by smallpox and malaria.[44] Populations that had never experienced such epidemics remained more vulnerable, as was especially evident in natives of North and South America when they were exposed to smallpox.[45] This, too, is culture-gene coevolution, in which culture provides the selective environment for genetic changes.

Cultural evolution became a driver of human biological evolution long before the agricultural revolution—in fact, long before humans as a species entered the African scene about 200,000 years ago. The advent of cooking was a particularly significant prehuman cultural innovation in functioning as a method for predigestion. Calories can be extracted more efficiently from both meat and vegetable foods when they are cooked. Estimates for the advent of cooking vary, but it is universally agreed that *Homo* was routinely cooking by 250,000 years ago, and probably a lot earlier. We should expect to find some evidence of biological adaptation to cooking in humans since then. Among the proposed adaptations to cooking are reductions in jaw muscles,[46] tooth enamel, and molar size.[47] Richard Wrangham claims that cooking also made possible a marked increase in human brain size[48] (see Appendix

15 on page 350). These biological responses to the cultural invention of cooking provide yet more examples of biological evolution driven by cultural evolution.

How does this dynamic of culturally driven biological evolution relate to our original question of the evolution of culturally modern humans and ultimately our rise to domesticator status? First, it suggests that whatever the role of natural selection on the relevant cognitive evolution, it was driven by cultural imperatives. That is, cultural evolution was the driving force and natural selection followed. Recently it was proposed that this "culture first, biology second" dynamic resulted in a degree of human self-domestication beyond that described in the previous chapter.

SELF-DOMESTICATION AS ADAPTATION TO THE HUMAN CULTURAL ENVIRONMENT

This particular self-domestication hypothesis is part of a broader coevolutionary framework developed by Michael Dor and Eva Jablonka, which also includes a theory of language evolution.[49] Dor and Jablonka begin with the premise that for the last few hundred thousand years, cultural conditions have provided the primary environment in which human cognitive and emotional evolution has occurred. Proto-language was one of the human cognitive capacities to emerge from cultural dynamics, and only secondarily did biological evolution stabilize and enhance our language skills, through a process akin to the Baldwin effect discussed in Chapter 5.

This view of language evolution differs markedly from the received view embraced by evolutionary psychologists such as Steven Pinker, according to which language is a specialized mental module.[50] This module supposedly arose in the human lineage from a mutation (or mutations), such as *FOXP2*. As for all mutations, it originated in a single individual and then rapidly spread through the human species. This language mutation was the basis for the elaboration of human culture to the culturally modern state, perhaps the basis for the putative great leap forward. As is typical of evolutionary psychology, this is a biology-first (or gene-first) view.

Dor and Jablonka basically turn this argument on its head. By their account, cultural evolution came first. Language was a secondary adaptation to the cultural environment. Dor and Jablonka emphasize that language is an irreducibly collective trait, not an individual trait. Therefore, a mutant individual with a capacity for language—one who could only talk to him- or herself—would have no selective advantage in a Darwinian dynamic. There must be an enabling preexisting cultural language environment for any language-enhancing capacity to thrive. In support of this view, Dor and Jablonka refer to Kanzi, the bonobo who could communicate through lexigraphs at the level of a three-year-old human, but only in a language already invented by humans. Kanzi could never have developed those capacities outside of the human cultural environment—not because bonobos and other apes lack the cognitive capacities, but because these apes lack the collective capacity for the cultural scaffolding of language acquisition. This scaffolding itself was acquired though the cumulative effects of cultural evolution.

From the perspective of Jablonka and Dor, language is much more like the rest of human cognition than it is a special-purpose adaptation.[51] Moreover, by their account language evolution and emotional evolution are closely linked. In particular, certain emotional preconditions were required for language evolution. At bottom, a high degree of emotional control and social sensitivity, expressed in such "higher" social emotions as shame, guilt, and embarrassment, was necessary to promote and stabilize cooperation to the point that linguistic information sharing was possible. Jablonka and Dor consider this additional, perhaps human-specific, emotional evolution to be a special form of self-domestication. Such self-domestication was, in turn, enhanced by the further evolution of our language capacities, which promoted further expansion of the human emotional world to include humor, social identity, and agency. Through the use of metaphors, our emotional responses were further refined in ways appropriate in ever more sophisticated social contexts.

These are certainly intriguing ideas, and refreshingly outside of the constrictive box that is evolutionary psychology. Unfortunately, as it stands, this language-emotion coevolution hypothesis suffers from the same defect as so much "theorizing" in evolutionary psychology: the "N of 1" problem. It is difficult to conceive of ways in which the hypothesis could be tested. Culture is not unique to humans, but the human level of culture is unprecedented in the history of life on earth. The one-off nature of human cultural evolution is a particular challenge when it comes to evaluating—using the standards of mainstream evolutionary biology—proposals such as that of Dor and Jablonka.

Nonetheless, Dor and Jablonka point to an interesting sense in which self-domestication is not only a cause, but also a consequence, of human cultural evolution. And with respect to language and our other higher cognitive faculties, the causal role of culture has been particularly pronounced over the last 200,000 years. Certainly, in explaining the rapid rise of humans to domesticator status, we should look to culture first. It is through our cultural resources that humans came to be the dominant large animal on earth, one indication of which is the evolutionary alterations in dogs, cats, aurochs, and the other mammals discussed in this book. Domestication is an example of how our dominion is less a matter of adaptation, as biologists typically understand the term, and more a matter of adapting our environment to our own ends.

DOMESTICATION AND THE HUMAN NICHE

"Niche" is an important construct in ecology, and one that is central to the evolutionary concept of adaptation. A niche is simply the suite of

environmental conditions—physical and biological—in which a particular species or population thrives. The physical environment may include
temperature, rainfall, elevation, soil conditions, and such; the biological
environment includes food resources, predators, competitors, and parasites. Traditionally, the physical and biological conditions that define
a niche are viewed as given (preexisting) conditions to which a species
must adapt. Adaptation is then a matter of accommodation to a niche.
Camouflage, as in the leafy sea dragon, is the epitome of adaptation from
this perspective.

Richard Lewontin has long dissented from this portrayal of organism-
environment relations in evolutionary biology.[52] First, there is the problem of defining a niche.[53] Any environment has innumerable physical
and biological variables, only a fraction of which are relevant to a given
organism. So an organism's niche cannot be defined independently of
the organism itself. Moreover, many creatures take a more active role in
creating their niche. For instance, many mammals, especially rodents,
live in burrows of their own construction. The environmental conditions within these burrows differ markedly from those without. By
burrowing, the burrowers have created—or constructed—a new niche
by altering their environment in ways that change the natural selection
regime that they must negotiate.

Niche construction is widespread and prevalent among living
things.[54] Among mammals, beavers are paradigmatic niche constructors, engineering their environment to their own ends. Social rodents,
such as prairie dogs, create complicated underground structures from
which they rarely stray far. Even aboveground, the prairie within prairie
dog "towns" differs markedly from the prairie outside of these towns,
because of changes in soil conditions, drainage, and such.

As prairie dogs illustrate, sociality can greatly increase the scale of
niche construction effects, one pinnacle of which has been achieved by
some social insects, such as termites. In tropical grasslands throughout
the world, their impressively engineered mounds are some of the most
notable structures. Temperature and humidity are carefully regulated,

as are oxygen and carbon dioxide levels. The entirely underground complexes of leaf-cutter ants are even more impressive, sometimes acres in extent.

The supreme niche constructors, though, are we humans. We are much less adapted to our environment than we are adaptors of our environment. The human niche was created by humans. The domestication of plants and animals was one of our seminal achievements in this regard.

AFTER THE REVOLUTION

Farming spread rapidly from its cradles, eventually engulfing all but the remoter traditional (hunter-gatherer) societies. Agriculture marginalized this formerly universal way of life, with enormous consequences, one of which was permanent settlements and eventually urban life, a novel development in the history of our species. Large human settlements caused enormous changes to religious life. Animism was superseded by more hierarchical representations and projections of the natural world and its assumed spiritual underpinning—the birth of gods. These religious developments were closely linked to political developments, specifically a more hierarchical organization of society, in which the supreme leader often functioned as a representative or manifestation of a god here on earth. No doubt, these religious changes were crucial in maintaining the social order in larger societies.[55]

The concentration of human minds in urban environments greatly quickened the pace of cultural evolution, including art and technology. While artistic innovations largely remained confined to the cities and courts, technological advances had more pervasive effects. And the human population rapidly grew. With population growth came more concentration of human minds and further technological advances—a feed-forward dynamic that continues to this day.

The rest of nature came to increasingly feel the impact of this dynamic. More and more, we humans have become the ecological context in which other species evolved, and many have been unable to adapt or construct niches in this new, human-perfused environment. Even as hunter-gatherers, humans have had an adverse impact on many other species. There was never a golden age in which the ecological scales were balanced. The arrival of humans in Australia[56] was followed by the extinction of many large mammals, as was the arrival of humans in North and South America.[57] Islands were particularly vulnerable, even large islands. The moas in New Zealand did not long persist once the Maori arrived, nor did the elephant birds of Madagascar once it was colonized. On the smaller islands of Polynesia, avian extinctions were much more rapid.

But after the agricultural revolution the human footprint became ever larger, and much of nature was squashed as a result. Agriculture requires land in which natural vegetation has been removed—ever more such land as human population growth accelerates. Technological innovations helped humans to become increasingly efficient at exploiting the creatures in the adjacent wild. But the indirect effects of our technology and population growth have even more pervasive consequences, culminating in climate change and the still very real possibility of a radiation catastrophe on a worldwide scale.

Today, no part of nature remains untouched: the remotest polar environments, the deepest rain forests, the barrenest deserts, every square inch of the vast oceans—all evidence the heavy human hand. One measure of the pervasive human influence is extinction rates, which now approach those reached only five other times in the history of life on earth (more than 2 billion years). Those five events are referred to as mass extinctions.[58] They were caused by immense geological events on earth, or collisions with extraterrestrial objects, or both. This, the sixth mass extinction, though, is all on us.[59]

We have become more than the dominant component of the biotic environment; we have become a major impactor on the physical environment on earth as well—even a geological force. It is now increasingly popular, in fact, to acknowledge our physical impact on the earth with a geological name. The Pleistocene is the term for the Ice Age (roughly 2.6 mya to 12,000 years ago). The succeeding post–Ice Age period is referred to as the Holocene. Now there is talk of renaming this period, the beginning of which coincides with the agricultural revolution, as the Anthropocene, the age of man.[60]

This has been a singular journey, from a creature that a hyperintelligent alien might not have even noticed 100,000 years ago to a force of nature so powerful as to define a geological epoch.

EPILOGUE

DOMESTICATION IS AN EVOLUTIONARY PROCESS. IN THAT regard, domestication is not particularly exceptional, except to the extent that it was consciously directed by humans. But as we have seen, much—perhaps most—of domestication occurred by means of garden-variety natural selection, when individuals of some species (most obviously cats and dogs) found it advantageous to occupy the human environment. In doing so, these pioneers altered the conditions for natural selection and genetic drift, to which they and their descendants were subject.

As Belyaev emphasized, the key to exploiting human resources was the ability to tolerate human proximity—in a word, tameness. Belyaev experimentally condensed the evolution of tameness through strong artificial selection on farmed foxes, a proof in principle of what happened much more gradually through natural selection.

For many of the species discussed here, it was only after an extended period of self-domestication that humans began to assert their influence through artificial selection. Cat self-domestication probably began over 10,000 years ago, but it is only within the last 100 years that a small minority has been subjected to much artificial selection. With the notable exception of some recently developed breeds of freaks like the Levkoy and latter-day Siamese, domestic cats closely resemble their wild progenitors except for their coloration and sociability.

The effects of artificial selection have been most pronounced on dogs. Cattle, sheep, some pig and horse breeds also evidence a heavy human hand, much less so goats. Reindeer and camels continue to be relatively uninfluenced by artificial selection, in part because they remain confined to the harsh environments of their ancestors.

For the ungulate species—pigs, cattle, sheep, goats, camels, reindeer, and horses—artificial selection probably began with selective culling of young males, which inadvertently diminished sexual selection and hence sex differences by reducing male-male competition. It was generally only late in the domestication process that humans took over the role of mate choice, further reducing sexual selection. Once the point was reached in which mate choice was a human prerogative, the evolutionary rate was greatly accelerated. This was how Belyaev, Trut, and their coworkers were able to achieve such remarkable behavioral alterations in their foxes in just 40-plus generations.

The correlated physical alterations, from shortened faces to altered coloration to floppy ears, were by-products of selection for tameness born of their developmental linkage. The fact that the same suite of characters tends to occur in many of the species described here, from dogs to pigs, is testimony to evolution's deep conservatism. Some of the conserved developmental processes, such as those influenced by selection for tameness, are mammal-wide; some, such as those involved in depigmentation, are vertebrate-wide; some, such as neuronal development, are animal-wide.

It is traditional, in discussing convergent evolution such as occurred in mammals under domestication, to emphasize the role of environmental similarities—in these cases, those of the human environment. But absent the conserved and hence shared developmental processes in mammals, such convergences are much less probable. Put another way, conserved developmental processes mean that natural selection has a lot less work to do to generate these similarities.

We have seen that a number of the convergent traits wrought by domestication are paedomorphic, often a combination of a generally slowed rate of development (neoteny) and accelerated sexual maturation (progenesis). But neoteny is not a universal feature of domestication, as evidenced by its absence in domesticated cavies and other rodents. The only universal feature of domestication is tameness. Neoteny is one avenue to tameness through the retention of juvenile behavior by means of decelerated development of the stress response and perhaps some key limbic neural circuits.

Which brings us to the question of human domestication. Are we humans self-domesticated? Does dog domestication, in particular, recapitulate, in certain essential ways, our own evolution since we diverged from chimpanzees? It's an intriguing idea, but one for which we cannot with any confidence provide an answer yet. It is certainly not enough to demonstrate that humans and/or bonobos are more prosocial than chimpanzees. Nor does evidence for human paedomorphosis—which is, in any case, equivocal—count as strong support for the self-domestication hypothesis. The strongest evidence for the self-domestication hypothesis, from an evo-devo perspective, would be evidence of convergent developmental alterations, unique to domestic mammals, in features of the human neuroendocrine traits that are shared by all mammals. Short of that, human self-domestication is merely a suggestive metaphor for recent human evolution.

Whatever its ultimate fate, the self-domestication hypothesis is valuable in reorienting our focus somewhat from our singular intelligence to our emotional constitution, which is every bit as singular. Our prosocial emotional tendencies are what afford human groups unrivaled capacities for coordinated action and, ultimately, our capacity for culture. Intelligence is secondary in this regard. Spock-like creatures, much more intelligent than we are, would never have achieved what we have, for lack of motivation.

The dawn of culture, which occurred long before our species first appeared in Africa, was momentous. With culture came a new evolutionary dynamic, with some features of evolution by natural selection but additional characteristics of its own, which contributed to cultural evolution's accelerated rate relative to biological evolution. We have never stopped evolving biologically, but the cultural evolutionary dynamic has increasingly come to make us what we are, including our unprecedented capacity, now on a geological scale, to shape our environment to our own ends. The domestication of plants and animals was an effect of our growing preeminence and a crucial cause of what we call civilization.

FROM THE MODERN SYNTHESIS TO AN EXTENDED SYNTHESIS?

THESE ARE HEADY DAYS FOR EVOLUTIONARY BIOLOGY, A period of great ferment. Within the last two decades, several new research areas have emerged, though their implications are in dispute. The controversy concerns the degree to which the framework for a consensus established in the 1930s and 1940s, called the modern synthesis, requires modification. The modern synthesis was indeed a monumental achievement deserving of deep respect. First and foremost it filled a gaping hole in Darwin's Darwinism—by providing a theory of inheritance. Darwin never developed an adequate theory to explain how traits are inherited, and he was completely unaware of the research of his contemporary, the Moravian monk Gregor Mendel, which provided the foundation for modern genetics. As a result, Darwinism and Mendelism developed independently and were often at odds throughout the first decades of the twentieth century. The modern synthesis cured that discrepancy through a seamless merger of the two; no wonder it became the received view, canonical Darwinism.

The modern synthesis also nicely integrated other areas of biology, such as paleontology and systematics (the study of genealogical [phylogenetic] relationships among living things). So the "synthesis" in the modern synthesis is in no way hyperbolic. But one huge field of biology was conspicuously omitted: developmental biology, the study of how a fertilized egg becomes a dog or horse or human, as the case may be. The reasons for this omission are themselves contested. Ernst Mayr, who was perhaps the foremost advocate of the modern synthesis—and certainly the most fervent—claimed that developmental biologists were welcomed to the table but declined the invitation. Most developmental biologists from this period had a quite different view of things; they complained that the invitation was grudging and full of preconditions. In any case, developmental biology and evolutionary biology continued to develop or evolve independently, despite the pleas of a few visionaries, notably Conrad Waddington, that integration was essential for both.[1] Since developmental biology played an important role in Darwin's Darwinism, its omission from the modern synthesis had important implications for the evolution of evolutionary biology, as Waddington so well understood.

Recently, Waddington's goal has been realized in the form of a synthetic research program called evolutionary developmental biology, more commonly referred to by its contraction "evo devo." The implications of evo devo to date are, as I indicated earlier, disputed. One common reaction is that evo devo is overhyped and poses no particular challenge for the modern synthesis. This is the position of some population geneticists, who have generally reserved for themselves the mantle of "theorists" in matters evolutionary. Jerry Coyne is a particularly vocal proponent of this attitude.[2]

On the opposite end of the spectrum are those, such as Lindsey Craig, who claim that a complete overhaul of the modern synthesis is required.[3] Occupying various points in the middle ground are a number of diverse evolutionists, such as Brian Hall, Gerd Muller, Sean Carroll, and Massimo Pigliucci, who reject both of the extreme views and call instead for an "extended synthesis."[4] Their goal is to loosen up what they see as an ossified framework, in light of evo devo and recent developments in molecular biology, such as genomics and epigenetics. To be candid, my sympathies lie here. In any case, evo devo will prove crucial in explicating the process of domestication.[5]

GENES AND PHENOTYPES

One common sentiment among those calling for an extended synthesis is that the theory of inheritance, which, as we saw, was a latecomer to evolutionary biology, has become the tail that wags the dog. Since the modern synthesis, evolutionary thinking certainly has become increasingly genocentric; it is now standard to define evolution as changes in gene frequencies, not changes in form. The apotheosis of genocentrism is the gene's-eye view of evolution developed by George Williams and popularized by Richard Dawkins, according to which evolution is, first and last, a competition between genes.[6] That is, genes—not individual organisms—are the only true evolutionary agents. Organisms are merely the vehicles through which individual genes conduct their increasingly circuitous battles through temporary alliances and confederations.

Critics of the genocentric approach, including some proponents of evo devo, complain that it leaves out much that concerned not only Darwin but also the architects of the modern synthesis, such as Mayr—most notably, what causes changes in the forms of organisms and in the complexity of these forms. This critique is based on the observation that an individual's genes (or genotype) do not map in any direct way to its phenotype, the suite of physical and behavioral traits upon which natural selection directly acts. Genes themselves are not generally visible to the selection process; rather, phenotypes are what evolution sees. Mary Jane West-Eberhard goes so far as to claim, "Phenotypes lead, genes follow."[7] By this she means that adaptation begins with existing developmental plasticity, which then channels subsequent genetic phenotypic alterations by natural selection.

The phenocentric view is bolstered by two incontrovertible facts that are problematic from the gene's-eye view of evolution. First, large changes in the phenotype often involve small changes in the genome. The Pekingese has diverged considerably from its wolf ancestors, but with only minute genetic alterations. Second, and conversely, much genetic evolution has no or little influence on the phenotype. Horseshoe crabs, for instance, belong to a lineage that has changed little over hundreds of millions of years, though their genomes have continued to evolve at a rate typical of other arthropods.[8]

GENOMICS

Since developmental biology was never a part of the modern synthesis, it is not so surprising that evo devo might provide challenges to the received view. Genomics, though, would seem to slide in smoothly, especially with the more recent genocentric versions. But genes, as conceived in the modern synthesis, are purely abstract entities; at the time, no one knew that genes are composed of DNA. And even as the modern synthesis became increasingly genocentric, the view of genes inherited from the 1930s has not been updated; genes largely remain disembodied abstractions in, for example, population genetics. The modern synthesis has never fully taken on board the material gene as revealed by molecular biology. But the material gene, the one that consists of DNA, has been fully embraced by evo devo, and the two research areas nicely complement each other.

GENOMICS AND THE TREE OF LIFE

WHOLE-GENOME COMPARISONS HAVE REVEALED TREMEN-dous heterogeneity in the rates at which various kinds of sequences evolve. This variability has proved useful in reconstructing the tree of life, including the genealogies of domestic breeds. Some particular kinds of genomic bits are highly conserved and hence evolve very slowly. Other kinds of sequences are poorly conserved and hence evolve relatively rapidly. The more conserved sequences in the genome are more suitable for reconstructing branches of the tree of life encompassing tens of millions of years of evolution; less conserved sequences are better for reconstructing branches representing tens of thousands of years of evolution.

The more highly conserved parts of the genome are the genes, the bits that code for proteins. These regions are subjected to high levels of "normalizing" or "purifying" selection, in which any mutation that affects the coded protein is generally eliminated.[1] The exceptions are the extremely rare mutations that result in a protein that confers an advantage over the nonmutated protein. In mammals, the protein-coding bits of the genome comprise only 1–2 percent of the total DNA; the rest of the genome is noncoding.[2]

Some bits of the noncoding genome actually evolve even more slowly than the coding bits do. These are called "ultraconserved regions."[3] Most of the noncoding bits, though, are much less conserved than the coding bits; they experience less purifying selection. But these vast, relatively unconserved noncoding regions vary greatly in how fast they evolve. Some of the relatively rapidly evolving noncoding bits are called transposons, which also have the capacity to move around the genome.[4] The most mutable genomic sequences are called microsatellites, which consist of highly repetitive base sequences.[5] Because they evolve so rapidly, microsatellites have been one of the most useful tools for determining the evolutionary history of dog breeds, for example.

The primary pregenomic DNA-based tool for such analyses is mitochondrial DNA. All animals actually have two genomes: the nuclear genome (just described) and a mitochondrial genome. The mitochondrion is a cellular component outside of the nucleus that produces most of the cell's energy. It has a genome of its own, but that genome is a fraction the size of the nuclear genome. This mitochondrial genome is also much less conserved than most parts of the nuclear genome, and as such it is useful for analyses of evolution over relatively short timescales, such as that of domestication. In fact, this has been the pregenomic method of choice for discerning the origin of domesticated mammals.

There is, however, a significant limitation in the usefulness of mitochondrial DNA. Unlike nuclear DNA, mitochondrial DNA is transmitted only through the female line. Therefore, it is insensitive to the genetic effects of, for example, the mating of male wolves with female dogs, which may have been common, especially during the early stages of domestication. And in fact, such sexual encounters between male wild types and female domestics were a fairly common occurrence among domesticated mammals generally.

Fortunately, there is a corrective, but one that could be fully exploited only recently as a result of genomics breakthroughs: the Y chromosome. The Y is an odd little chromosome, a fraction the size of the next-smallest chromosome. What makes the Y chromosome odd, though, is the fact that it has very few genes, even for its diminutive size. The vast majority of the Y chromosome consists of truly junky junk DNA. This junk can mutate willy-nilly because it doesn't affect the phenotype; it is largely invisible to natural selection. Therefore, mutations in the Y chromosome accumulate at a rapid rate, similar to the rate at which mutations accumulate in mitochondrial DNA.

Much has been learned about the history of domesticated mammals through a combination of analyses based on mitochondrial DNA and analyses based on Y-chromosome DNA. But the recent influx of whole-genome information has greatly expanded our knowledge of domestication with respect to both historical reconstructions and the DNA modifications involved.

FROM LANDRACES TO BREEDS

AS DOMESTIC CATTLE EXPANDED THEIR RANGE UNDER human auspices, they began to genetically differentiate along geographic lines, such that by 2000 BP, there were a number of distinct landraces of both the zebu and taurine subspecies. Some of this differentiation may be attributable to interbreeding with what remained of local wild aurochs, especially in zebu cattle. But most was due to genetic drift and adaptation to the local environment, both cultural and physical. These landraces, or proto-breeds would later be the raw material from which true breeds were developed in the nineteenth century.

Because the breed concept was first developed in Europe, the transition from landraces to breeds was better defined there than elsewhere in the world. Many geographically distinct zebu cattle remained at the landrace stage until quite recently, as did taurine cattle of eastern Asia. There are also a few taurine landraces of European origin remaining to this day elsewhere in the world. The criollo-type cattle of the Americas, for instance, are actually landraces. These include the Corriente cattle so popular in rodeos (roping and steer wrestling), the Florida Cracker and Pineywoods cattle of the Gulf Coast region, Crioulo Lageano of Brazil, and Texas Longhorns. All of these are descendants of free-ranging—primarily Spanish—cattle imported during the conquest of the New World, beginning in the fifteenth century. They became adapted to the new local, often marginal conditions under varying degrees of human control.

Criollo landraces have a large European component, as you would expect—mainly southern European landraces that arrived in Europe via the Mediterranean route and populated much of southern Europe.[1] In general, these southern European cattle were less intensively managed than those that came via the northern route. And there are more so-called primitive breeds in southern Europe, including some, like Tudanca, Sayaguesa, Pajuna (Spain), and Maronesa (Portugal) that most closely resemble wild aurochs.

It is noteworthy as well that Spanish Fighting Bull cattle—as well as Camargue cattle, which were originally bred for the same purpose—are also quite auroch-like in horns and build, albeit much smaller. Other, less auroch-like, but older southern European breeds include the Chianina and Marchigiana of Italy. Both of these breeds show evidence of zebu introgression, which is also evident, to varying degrees, in creole breeds derived from Iberian cattle.[2] There is also evidence of a taurine African influence on the original imports to the New World;[3] some creole breeds may also have a genetic component derived from taurine African cattle imported to the New World directly from Africa.[4] Most recently, some creole breeds, notably the Texas Longhorn, have experienced admixture from British breeds.[5]

In northern Europe, the human influence weighs more heavily on even the oldest breeds, especially the dairy cattle. Icelandic cattle, for instance, derive from a landrace originally transported there by the Vikings over a thousand years ago and isolated on that island ever since.[6] Yet, aside from their distinctive coloration, Icelandic cattle look like fairly typical dairy cows. The Jersey and Guernsey, from the islands for which they are named, have also long been isolated but are nevertheless rather typical dairy cows. Dairy cattle generally have more of the auroch removed from them than do beef, draft, and multipurpose breeds. Other relatively ancient northern European cattle types derive from landraces adapted to marginal habitats such as mountains. These include the

Braunvieh of the Swiss Alps, Harz Red from Germany, Highland from Scotland, Blonde d'Aquitaine from the Pyrenees, and Irish Mountain Cattle/Dexter from Ireland.[7]

Most zebu breeds are more recently derived from local landraces; their breed names usually reflect their geographic origin—for example, Gir, Guzerat, Red Sindhi, Ongole, and Kankrej. In general, there was less breed specialization for dairy, draft, and, in particular, beef among zebu breeds.[8] A greater percentage of zebu breeds are multipurpose. The Red Sindhi and Sahiwal are among the few zebu breeds specialized primarily for dairying; most milking breeds are also used extensively as draft animals.[9] The dung of all zebu cattle in India is also of critical importance as fuel in deforested areas.[10]

The development of hybrid African breeds followed yet a different course. Until quite recently, most cattle were owned by wandering pastoralists, rather than sedentary farmers; hence, breeds tend to be more related to tribe rather than geography, though the two factors often coincide. For many pastoralists, cattle function at least partly as a form of currency. As such, there is a disincentive to eat them. But for breeds like the Ankole, the symbolic function supersedes even dairying or any other use. Bulls, in particular, are primarily status symbols.[11]

Pregenomic studies identified two basic breed groups: taurine and zebu (Figure 7A.1). One recent study also identified a third major branch of cattle breeds from Africa, identified as African taurine.[12] This third group may reflect introgression from native North African aurochs, or simply long-standing isolation from Eurasian taurine breeds. Zebu-taurine hybrids, including Sanga cattle, also form a distinct breed cluster.[13]

There is a clear geographic component in the genealogy of zebu breeds in India.[14] The geographic signal is not as strong among European taurine breeds, because of human-mediated dispersal. Nonetheless, northern European, central European, and Iberian breeds form distinct clumps (Figure 7A.2). Another breed clump represents cattle from the Podolian steppe region of eastern Europe.[15] Podolian cattle show genetic

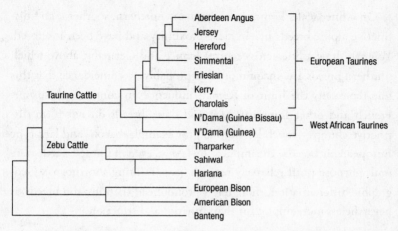

FIGURE 7A.1   Cattle phylogeny, showing the taurine-zebu split and a few representative breeds. (Adapted from MacHugh et al. 1997, 1079, fig. 3.)

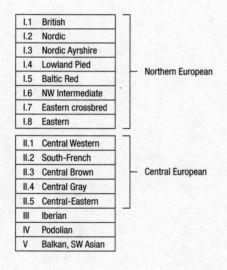

FIGURE 7A.2   Five breed clusters of European cattle: northern European, central European, Iberian, Podolian, and Balkan. (Adapted from Felius et al. 2011, 685.)

affinities with cattle from the Near East[16] and central Asia.[17] Some Podolian breeds, including the Maremmana, made their way to central Italy perhaps by way of the Etruscan migration,[18] or via trade during the Roman period.[19]

On a finer scale, France is where many northern, southern, and distinctive alpine breeds live in close proximity and have been interbred. You can draw a line across central France and Germany, above which southern breeds are uncommon. It is probably no coincidence that this line represents the limit of Roman influence on continental Europe. French and other European beef and dairy breeds do not form distinctive clumps—much less so than, for example, bacon- and lard-type European pig breeds. The implication is that, even in Europe, cattle were multipurpose until relatively recently (e.g., milking shorthorns). Geographic differentiation, though less pronounced than in zebu breeds, is nevertheless more important than functional distinction.[20]

WHAT'S IN A GAIT

CAMELS ARE UNIQUE AMONG UNGULATES IN THEIR COM-mitment to the pacing gait. Camels pace when they walk slowly, when they amble, and when they are raced at top speed.[1] What is it about pacing, of all possible gaits, that suits camels so? Is there something about camels that makes pacing more energetically efficient? According to one hypothesis, there is, and that something is their long legs.[2] Advocates of this view point to the fact that cheetahs pace, as do long-legged dog breeds such as the Afghan hound. But cheetahs and Afghans pace only when they walk. It has also been claimed, erroneously, that giraffes pace.[3] In fact, though, giraffes neither pace nor trot; they have a unique gait of their own. There seems to be more to pacing in camels than just long legs.

Horses, another long-legged species, may be instructive in this regard. Horses have different natural gaits for different speeds: walk, trot, canter, gallop. Moreover, horses can use multiple gaits for all but the highest speeds. Of particular note is the fact that at trot speeds, some horses pace. Among all horses of all breeds, the vast majority trot, but for the Icelandic pony breed, pacing is the most prevalent, suggesting a genetic component to horse gaits.[4] (But both trotting and pacing horses naturally gallop at higher speeds. Standard-bred racers, both trotters and pacers, must be intensively trained not to gallop in races.) And indeed, trotters tend to beget trotters, and pacers tend to beget pacers.[5] Recently, a genetic difference between trotters and pacers was identified—one that affects patterns of neural impulses from the spine to the limbs.[6]

The horse research on gait heritability suggests an alternative explanation for camel pacing: perhaps the camel's peculiar gait simply reflects the retention of an ancestral trait. Fossil camel tracks are instructive in this regard. They clearly indicate that millions of years before either Bactrian or dromedary camels evolved, their ancestors were pacing.[7] It seems that pacing is a conservative feature of camel evolution. If indeed the camel's wide footpads did evolve to stabilize the pacing gait, it wouldn't be the first time that a behavioral trait channeled subsequent morphological evolution.[8] And once the morphological change occurs, the behavior becomes harder to tinker with. Should humans ever decide to make miniature or toy camels, they may still pace.

HORSE EVOLUTION

THE FIRST EQUIDS FOR WHICH WE HAVE FOSSIL EVIDENCE were about the size of a fox and had the dentition typical of an omnivore.[1] The teeth, which included three incisors, one canine, four premolars, and three molars on each jaw quadrant, were all low crowned, like ours. These early equids had already departed from the primitive mammalian condition of five digits, having lost the first. These modest creatures inhabited the undergrowth of tropical forests, where they foraged for fruit and soft vegetation, and perhaps insects and other small creatures.

About 50 mya, some of these first equines began to specialize in browsing tougher plant material, and their dentition changed accordingly, most notably in the loss of one premolar and the transformation of the rearmost premolar into a molar-like tooth.[2] This next wave of equines showed adaptations for jumping and lost another digit on the hind legs.[3] It should be emphasized that this wave of equines did not replace the first equines, which continued with only subtle modifications for another 20 million years.

Over the next 14 million years (Eocene and Oligocene), many new equid species appeared on the scene, some showing increased dental modifications suitable for tough vegetation.[4] Of particular significance was a general but by no means universal trend toward longer, or higher

Mya Approximate

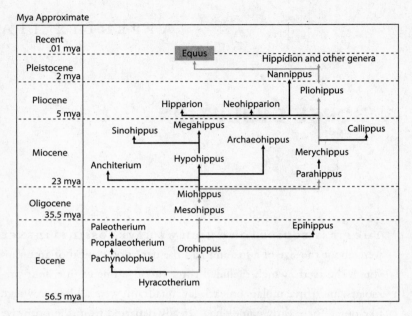

FIGURE 11A.1   Equid family tree. Note the branching pattern.

crowned, teeth and molars with extra ridges and grinding surfaces. There was also a general but by no means universal trend toward longer legs.[5] About 24 mya (Oligocene-Miocene boundary), newly evolved temperate grasses began to predominate in the unforested areas, which increased because of climate change.[6] This new habitat seems to have triggered rapid speciation in the horse family, as it did in several artiodactyl families. Some of the new species became increasingly specialized on this new vegetation, though many remained forest creatures at this time. Those that specialized on grasses exhibited more increases in tooth crown height, leg length, and overall size[7] (Figure 11A.1).

Of the species that took to the prairies, one at least (*Parahippus*) approached the size of a pony.[8] It also had an elongated horselike skull with elaborate high-crowned molars and molar-like premolars specialized for grinding grasses. Though it was still three-toed, most of its weight was born by the middle (third) toe with hoof-like modifications.

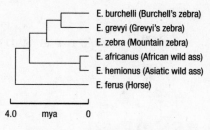

FIGURE 11A.2   Genealogical relationships among members of the genus *Equus*. (Adapted from MacFadden 1999.)

As grasslands expanded throughout the Miocene, several lineages of horse species showed trends toward longer legs, reduction in all but the third digit, and dentition increasingly specialized for a grass diet.[9] By 5 mya (at the onset of the Pliocene epoch), there were numerous such horse species in North America, and by 3.5 mya some genera were the size of a modern horse.[10]

One of these was *Equus*, to which the modern horses, asses, and zebras belong. The evolutionary relationships of modern equines are disputed.[11] Here I will use the tree developed by Steiner and colleagues in 2012, but this should not be taken as the last word on the subject (Figure 11A.2). In their account, the first branching point separates the three zebra species from all other members of the genus. The final branching point separates the true horses (caballines) from the African asses (Nubian and Somali) and Asiatic asses (e.g., onagers, kunluns [hermiones]).

GENEALOGY OF HORSE BREEDS

AS IN ALL DOMESTIC ANIMALS, MITOCHONDRIAL DNA (mtDNA) has been the workhorse in establishing genealogical relationships among horse landraces and breeds. The Y chromosome has also proved useful, but in a much more limited way. It was analysis of mtDNA that enabled investigators to trace the area of origin for domestic horses and their subsequent dispersal to Europe, China, and Egypt, as well as the fact that during this range expansion there was much interbreeding with wild mares. Analysis of Y chromosomes demonstrated no contribution from wild stallions during the range expansion and only extremely restricted contribution of wild stallions to the original domestic stock. For breed differentiation, though, scientists need to use the faster-evolving microsatellite DNA, which is distributed throughout the nuclear genome, as well as whole-genome data. This research is in a very preliminary state.

As in other domestic animals, hybridization complicates efforts to construct genealogies for horse breeds. For example, Thoroughbreds have contributed to the construction of several draft and riding breeds. Nonetheless, this horizontal dimension does not completely obscure the historical signal as manifest in breed clusters and ultimately genealogical trees. Currently, the best we can manage is breed clusters, of which four could be distinguished in one recent study[1] (Figure 11B.1).

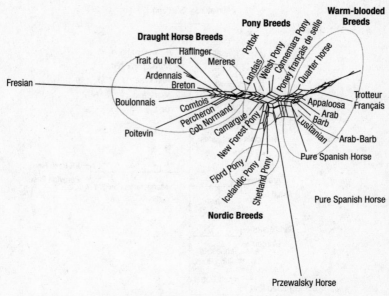

FIGURE 11B.1   Genealogical relationships among some French horse breeds, based on microsatellite data. (Adapted from Leroy et al., *Genetics Selection Evolution*, 2009, 6, fig. 1.)

One cluster consisted of Nordic landraces/breeds, including the Icelandic pony, Fjord horse, and Shetland pony. A second distinct cluster was composed of pony breeds from the rest of Europe, including the Welsh pony, Connemara pony, New Forest pony, and French Saddle Pony. The separation of these two groups of small horses indicates the importance of geography. A third cluster was composed of draft breeds, including the Ardennais, Breton, and Comtois. (As it happens, all of the draft breeds sampled were French.) The final cluster consisted primarily of the so-called warm-blooded breeds used primarily for riding and formerly carriage transportation. These included the Arabian, Quarter Horse, Lusitano, Appaloosa, and French Trotter.

More recently, a breed tree was constructed from genome-wide association studies (GWASs) of more diverse breeds with a focus on single nucleotide polymorphisms (SNPs)[2] (Figure 11B.2). Among the findings of note was a cluster that included two steppe breeds—the Tuva and

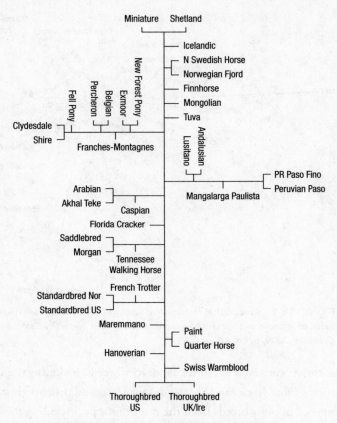

FIGURE 11B.2 Genealogical relationships among diverse horse breeds from around the world, based on SNPs identified from genome-wide association studies (GWAS). (Adapted from Petersen et al. 2013, 7, fig. 2.)

Mongolian—considered to be primitive (basal) breeds, and several Nordic breeds, including the Norwegian Fjord, Finnhorse, Shetland, and Icelandic, all of which experienced long histories of isolation. Another breed cluster includes two Iberian breeds—Andalusian and Lusitano—and several New World breeds, but not the Florida Cracker, a Criollo landrace. Again, there was also a cluster of western European breeds, including large draft breeds—Percheron, Shire, and Clydesdale—as well as two small breeds: Exmoor and New Forest ponies. Finally, one breed cluster reflected hybridization with Thoroughbreds, including the Italian Maremmano and the Hanoverian (originally German), as well as the Morgan and Quarter Horse from the United States.

These breed trees should be considered very preliminary—first approximations at reconstructing genealogical relationships among horses. It seems obvious that geography will loom large. So, too, will human history, especially as it relates to conquest and trade. Functional attributes—as reflected by, for example, size and conformation—appear to be less important, since similarly altered traits evolved independently in different regions from more generic landraces or through hybridization.

THE EPIGENETIC DIMENSION

THE BASE SEQUENCE THAT CONSTITUTES THE GENETIC code is one dimension of DNA, but like all biochemicals, DNA is three-dimensional. And the three-dimensional properties of DNA are what determine which genes are "expressed"—that is, which genes are actively engaged in making proteins. Epigenetics is the relatively new field of biology in which the goal is to understand the long-term control of gene expression, in the form of chemical attachments to DNA or the histone proteins with which DNA is inextricably intertwined in cells.[1]

In mammals, much of the epigenetic action occurs before birth. Many epigenetic alterations are environmentally induced. That is, some specific epigenetic alterations to particular genes occur predictably in response to the environment, both physical and social. Other epigenetic alterations, though, are essentially random. And it is this random epigenetic variation that might explain the large amount of phenotypic variation in isogenic mice.[2]

Epigenetics has had a huge impact on the study of development. For example, though the hair cells, blood cells, neurons, and other cells in your body are genetically identical, they are phenotypically quite different. What makes them phenotypically different is their epigenetic differences. But is epigenetics relevant to an understanding of evolution? There is increasing interest in this question, especially in the evo-devo community. Marion Lamb and Eva Jablonka, who have pioneered this attempt at integrating epigenetics and evolution, argue emphatically that epigenetics is essential for evolutionary biology.[3] Guardians of Darwinian orthodoxy, such as Jerry Coyne, are equally emphatic, sometimes apoplectically so, that this is nonsense.[4] Things are trending toward Jablonka and Lamb.[5]

The main point of dispute centers on epigenetic inheritance: how common it is, and whether it is evolutionarily relevant. Epigenetic inheritance is the inheritance of epigenetic states and hence degrees of gene expression. It is well known that epigenetic states are inherited during cell division; once an epigenetic mark is established in one cell, all of its descendant cells (called daughter cells) will inherit that epigenetic state, which is why blood cell precursors don't give rise to neurons. This is called somatic epigenetic inheritance. But does epigenetic inheritance also sometimes extend across generations, from parent to offspring, like the one-dimensional DNA base sequences? For this to occur, the epigenetic mark must be present in the only cells that make this passage: sperm and egg, which are collectively called "germ (as in "germinate") cells." Furthermore, this epigenetic mark must be transmitted to the fertilized egg. This is called germ-line epigenetic inheritance or transgenerational epigenetic inheritance.

There is ample evidence for transgenerational epigenetic inheritance in diverse plants and animals, including farm foxes. The white blaze—called the star pattern—that frequently appears on the foreheads of farm foxes, horses, and many other domestic animals is inherited in a way that cannot be explained by standard Mendelian genetics. The evidence now points to epigenetic inheritance.[6]

But for epigenetic inheritance to be evolutionarily relevant, there are a couple of prerequisites. First, there must be ample epigenetic variation; second, this epigenetic variation must occur independently of any genetic variation. The isogenic mice provide evidence for both. Moreover, there has been successful selection on such epigenetic variation in isogenic mice to increase the frequency of certain coat color variants.[7] Nonetheless, the evidence to date suggests that though epigenetic inheritance is common in plants and simple organisms, it is much rarer in mammals and other vertebrates.[8]

EVOLUTIONARY BIOLOGY, EVOLUTIONARY ANTHROPOLOGY, AND EVOLUTIONARY PSYCHOLOGY

THE COUNSEL OF CAUTION REGARDING OUR ABILITY TO establish human mind/brain adaptations goes unheeded by those who call themselves "evolutionary psychologists"; they are not at all chary of claiming to have demonstrated all manner of human cognitive adaptations on what for mainstream evolutionary biology is very weak evidence. But evolutionary psychology was born and nurtured outside of the mainstream of evolutionary biology. It was developed by psychologists who sought a unifying monolithic framework for psychology to replace that of a discredited behaviorism.[1]

It is not surprising, therefore, that the "evolution" in evolutionary psychology runs exceedingly shallow.[2] For instance, the developmental turn in mainstream evolutionary biology occurred before the advent of evolutionary psychology but has never been so much as acknowledged by evolutionary psychologists. Their use of genomic evidence is extremely limited, and their awareness of the increasingly sophisticated methods and tools for genealogical (phylogenetic) reconstructions seems completely absent. In this, evolutionary psychology stands in marked contrast to evolutionary anthropology, for which evo devo, genomics, and genealogical techniques essential for applying the comparative method have become increasingly central.[3]

There are several problematic foundational assumptions in evolutionary psychology, including that of the "environment of evolutionary adaptation," according to which the human mind evolved to solve problems posed by the Pleistocene, the so-called environment of evolutionary adaptation (EEA). Because of an inevitable evolutionary lag, the human mind has not had time to catch up with the problems posed by our current environment.[4] There are three problems—one methodological and two factual—with this assumption.

The methodological problem is that the EEA assumption has the effect of shielding evolutionary psychology from negative evidence. If the data don't fit the reverse-engineering prediction, it is because of a mismatch between the EEA and the present environment; if the data tend to support the prediction, it is because of the remaining similarities in these two environments. The EEA assumption renders much evolutionary psychology resistant to empirical access, in principle. The first factual problem is that the proposed EEA was the Pleistocene, which was notorious for its wild fluctuations.[5] You can't just average these wildly varied conditions, as evolutionary psychologists do[6] to determine the conditions to which the human mind was adapted. The second factual problem with the EEA assumption is that there has been a lot of human evolution since the Pleistocene;[7] in fact, as we shall see in chapter 15, there has been a lot of human evolution since the agricultural revolution 12,000 years ago. There is no good reason to believe that the evolution of the human mind/brain came to a halt during the Pleistocene.

Evolutionary psychologists have also chosen quite selectively from the vast literature in evolutionary biology in assembling their "foundational texts."[8] Moreover, they have somehow managed to overlook a central tenet of one of those texts, *Adaptation and Natural Selection*, by George C. Williams: that adaptation (the product of "selection for") is an "onerous" concept.[9] The term "onerous" is not one you often see in scientific discourse, so Williams, whose command of the English language was exceptional, would not have used it lightly.

Let's take a closer look at what Williams meant by this curious term.

Some of the primary connotations of "onerous" include "burdensome," "oppressive," "exacting," "stringent," even "agonizing." Antonyms include "comfortable," "moderate," "restful," and "agreeable." Clearly, Williams was aware of the obstacles to adaptive inference already described. He was particularly keen to guard against the then (1960s) prevalent temptation for facile claims of adaptation through group selection. These claims were often based on a perceived—or imagined—benefit for a group trait, such as a stable population size.

For Williams, adaptation is easy to claim but hard to demonstrate. The onerous burden of proof falls on those who make such claims. Adaptation should certainly not be assumed from the outset. Evolutionary psychologists, while worshipping Williams, somehow manage to ignore that message and, quite to the contrary, presume the adaptiveness (based on reverse engineering from putative conditions in the EEA) of any cognitive trait or behavioral disposition that they can distinguish— for example, our war-making proclivities or rape or putative cognitive sex differences.[10] Unconstrained by normal standards of inference in evolutionary biology, evolutionary psychology is prone to the construction of what Gould and Lewontin have labeled "just-so stories," after the famous collection by Rudyard Kipling.[11]

Some of the most egregious just-so stories to emerge from evolutionary psychology concern human sex differences. A typical evolutionary psychological account would lead you to believe that male and female humans belong to different species, so great are our sex differences. Yet, as is clear from the previous chapter, there has been a notable diminution of sex differences, or sexual dimorphism, in the hominid lineage—a trend that began over 3 mya. Humans are at the low end of the sexual dimorphism scale, by mammal standards, by primate standards, and by hominid standards. That fact is consistent with, but by no means convincing evidence for, the self-domestication hypothesis; it should also give pause to those prone to rampant speculation based on "adaptationist logic" concerning human sex differences, including psychological sex differences.[12]

THE CONTROL OF FIRE AND ITS CONSEQUENCES

IT IS UNIVERSALLY ACCEPTED THAT HUMANS HAD achieved the control of fire by 250,000 BP. Many anthropologists would push the origin back further, into the range of 300,000–500,000 years ago. Much more controversially, Richard Wrangham has proposed that the human use of controlled fire extends as far back as 1.5 million years, when *Homo erectus* and *Homo ergaster* roamed Africa and Eurasia.[1]

On Wrangham's view, cooking had enormous consequences for human biological evolution, including reduced tooth size, and a general reduction in the digestive tract. The energy formerly devoted to digestion could then be put to other uses, such as brain growth. Indeed, Wrangham attributes much of the increase in brain size of hominins to the invention of cooking. In this respect he builds on an older explanation for brain size increase called the "expensive tissue hypothesis."[2]

The expensive tissue hypothesis posits a trade-off between energy devoted to the gastrointestinal tract and energy devoted to the brain, because digestive tissue is second only to brain tissue in energy consumption. Advocates also note an inverse correlation between the size of the GI tract and the size of the brain—relative to body mass—in primates and other mammals, which for them indicates that you can't have both a large brain and a large gastrointestinal system.

It has long been speculated that an increasingly meat-heavy diet made gastrointestinal diminishment possible because plant materials require the most digestive work, as in the four-chambered stomachs of cows. Proponents of the expensive tissue hypothesis see this meat-based diet as enabling brain growth through diversion of resources from the GI system. According to Wrangham, meat was not enough; it had to be cooked. Furthermore, the truly salient dietary change was the addition of tubers and roots (potatoes, yams, cassava, manioc, turnips, etc.), the complex starches of which could be made digestible only through cooking.[3] It was the addition of calories from cooked tubers that made brain expansion in hominins possible, according to Wrangham.

ACKNOWLEDGMENTS

I AM DEEPLY GRATEFUL TO A NUMBER OF SCIENTISTS AND scholars for their comments and suggestions: Anna Kukekova for Chapter 1 (Foxes) and Chapter 2 (Dogs); Raymond Coppinger for Chapter 2 (Dogs); Greger Larson for Chapter 2 (Dogs) and Chapter 6 (Pigs); Dennis A. Turner for Chapter 3 (Cats) and Chapter 4 (Other Predators); Massimo Pigliucci for Chapter 5 (Evolutionary Interlude); David W. Burt for Chapter 7 (Cattle); James W. Kijas for Chapter 8 (Sheep and Goats); Knut H. Roed for Chapter 9 (Reindeer); Robin Bendrey for Chapter 10 (Camels); David W. Anthony for Chapter 11 (Horses); Steven N. Austad for Chapter 12 (Rodents); Angel Spotorno for Chapter 12 (Rodents); and Jeffrey Schwartz for Chapter 13 (Humans Part 1: Evolution) and Chapter 14 (Humans Part II: Sociality). Their efforts considerably aided the execution of this book.

Thanks as well to Jack Repcheck, my editor at W. W. Norton, for his enthusiastic support of this project from its inception as well as his light touch in applying his unerring judgment. Jack's assistant, Theresia Kowara, is a jewel to be treasured for her diligence and unflagging bonhomie. I also benefited from the ministrations of an exceptional copyeditor, Stephanie Hiebert.

I also want to acknowledge the support of some dear friends. Lori Cannatella, unabashed hippy chick, for some memorable *Jeopardy!*

nights and stimulating conversations that had nothing to do with this book. And Jeanne and Michael Williams, whose hospitality was well beyond generous. Their idyllic guesthouse in the hills of Lafayette, California, was my abode during the last phase of the book. I was also the beneficiary of Jeanne's considerable culinary skills.

Finally, special thanks to Pauline Burns (aka "Puppet Master"), without whom my stay in California would not have been possible.

Notes and References

Chapter 1: HOUSE FOX

NOTES

1. Gibbs 2002.
2. There were two versions of this tableau. The first, from 1989, displayed at the Georges Pompidou Centre, Paris, had the reverse color scheme (gray overall, with red foxes), white tablecloths, and a few people in the background. This is the Denver version.
3. Lewontin and Levins 1976; reprinted in Levins and Lewontin 1985.
4. The environmental manipulation involved was referred to as "vernalization."
5. See Burkhardt 1995. The long-standing antipathy, among English-speaking evolutionists, to Lamarck—who was, after all, the first to propose a well-considered evolutionary theory—is difficult to fathom. Even Ernst Mayr, as orthodox a Darwinian as ever was, found it perplexing (Mayr 1972). In part this antipathy stems from the Lysenko affair; in part it is a matter of English cultural nationalism. But there is more here for a historian of science to investigate.
6. Vavilov is generally associated with his law of homologous variation (see Vavilov 1922): the potential for subsequent genetic variability of a genotype is constrained in important ways and is therefore broadly predictable. Vavilov also made important contributions to the study of the domestication of plants (Vavilov 1992, a translation of the 1956 Russian original). See Levina, Yesakov, and Kisselev 2005 for an appreciation.
7. Schmalhausen 1949; see Hall 2001 and Gilbert 2003 for Schmalhausen's contributions to evo devo.
8. Drake and Klingenberg 2010.
9. This important distinction was first formulated by the philosopher (of science) Eliot Sober (in Sober 1984). The failure to make this distinction is acute among adaptationists.

10. Belyaev 1979; Trut 1999; Trut, Plyusnina, and Oskina 2004; Trut, Oskina, and Kharlamova 2009.
11. These farmed foxes ultimately derived from wild silver foxes from Prince Edward Island, Canada (Statham et al. 2011).
12. Trut 1999.
13. Ibid.
14. Trut, Oskina, and Kharlamova 2009.
15. Hare et al. 2005; Trut, Oskina, and Kharlamova 2009.
16. Trut 1999.
17. Ibid.
18. Ibid.
19. Belyaev 1979; Trut 1998.
20. Schmalhausen 1949; see also Pigliucci 2005 for a good discussion.
21. The norm-of-reaction concept originated with Woltereck 1909.
22. Waddington 1942, 1959.
23. Scott 1965.
24. Trut, Plyusnina, and Oskina 2004. The pituitary hormone is adrenocorticotropic hormone (ACTH), also known as corticotropin.
25. Ibid.
26. The term "heterochrony" was introduced by Ernst Haeckel and popularized by Stephen Jay Gould (1977). See Smith 2001 for an updated account.
27. Trut, Plyusnina, and Oskina 2004.
28. See Kukekova et al. 2004, 2007, 2008, 2011, and 2012 for preliminary results. Among the highlights: A gene promoting nonaggressiveness was mapped to a homologous region in the dog genome, but it interacts in complex—epistatic—ways with a number of other genes. The tame and aggressive foxes show evidence of many (335) differences in gene expression in the prefrontal cortex.
29. The precursor peptide is called pro-opiomelanocortin (POMC). One reason the gene involved is so massively pleiotropic is that POMC is alternately spliced; that is, depending on the type of the cell where it is expressed, it can be made into different peptides. One peptide is ACTH, another is an endogenous opiate, and still another is involved in melanin metabolism in the epidermis. See Gulevich et al. 2004.
30. Trut, Plyusnina, and Oskina 2004; Trut, Oskina, and Kharlamova 2009.
31. Trut 1999, 2001.
32. Wilkins, A. S., Wrangham, R. W., and Fitch, W. T. (2014).
33. Diamond 2002.
34. Hare and Tomasello 2005; Brüne 2007.

REFERENCES

Belyaev, D. K. (1979). Destabilizing selection as a factor in domestication. *Journal of Heredity* 70 (5): 301–8.
Brüne, M. (2007). On human self-domestication, psychiatry, and eugenics. *Philosophy, Ethics, and Humanities in Medicine* 2: 21.
Burkhardt, R. W. (1995). *The Spirit of System: Lamarck and Evolutionary Biology: Now with "Lamarck in 1995."* Cambridge, MA: Harvard University Press.

Diamond, J. (2002). Evolution, consequences and future of plant and animal domestication. *Nature* 418: 700–707.

Drake, A. G., and C. P. Klingenberg. (2010). Large-scale diversification of skull shape in domestic dogs: Disparity and modularity. *American Naturalist* 175 (3): 289–301.

Gibbs, L., trans. (2002). *Aesop's Fables*. Oxford: Oxford University Press.

Gilbert, S. F. (2003). The morphogenesis of evolutionary developmental biology. *International Journal of Developmental Biology* 47 (7/8): 467–78.

Gould, S. J. (1977). *Ontogeny and Phylogeny*. Cambridge, MA: Harvard University Press.

Gulevich, R., I. Oskina, S. Shikhevich, E. Fedorova, and L. Trut. (2004). Effect of selection for behavior on pituitary-adrenal axis and proopiomelanocortin gene expression in silver foxes (*Vulpes vulpes*). *Physiology & Behavior* 82: 513–18.

Hall, B. K. (2001). Organic selection: Proximate environmental effects on the evolution of morphology and behaviour. *Biology and Philosophy* 16 (2): 215–37.

Hare, B., I. Plyusnina, N. Ignacio, O. Schepina, A. Stepika, R. Wrangham, and L. Trut. (2005). Social cognitive evolution in captive foxes is a correlated by-product of experimental domestication. *Current Biology: CB* 15 (3): 226–30.

Hare, B., and M. Tomasello. (2005). Human-like social skills in dogs? *Trends in Cognitive Sciences* 9 (9): 439–44.

Kukekova, A. V., S. V. Temnykh, J. L. Johnson, L. N. Trut, and G. M. Acland. (2012). Genetics of behavior in the silver fox. *Mammalian Genome* 23 (1–2): 164–77.

Kukekova, A. V., L. N. Trut, K. Chase, A. V. Kharlamova, J. L. Johnson, S. V. Temnykh, . . . and K. Lark. (2011). Mapping loci for fox domestication: Deconstruction/reconstruction of a behavioral phenotype. *Behavior Genetics* 41 (4): 593–606.

Kukekova, A. V., L. N. Trut, K. Chase, D. V. Shepeleva, A. V. Vladimirova, A. V. Kharlamova, . . . and G. M. Acland. (2008). Measurement of segregating behaviors in experimental silver fox pedigrees. *Behavior Genetics* 38 (2): 185–94.

Kukekova, A. V., L. N. Trut, I. N. Oskina, J. L. Johnson, S. V. Temnykh, A. V. Kharlamova, . . . and G. M. Acland. (2007). A meiotic linkage map of the silver fox, aligned and compared to the canine genome. *Genome Research* 17 (3): 387–99.

Kukekova, A. V., L. N. Trut, I. N. Oskina, A. V. Kharlamova, S. G. Shikhevich, E. F. Kirkness, . . . and G. M. Acland. (2004). A marker set for construction of a genetic map of the silver fox (*Vulpes vulpes*). *Journal of Heredity* 95 (3): 185–94.

Levina, E. S., V. D. Yesakov, and L. L. Kisselev. (2005). Nikolai Vavilov: Life in the cause of science or science at a cost of life. *Comprehensive Biochemistry* 44: 345–410.

Levins, R. L., and S. R. Lewontin. (1985). *The Dialectical Biologist*. Cambridge, MA: Harvard University Press.

Lewontin, R., and R. Levins. (1976). "The Problem of Lysenkoism." In *The Radicalisation of Science: Ideology of/in the Natural Sciences*, edited by H. Rose and S. Rose, 32–64. London: Macmillan.

Mayr, E. (1972). Lamarck revisited. *Journal of the History of Biology* 5 (1): 55–94.

Pigliucci, M. (2005). Evolution of phenotypic plasticity: Where are we going now? *Trends in Ecology & Evolution* 20 (9): 481–86.

Schmalhausen, I. I. (1949). *Factors of Evolution: The Theory of Stabilizing Selection*. Translated by Isadore Dordick. Edited by Theodosius Dobzhansky. Philadelphia: Blakiston.

Scott, J. P. (1965). *Genetics and the Social Behavior of the Dog.* Chicago: University of Chicago Press.

Smith, K. K. (2001). Heterochrony revisited: The evolution of developmental sequences. *Biological Journal of the Linnean Society* 73 (2): 169–86.

Sober, E. (1984). *The Nature of Selection.* Cambridge, MA: MIT Press.

Statham, M. J., L. N. Trut, B. N. Sacks, A. V. Kharlamova, I. N. Oskina, R. G. Gulevich, . . . and A. V. Kukekova. (2011). On the origin of a domesticated species: Identifying the parent population of Russian silver foxes (*Vulpes vulpes*). *Biological Journal of the Linnean Society* 103 (1): 168–75.

Trut, L. N. (1998). The evolutionary concept of destabilizing selection: *status quo* In commemoration of D. K. Belyaev. *Journal of Animal Breeding and Genetics* 115 (6): 415–31.

Trut, L. N. (1999). Early canid domestication: The farm-fox experiment. *American Scientist* 87: 160–69.

Trut, L. N. (2001). "Experimental Studies of Early Canine Domestication." In *Genetics of the Dog*, edited by A. Ruvinsky and J. Sampson, 15–43. Wallingford, UK: CABI.

Trut, L., I. Oskina, and A. Kharlamova. (2009). Animal evolution during domestication: The domesticated fox as a model. *BioEssays* 31 (3): 349–60.

Trut, L. N., I. Z. Plyusnina, and I. N. Oskina. (2004). An experiment on fox domestication and debatable issues of evolution of the dog. *Russian Journal of Genetics* 40 (6): 644–55.

Vavilov, N. I. (1922). The law of homologous series in variation. *Journal of Genetics* 12 (1): 47–89.

Vavilov, N. I. (1992). *Origin and Geography of Cultivated Plants.* Cambridge: Cambridge University Press.

Waddington, C. H. (1942). Canalization of development and the inheritance of acquired characters. *Nature* 150 (3811): 563–65.

Waddington, C. H. (1959). Canalization of development and genetic assimilation of acquired characters. *Nature* 183 (4676): 1654–55.

Wilkins, A. S., Wrangham, R. W., and Fitch, W. T. (2014). The "domestication syndrome" in mammals: a unified explanation based on neural crest cell behavior and genetics. *Genetics* 197 (3): 795–808.

Woltereck, R. (1909). Weitere experimentelle Untersuchungen über Artveränderung, speziell über das Wesen quantitativer Artunterscheide bei Daphniden. *Verhandlungen der Deutschen Zoologischen Gesellschaft* 19: 110–72.

Chapter 2: DOGS

NOTES

1. Williams 1974.
2. X. Wang et al. 2004.
3. Unlike Asiatic and African wild dogs, which almost always occur in social groups, wolves can be solitary. In some areas, such as India, most wolves are solitary for much of the year.
4. Species status among members of genus *Canis* (coyote, red wolf, gray wolf) is a vexed issue; see, for example, Coppinger, Spector, and Miller 2010.
5. Germonpré et al. 2009; Galibert et al. 2011.

6. Greger Larson, e-mail, February 12, 2014.
7. Savolainen et al. 2002.
8. Niskanen et al. 2013.
9. VonHoldt et al. 2010.
10. Thalmann et al. 2013.
11. Vilà et al. 1997; Crockford 2000; Larson et al. 2012.
12. Clutton-Brock 1995; Coppinger and Coppinger 2001.
13. Coppinger and Coppinger 2001.
14. But see Coppinger and Coppinger 2001.
15. Nobis 1979.
16. Morey 2006; Losey et al. 2011.
17. Davis and Valla 1978.
18. Clutton-Brock 1995.
19. Larson et al. 2012.
20. Ibid.
21. Morey 1992, 1994.
22. At Coxcatlan Cave.
23. Gautier 2002.
24. Larson et al. 2012.
25. Prates, Prevosti, and Berón 2010.
26. Sutter et al. 2007; Gray et al. 2010: mutation derived from Near Eastern wolves.
27. G. D. Wang et al. 2013.
28. Gould 1969; Clutton-Brock 1995.
29. Salima Ikram, American University in Cairo, quoted in Lobell and Powell 2010.
30. This mummy was discovered in 1902 at Abydos and dated to the fourth century BCE; it was CT-scanned at the University of Pennsylvania by Janet Monge.
31. Homer, *Odyssey* book 17, quote beginning "As they were talking."
32. MacKinnon 2010.
33. Collie 2013.
34. MacKinnon 2010; Driscoll and Macdonald 2010.
35. Robert Rosenswig, quoted in Lobell and Powell 2010.
36. Dog consumption was not confined to the New World. In Europe, dogs were consumed in Iron Age France. There is even evidence of dog farming at Levroux. Evidence from the Celtic village of La Tene in northern Slovakia suggests that dog consumption in Europe continued at least through the Roman period (Chrószcz et al. 2013).
37. Tito et al. 2011. The discovery was made by Samuel Belknap III, then a graduate student at the University of Maine. These bones were dated to 9400 BP.
38. Lobell and Powell 2010.
39. Clutton-Brock and Hammond 1994; White et al. 2001.
40. The Aztec breed is the Xoloitzcuintli, which still exists today.
41. Rosenswig 2006.
42. Zimmerman 2001.
43. Rangel was De Soto's personal secretary; he wrote, "The Indians came forth in peace and gave them corn, although little and many hens, and a few little dogs, which are good food. These are little dogs that do not bark, and they rear them in the house for food."

("Rodrigo Rangel's Account, Part 2," FloridaHistory.com, http://www.floridahistory.com/rangel-2.html.)

44. Serpell 1995, 2009.

45. Craig Skehan, "Dog-Meat Mafia: Inside Thailand's Smuggling Trade," *Globe and Mail*, May 20, 2013. Most of the dogs eaten are feral.

46. The Carolina dog may be a remnant North American landrace, but in its current form it is a largely reconstructed breed.

47. Raymond Coppinger fervently advocates that all dog domestication came by way of village dogs (Coppinger and Coppinger 2001).

48. Macdonald and Carr 1995; Boitani and Ciucci 1995; Boitani, Ciucci, and Ortolani 2007; Lord et al. 2013.

49. Brown et al. 2011.

50. Boyko et al. 2009.

51. Savolainen et al. 2004; Oskarsson et al. 2012.

52. But Savolainen et al. 2004 estimate that dingoes arrived in Australia around 5000 BP.

53. Corbett 1995.

54. Newsome, Corbett, and Carpenter 1980.

55. B. Smith and Litchfield 2010.

56. At the Prospect Park Zoo in Brooklyn, for example, the sign for the dingo exhibit suggests that dingoes are something like wolf-dogs—half wolf, half dog.

57. Parker et al. 2004.

58. See Morris 2002 for recent breed reconstruction of the Shiba Inu and Finnish spitz.

59. That is, they are artifacts of the methods used for constructing breed lineages on the basis of DNA (Larson et al. 2012).

60. Parker et al. 2004; VonHoldt et al. 2010.

61. Larson et al. 2012; but see Dayan 1994 for proposed West Asian origin.

62. Ashdown and Lea 1979.

63. Fuller 1956.

64. See Larson et al. 2012 for evidence that the Afghan was extinct in Europe after World War II, after which it was reestablished through three imported dogs.

65. Vilà et al. 2003; VonHoldt et al. 2013.

66. Only the Akita is basal in Larson et al. 2012.

67. John Henry Walsh, using the pseudonym "Stonehenge," adapted standards from pigeon fanciers. The first dog to serve as a breed standard was a pointer named "Major." Some refer to Major as the first modern dog.

68. VonHoldt et al. 2010. This correlation between function and genealogy holds only for western Europe, and particularly Great Britain (and its former colonies). Geography becomes more important for genealogical relationships over larger geographic scales—Eurasia, for example.

69. Much of this discussion on collie landraces comes from "The Collie Spectrum: Understanding the Scotch Landrace," Old-Time Farm Shepherd, http://www.oldtimefarmshepherd.org/current-collie-articles/the-collie-spectrum-understanding-the-scotch-landrace.

70. VonHoldt et al. 2010.

71. Vilà, Seddon, and Ellegren 2005. This is actually true for members of the genus *Canis* generally. Hybridization is rampant—hence the difficulty of delineating species using the

biological species concept. For example, during the recent expansion of coyotes eastward, there was much interbreeding with eastern timber wolves. As a result, eastern coyotes are much larger and more wolflike in behavior than western coyotes (Kays, Curtis, and Kirchman 2010).

72. The insertion or deletion of a base also counts as a point mutation. C = cytosine; G = guanine; A = adenine; T = thymine.

73. Tandem repeats could be considered a subcategory of indels, but the latter term is usually reserved for additions or deletions of a larger base sequence that are one-offs.

74. Fondon et al. 2008.

75. Fondon and Garner 2004.

76. Nicholas et al. 2011; Alvarez and Akey 2011.

77. Sutter et al. 2007; Gray et al. 2010.

78. Parker et al. 2009.

79. Cadieu et al. 2009.

80. Reviewed in Parker, Shearin, and Ostrander 2010.

81. Wayne and VonHoldt 2012.

82. Cardieu et al. 2009.

83. Boyko et al. 2010.

84. L. Smith et al. 2008; Bannasch et al. 2010.

85. See Boyko et al. 2010; Vaysse et al. 2011.

86. See Parker, Shearin, and Ostrander 2010 for a review.

87. Kraus, Pavard, and Promislow 2013.

88. My beloved toy terrier–Chihuahua mix, Gigi, survived to age 18.

89. Careau et al. 2010.

90. Galis et al. 2007.

91. See Van Noort, Snel, and Huynen 2003 for a general account of genetic architecture.

92. Drake and Klingenberg 2010.

93. See, for example, Raff and Raff 2000; Mitteroecker and Bookstein 2007; and Klingenberg 2008.

94. Drake and Klingenberg 2010.

REFERENCES

Alvarez, C. E., and J. M. Akey. (2011). Copy number variation in the domestic dog. *Mammalian Genome* 23 (1–2): 1–20.

Ashdown, R. R., and T. Lea. (1979). The larynx of the Basenji dog. *Journal of Small Animal Practice* 20 (11): 675–79.

Bannasch, D., A. Young, A. Myers, K. Truvé, P. Dickinson, J. Gregg, . . . and N. Pedersen. (2010). Localization of canine brachycephaly using an across breed mapping approach. *PLoS One* 5 (3): e9632.

Boitani, L., and P. Ciucci. (1995). Comparative social ecology of feral dogs and wolves. *Ethology, Ecology & Evolution* 7 (1): 49–72.

Boitani, L., P. Ciucci, and A. Ortolani. (2007). Behaviour and social ecology of free-ranging dogs. In *The Behavioural Biology of Dogs*, edited by Per Jensen, 147–65. Wallingford, UK: CABI International.

Boyko, A., R. Boyko, C. Boyko, H. Parker, M. Castelhano, L. Corey, . . . and C. Bustamante.

(2009). Complex population structure in African village dogs and its implications for inferring dog domestication history. *Proceedings of the National Academy of Sciences USA* 39: 13903–8.

Boyko, A., P. Quignon, L. Li, J. Schoenebeck, J. Degenhardt, K. Lohmueller, . . . and E. Ostrander. (2010). A simple genetic architecture underlies morphological variation in dogs. *PLoS Biology* 8: e1000451.

Brown, S. K., N. C. Pedersen, S. Jafarishorijeh, D. L. Bannasch, K. D. Ahrens, J.-T. Wu, . . . and B. N. Sacks. (2011). Phylogenetic distinctiveness of Middle Eastern and Southeast Asian village dog Y chromosomes illuminates dog origins. *PLoS One* 6 (12): e28496.

Cadieu, E., M. Neff, P. Quignon, K. Walsh, K. Chase, H. Parker, . . . and E. Ostrander. (2009). Coat variation in the domestic dog is governed by variants in three genes. *Science* 39 (5949): 150–53.

Careau, V., D. Réale, M. M. Humphries, and D. W. Thomas. (2010). The pace of life under artificial selection: Personality, energy expenditure, and longevity are correlated in domestic dogs. *American Naturalist* 175 (6): 753–58.

Chrószcz, A., M. Janeczek, Z. Bielichová, T. Gralak, and V. Onar. (2013). Cynophagia in the Púchov (Celtic) Culture Settlement at Liptovská Mara, Northern Slovakia. *International Journal of Osteoarchaeology*, May 26.

Clutton-Brock, J. (1995). "Origins of the Domestic Dog: Domestication and Early History." In *The Domestic Dog: Its Evolution, Behaviour, and Interactions with People*, edited by J. Serpell, 7–20. Cambridge: Cambridge University Press.

Clutton-Brock, J., and N. Hammond. (1994). Hot dogs: Comestible canids in Preclassic Maya culture at Cuello, Belize. *Journal of Archaeological Science* 21 (6): 819–26.

Collie, N. (2013). "Ritualising Encounters with Subterranean Places: An Investigation of Urban Depositional Practices of Roman Britain." PhD diss., University of Tasmania.

Coppinger, R., and L. Coppinger. (2001). *Dogs: A Startling New Understanding of Canine Origin, Behavior and Evolution*. New York: Scribner.

Coppinger, R., L. Spector, and L. Miller. (2010). "What, If Anything, Is a Wolf?" In *The World of Wolves: New Perspectives on Ecology, Behaviour, and Management*, edited by M. Musiani, L. Boitani, and P. C. Paquet, 1–52. Energy, Ecology, and the Environment Series 3. Calgary, Alberta: University of Calgary Press.

Corbett, L. K. (1995). *Dingo in Australia and Asia*. Sydney, Australia: UNSW Press.

Crockford, S. J. (2000). A commentary on dog evolution: Regional variation, breed development and hybridization with wolves. *BAR International Series* 889: 295–312.

Davis, S., and F. Valla. (1978). Evidence for domestication of dog 12,000 years ago in Natufian of Israel. *Nature* 276: 608–10.

Dayan, T. (1994). Early domesticated dogs of the Near East. *Journal of Archaeological Science* 21 (5): 633–40.

Drake, A. G., and C. P. Klingenberg. (2010). Large-scale diversification of skull shape in domestic dogs: Disparity and modularity. *American Naturalist* 175 (3): 289–301.

Driscoll, C., and D. Macdonald. (2010). Top dogs: Wolf domestication and wealth. *Journal of Biology* 9 (2): 10.

Fondon, J. W., III, and H. R. Garner. (2004). Molecular origins of rapid and continuous morphological evolution. *Proceedings of the National Academy of Sciences USA* 101 (52): 18058–63.

Fondon, J. W., III, E. A. D. Hammock, A. J. Hannan, and D. G. King. (2008). Simple sequence repeats: Genetic modulators of brain function and behavior. *Trends in Neurosciences* 31 (7): 328–34.

Fuller, J. L. (1956). Photoperiodic control of estrus in the Basenji. *Journal of Heredity* 47 (4): 179–80.

Galibert, F., P. Quignon, C. Hitte, and C. André. (2011). Toward understanding dog evolutionary and domestication history. *Comptes Rendus Biologies* 334 (3): 190–96.

Galis, F., I. Van Der Sluijs, T. J. M. Van Dooren, J. A. J. Metz, and M. Nussbaumer. (2007). Do large dogs die young? *Journal of Experimental Zoology. Part B: Molecular and Developmental Evolution* 308B (2): 119–26.

Gautier, A. (2002). "The Evidence for the Earliest Livestock in North Africa: Or Adventures with Large Bovids, Ovicaprids, Dogs and Pigs." In *Droughts, Food and Culture: Ecological Change and Food Security in Africa's Later Prehistory*, edited by F. A. Hassan, 195–207. New York: Kluwer Academic/Plenum.

Germonpré, M., M. V. Sablin, R. E. Stevens, R. E. M. Hedges, M. Hofreiter, M. Stiller, and V. R. Després. (2009). Fossil dogs and wolves from Palaeolithic sites in Belgium, the Ukraine and Russia: Osteometry, ancient DNA and stable isotopes. *Journal of Archaeological Science* 36 (2): 473–90.

Gould, R. A. (1969). Subsistence behaviour among the Western Desert Aborigines of Australia. *Oceania* 39: 253–74.

Gray, M., N. Sutter, E. Ostrander, and R. Wayne. (2010). The IGF1 small dog haplotype is derived from Middle Eastern grey wolves. *BMC Biology* 8 (1): 16.

Kays, R., A. Curtis, and J. J. Kirchman. (2010). Rapid adaptive evolution of northeastern coyotes via hybridization with wolves. *Biology Letters* 6 (1): 89–93.

Klingenberg, C. P. (2008). Morphological integration and developmental modularity. *Annual Review of Ecology, Evolution, and Systematics* 39: 115–32.

Kraus, C., S. Pavard, and D. E. Promislow. (2013). The size-life span trade-off decomposed: Why large dogs die young. *American Naturalist* 181 (4): 492–505.

Larson, G., E. K. Karlsson, A. Perri, M. T. Webster, S. Y. W. Ho, J. Peters, . . . and K. Lindblad-Toh. (2012). Rethinking dog domestication by integrating genetics, archeology, and biogeography. *Proceedings of the National Academy of Sciences USA* 109 (23): 8878–83.

Lobell, J. A., and E. A. Powell. (2010). More than man's best friend. *Archaeology* 63 (5): 26–35.

Lord, K., M. Feinstein, B. Smith, and R. Coppinger. (2013). Variation in reproductive traits of members of the genus *Canis* with special attention to the domestic dog (*Canis familiaris*). *Behavioural Processes* 92: 131–42.

Losey, R. J., V. I. Bazaliiskii, S. Garvie-Lok, M. Germonpré, J. A. Leonard, A. L. Allen, . . . and M. V. Sablin. (2011). Canids as persons: Early Neolithic dog and wolf burials, Cis-Baikal, Siberia. *Journal of Anthropological Archaeology* 30 (2): 174–89.

Macdonald, D., and G. Carr. (1995). "Variation in Dog Society: Between Resource Dispersion and Social Flux." In *The Domestic Dog, Its Evolution, Behaviour, and Interactions with People*, edited by J. Serpell, 199–216. Cambridge: Cambridge University Press.

MacKinnon, M. (2010). "Sick as a dog": Zooarchaeological evidence for pet dog health and welfare in the Roman world. *World Archaeology* 42 (2): 290–309.

Mitteroecker, P., and F. Bookstein. (2007). The conceptual and statistical relationship between modularity and morphological integration. *Systematic Biology* 56 (5): 818–36.

Morey, D. F. (1992). Size, shape and development in the evolution of the domestic dog. *Journal of Archaeological Science* 19 (2): 181–204.

Morey, D. F. (1994). The early evolution of the domestic dog. *American Scientist* 82 (4): 336–47.

Morey, D. F. (2006). Burying key evidence: The social bond between dogs and people. *Journal of Archaeological Science* 33 (2): 158–75.

Morris, D. (2002). *Dogs: The Ultimate Dictionary of Over 1,000 Dog Breeds*. North Pomfret, VT: Trafalgar Square.

Newsome, A., L. Corbett, and S. Carpenter. (1980). The identity of the dingo I. Morphological discriminants of dingo and dog skulls. *Australian Journal of Zoology* 28 (4): 615–25.

Nicholas, T., C. Baker, E. Eichler, and J. Akey. (2011). A high-resolution integrated map of copy number polymorphisms within and between breeds of the modern domesticated dog. *BMC Genomics* 12 (1): 414.

Niskanen, A. K., E. Hagstrom, H. Lohi, M. Ruokonen, R. Esparza-Salas, J. Aspi, and P. Savolainen. (2013). MHC variability supports dog domestication from a large number of wolves: High diversity in Asia. *Heredity* 110 (1): 80–85.

Nobis, G. (1979). "Problems of the Early Husbandry of Domestic Animals in Northern Germany and Denmark." In *Proceedings of the 18th International Symposium on Archaeometry and Archaeological Prospection, Bonn, 14–17 March 1978*. Archaeo-Physika 10. Cologne, Germany: Rheinland-Verlag.

Oskarsson, M. C. R., C. F. C. Klütsch, U. Boonyaprakob, A. Wilton, Y. Tanabe, and P. Savolainen. (2012). Mitochondrial DNA data indicate an introduction through Mainland Southeast Asia for Australian dingoes and Polynesian domestic dogs. *Proceedings of the Royal Society. B: Biological Sciences* 279 (1730): 967–74.

Parker, H. G., L. V. Kim, N. B. Sutter, S. Carlson, T. D. Lorentzen, T. B. Malek, . . . and L. Kruglyak. (2004). Genetic structure of the purebred domestic dog. *Science* 304 (5674): 1160–64.

Parker, H. G., A. L. Shearin, and E. A. Ostrander. (2010). Man's best friend becomes biology's best in show: Genome analyses in the domestic dog. *Annual Review of Genetics* 44 (1): 309–36.

Parker, H. G., B. M. VonHoldt, P. Quignon, E. H. Margulies, S. Shao, D. S. Mosher, . . . and E. A. Ostrander. (2009). An expressed *Fgf4* retrogene is associated with breed-defining chondrodysplasia in domestic dogs. *Science* 325 (5943): 995–98.

Prates, L., F. J. Prevosti, and M. Berón. (2010). First records of prehispanic dogs in southern South America (Pampa-Patagonia, Argentina). *Current Anthropology* 51 (2): 273–80.

Raff, E., and R. Raff. (2000). Dissociability, modularity, evolvability. *Evolution and Development* 2: 235–37.

Rosenswig, R. M. (2006). Sedentism and food production in early complex societies of the Soconusco, Mexico. *World Archaeology* 38 (2): 330–55.

Savolainen, P., T. Leitner, A. N. Wilton, E. Matisoo-Smith, and J. Lundeberg. (2004). A detailed picture of the origin of the Australian dingo, obtained from the study of mitochondrial DNA. *Proceedings of the National Academy of Sciences USA* 101 (33): 12387–90.

Savolainen, P., Y. Zhang, J. Luo, J. Lundeberg, and T. Leitner. (2002). Genetic evidence for an East Asian origin of domestic dogs. *Science* 298: 1610–13.

Serpell, J., ed. (1995). *The Domestic Dog: Its Evolution, Behaviour and Interactions with People*. Cambridge: Cambridge University Press.

Serpell, J. A. (2009). Having our dogs and eating them too: Why animals are a social issue. *Journal of Social Issues* 65 (3): 633–44.

Smith, B., and C. Litchfield. (2010). Dingoes (*Canis dingo*) can use human social cues to locate hidden food. *Animal Cognition* 13 (2): 367–76.

Smith, L., D. Bannasch, A. Young, D. Grossman, J. Belanger, and A. Oberbauer. (2008). Canine fibroblast growth factor receptor 3 sequence is conserved across dogs of divergent skeletal size. *BMC Genetics* 9 (1): 67.

Sutter, N. B., C. D. Bustamante, K. Chase, M. M. Gray, K. Zhao, L. Zhu, . . . and E. A. Ostrander. (2007). A single IGF1 allele is a major determinant of small size in dogs. *Science* 316 (5821): 112–15.

Thalmann, O., B. Shapiro, P. Cui, V. J. Schuenemann, S. K. Sawyer, D. L. Greenfield, . . . and R. K. Wayne. (2013). Complete mitochondrial genomes of ancient canids suggest a European origin of domestic dogs. *Science* 342 (6160): 871–74.

Tito, R. Y., S. L. Belknap, K. D. Sobolik, R. C. Ingraham, L. M. Cleeland, and C. M. Lewis. (2011). Brief communication: DNA from early Holocene American dog. *American Journal of Physical Anthropology* 145 (4): 653–57.

Van Noort, V., B. Snel, and M. Huynen. (2003). Predicting gene function by conserved co-expression. *Trends in Genetics* 19: 238–42.

Vaysse, A., A. Ratnakumar, T. Derrien, E. Axelsson, G. R. Pielberg, S. Sigurdsson, . . . and C. T. Lawley. (2011). Identification of genomic regions associated with phenotypic variation between dog breeds using selection mapping. *PLoS Genetics* 7 (10): e1002316.

Vilà, C., P. Savolainen, J. E. Maldonado, I. R. Amorim, J. E. Rice, R. L. Honeycutt, . . . and R. K. Wayne. (1997). Multiple and ancient origins of the domestic dog. *Science* 276 (5319): 1687–89.

Vilà, C., J. Seddon, and H. Ellegren. (2005). Genes of domestic mammals augmented by back-crossing with wild ancestors. *Trends in Genetics* 21 (4): 214–18.

Vilà, C., C. Walker, A. K. Sundqvist, Ø. Flagstad, Z. Andersone, A. Casulli, . . . and H. Ellegren. (2003). Combined use of maternal, paternal and bi-parental genetic markers for the identification of wolf-dog hybrids. *Heredity* 90 (1): 17–24.

VonHoldt, B., J. Pollinger, D. Earl, H. Parker, E. Ostrander, and R. Wayne. (2013). Identification of recent hybridization between gray wolves and domesticated dogs by SNP genotyping. *Mammalian Genome* 24 (1–2): 80–88.

VonHoldt, B., J. Pollinger, K. Lohmueller, E. Han, H. Parker, P. Quignon, . . . and W. Huang. (2010). Genome-wide SNP and haplotype analyses reveal a rich history underlying dog domestication. *Nature* 39: 898–902.

Wang, G. D., W. Zhai, H.-C. Yang, R.-X. Fan, X. Cao, L. Zhong, . . . and Y.-P. Zhang. (2013). The genomics of selection in dogs and the parallel evolution between dogs and humans. *Nature Communications* 4: 1860.

Wang, X., R. Tedford, B. Van Valkenburgh, and R. Wayne. (2004). "Ancestry-Evolutionary History, Molecular Systematics, and Evolutionary Ecology of Canidae." In *The Biology and Conservation of Wild Canids*, edited by D. W. Macdonald and C. Sillero-Zubiri, 38–54. Oxford: Oxford University Press.

Wayne, R., and B. vonHoldt. (2012). Evolutionary genomics of dog domestication. *Mammalian Genome* 23 (1): 3–18.

White, C. D., M. E. D. Pohl, H. P. Schwarcz, and F. J. Longstaffe. (2001). Isotopic evidence for

Maya patterns of deer and dog use at preclassic Colha. *Journal of Archaeological Science* 28 (1): 89–107.

Williams, C. (1974). *Chinese Symbolism and Art Motifs: An Alphabetical Comprehensive Handbook on Symbolism in Chinese Art through the Ages*, 4th rev. ed. North Clarendon, VT: Tuttle.

Zimmerman, L. (2001). "Northern Plains Village." In *Encyclopedia of Prehistory*, edited by P. Peregrine and M. Ember, 377–88. New York: Kluwer Academic/Plenum.

Chapter 3: CATS

NOTES

1. For example, because Smoke was the only female in a litter of four, she received extra attention. Sylvester was the smallest of the three males, which didn't help in his competition with his brothers, and he was clearly dominated by them. Largely for these reasons, Sylvester was the kitten that the mother cat's owners most wanted to be rid of. They were much more reluctant to part with Smoke and her two other brothers. For an account of the formative role of early experience, see Bateson (2000) and Mendl and Harcourt (2000).

2. Among carnivores and other mammalian orders, such as primates, the more plant material there is in the diet, the longer the intestine is, because plant material requires more digestive effort. Members of the cat family (Felidae) have the shortest intestines among terrestrial carnivore families. Their high-protein diet is also reflected in their sense of taste. Felids are largely insensitive to the taste of salt, because they get salt as a by-product of animal protein; they have no sense of sweetness whatsoever.

3. Van Valkenburgh and Ruff 1987.

4. Generally, in coalitions of males (Caro and Collins 1987).

5. Johnson et al. 2006; O'Brien and Johnson 2007; O'Brien et al. 2008.

6. O'Brien et al. 2008.

7. Driscoll et al. 2007, 2009.

8. Cameron-Beaumont, Lowe, and Bradshaw 2002; Goldberg 2003.

9. Ginsburg et al. 1991; Clutton-Brock 1999.

10. Driscoll et al. 2009.

11. Zeuner 1958.

12. Vigne et al. 2004.

13. Serpell 2000; Driscoll et al. 2009.

14. Linseele, Van Neer, and Hendrickx 2007.

15. Clutton-Brock 1992, 1999.

16. Málek 1997.

17. Serpell 2000.

18. Driscoll et al. 2009.

19. Bateson and Turner 2014. They speculate that this is when the tail-up display may have evolved as a consequence of selection for sociability.

20. Serpell 2000. Citing Ahmad, Blumenberg, and Chaudhary 1980, Serpell describes what would have been a much earlier export of cats from Egypt to the Indus valley Harappan

civilization (4500–4100 BP). But that would be prior to the period of their prevalence in Egypt; so if these "urban cats" were in fact domesticated, they probably came from West Asia.

21. Driscoll et al. 2009. Serpell 2000, however, claims that both Greeks and Romans preferred domestic polecats and ferrets for rodent control, so some of the shipboard cats may have been stowaways (see also Beadle 1979). In any case, cats adapt well to life aboard ships, and ship transport was an important source of global cat dispersal (Todd 1977). For example, the sex-linked mutation found in orange tabby males, and calico/tortoiseshell females, seems to have originated in western Turkey, from which it was transported on Viking ships to Brittany, Britain, and Scandinavia.

22. Zeuner 1963.

23. Driscoll et al. 2009.

24. Driscoll et al. 2007.

25. Wastlhuber 1991.

26. Morris 1999.

27. Morris 1999.

28. Taylor 1989.

29. To my eyes the Levkoy is, aesthetically, the feline equivalent of ET.

30. Menotti-Reymond et al. 2007; Stephens 2001. G. Sutton, "Scottish Fold," Cat Fanciers' Association, http://www.cfa.org/Breeds/BreedsSthruT/ScottishFold/SFArticle.aspx. The ear deformation in Scottish folds is but one of several cartilage-related defects, which include stunted limb growth, or osteochondrodysplasia (Gunn-Moore, Bessant, and Malik 2008).

31. Zhigachev and Vladimirova 2002.

32. Gunn-Moore, Bessant, and Malik 2008; Stephens 2001.

33. Albuisson et al. 2011.

34. Polydactyls are not recognized by the British Cat Fanciers' Association (Dennis Turner, e-mail, February 5, 2014) for animal cruelty reasons, which is encouraging.

35. The condition is more technically referred to as radial hypoplasia.

36. See Gandolfi et al. 2010 for the mutation involved.

37. O'Brien et al. 2008. See Kehler et al. 2007 for mutations in $Fgf5$ related to hair length.

38. I'm not sure how these last two crosses were accomplished. I'm assuming that the domestic cat was the female in both cases, because no self-respecting female serval or caracal would let herself be mounted by a tomcat, no matter how brazen and brawny. But these out-of-genus crosses are problematic on other practical grounds; the gestation periods are different in the two species, and a suite of well-known defects is associated with interspecies hybridizations. So, many of the fetuses are aborted at various stages of development, and sterility is common, particularly in males. These impediments haven't deterred ambitious breeders or The International Cat Association (TICA), which provides official sanction for new breeds. Indeed, there is now a new breed, called Toyger, from a second-order cross of Bengal (leopard cat × domestic cat) and a domestic tabby. The Toyger is so called because it has stripes like a tiger. With the advent of artificial insemination we might expect the creation of a tiger-striped domestic cat from crosses with actual tigers (*Panthera tigris*), but I'm not sure it would make a good pet—maybe a "watchcat." For that matter, you might want to steer your toddlers away from a Savannah or a Caracat.

39. See, for example, Lynch 1991.
40. Thai cats are actually becoming quite rare in Thailand. The ones I observed were in small "zoos" exhibiting local fauna.
41. "Siam: America's First Siamese Cat," Rutherford B. Hayes Presidential Center, http://www.rbhayes.org/hayes/manunews/paper_trail_display.asp?nid=65&subj=manunews.
42. Jennifer Copley, "Naturally Occurring Cat Breeds, Cross-Breeds, and Recent Mutations," *Metaphorical Platypus* (blog), July 13, 2010, http://www.metaphoricalplatypus.com/articles/animals/cats/cat-facts/naturally-occurring-cat-breeds-cross-breeds-and-recent-mutations.
43. Pontius et al. 2007.
44. O'Brien and Nash 1982; Menotti-Raymond et al. 1999.
45. Kehler et al. 2007.
46. Geibel et al. 1991; Lyons et al. 2005.
47. Schmidt-Küntzel et al. 2005; Linderholm and Larson 2013.
48. Schmidt-Küntzel et al. 2005.
49. Albuisson et al. 2011; Lange, Nemeschkal, and Müller 2013.
50. See, for example, Ingham and Placzek 2006.
51. This cis-regulatory element is somewhat odd in its considerable distance from the gene (*shh*) it regulates (Lettice et al. 2008; Albuisson et al. 2011).
52. O'Brien et al. 1999; Menotti-Raymond et al. 2009.
53. For example, see Menotti-Raymond et al. 2007, 2010 for the cat as model for progressive retinal degeneration.
54. Zatta and Frank 2007.
55. Courchamp, Say, and Pontier 2000.
56. Macdonald 1983; Natoli and de Vito 1991.
57. Bradshaw et al. 1999; Fitzgerald and Karl 1986.
58. Macdonald 1983; Crowell-Davis 2005.
59. Bradshaw and Cameron-Beaumont 2000; Cafazzo and Natoli 2009.
60. Bradshaw and Cameron-Beaumont 2000.
61. Bradshaw and Cameron-Beaumont 2000; Bradshaw, Casey, and Brown 2012.
62. Nicastro 2004.
63. Sylvester's meow is weak compared to that of the late Misty, she of the casual attitude toward raccoons described in Chapter 5. Misty's meow was astoundingly and gratingly loud, which was especially irritating during phone conversations. And phone conversations never failed to elicit her best efforts in that regard, to the point of exasperation for those on both ends of the line.

REFERENCES

Ahmad, M., B. Blumenberg, and M. F. Chaudhary. (1980). Mutant allele frequencies and genetic distance in cat populations of Pakistan and Asia. *Journal of Heredity* 71 (5): 323–30.

Albuisson, J., B. Isidor, M. Giraud, O. Pichon, T. Marsaud, A. David, . . . and S. Bezieau. (2011). Identification of two novel mutations in Shh long-range regulator associated with familial pre-axial polydactyly. *Clinical Genetics* 79 (4): 371–77.

Bateson, P. P. G. (2000). "Behavioural Development in the Cat." In *The Domestic Cat: The*

Biology of Its Behaviour, edited by D. C. Turner and P. P. G. Bateson, 9–22. Cambridge: Cambridge University Press.

Bateson, P. and Turner, D. C. (2014). "Postscript: Questions and Some Answers." In *The Domestic Cat: The Biology of Its Behaviour*, 3rd ed., edited by D. C. Turner and P. Bateson, 231–40. Cambridge: Cambridge University Press.

Beadle, M. (1979). *Cat: A Complete Authoritative Compendium of Information about Domestic Cats*. New York: Simon and Schuster.

Bradshaw, J., and C. Cameron-Beaumont. (2000). "The Signalling Repertoire of the Domestic Cat and Its Undomesticated Relatives." In *The Domestic Cat: The Biology of Its Behaviour*, 2nd ed., edited by D. C. Turner and P. Bateson, 67–93. Cambridge: Cambridge University Press.

Bradshaw, J. W., R. A. Casey, and S. L. Brown. (2012). *The Behaviour of the Domestic Cat*. Wallingford, UK: CABI.

Bradshaw, J. W. S., G. F. Horsfield, J. A. Allen, and I. H. Robinson. (1999). Feral cats: Their role in the population dynamics of *Felis catus*. *Applied Animal Behaviour Science* 65 (3): 273–83.

Cafazzo, S., and E. Natoli. (2009). The social function of tail up in the domestic cat (*Felis silvestris catus*). *Behavioural Processes* 80 (1): 60–66.

Cameron-Beaumont, C., S. Lowe, and J. Bradshaw. (2002). Evidence suggesting preadaptation to domestication throughout the small Felidae. *Biological Journal of the Linnean Society* 75: 361–66.

Caro, T. M., and D. A. Collins. (1987). Male cheetah social organization and territoriality. *Ethology* 74 (1): 52–64.

Clutton-Brock, J. (1992). The process of domestication. *Mammal Review* 22 (2): 79–85.

Clutton-Brock, J. (1999). *A Natural History of Domesticated Mammals*. Cambridge: Cambridge University Press.

Courchamp, F., L. Say, and D. Pontier. (2000). Transmission of feline immunodeficiency virus in a population of cats (*Felis catus*). *Wildlife Research* 27 (6): 603–11.

Crowell-Davis, S. (2005). "Cat Behaviour: Social Organization, Communication and Development." In *The Welfare of Cats*, edited by I. Rochlitz, 1–22. Animal Welfare 3. Dordrecht, Netherlands: Springer.

Driscoll, C. A., J. Clutton-Brock, A. C. Kitchener, and S. J. O'Brien. (2009). The taming of the cat: Genetic and archaeological findings hint that wildcats became house cats earlier—and in a different place—than previously thought. *Scientific American* 300 (6): 68–75.

Driscoll, C. A., M. Menotti-Raymond, A. L. Roca, K. Hupe, W. E. Johnson, E. Geffen, . . . and D. W. Macdonald. (2007). The Near Eastern origin of cat domestication. *Science* 317: 519–23.

Fitzgerald, B., and B. Karl. (1986). Home range of feral house cats (*Felis catus* L.) in forest of the Orongorongo Valley, Wellington, New Zealand. *New Zealand Journal of Ecology* 9: 71–82.

Gandolfi, B., C. A. Outerbridge, L. G. Beresford, J. A. Myers, M. Pimentel, H. Alhaddad, . . . and L. A. Lyons. (2010). The naked truth: Sphynx and Devon Rex cat breed mutations in KRT71. *Mammalian Genome* 21 (9–10): 509–15.

Geibel, L. B., R. K. Tripathi, R. A. King, and R. A. Spitz. (1991). A tyrosinase gene missense mutation in temperature-sensitive type I oculocutaneous albinism. A human homologue to the Siamese cat and the Himalayan mouse. *Journal of Clinical Investigation* 87 (3): 1119.

Ginsburg, L., G. Delibrias, A. Minaut-Gout, H. Valladas, and A. Zivie. (1991). On the Egyptian origin of the domestic cat. *Bulletin du Museum National d'Histoire Naturelle. Section C: Sciences de la Terre Paleontologie Geologie Mineralogie* 13: 107–14.

Goldberg, J. (2003). Domestication and behaviour. *Domestication et Comportement* 128 (4): 275–81.

Gunn-Moore, D., C. Bessant, and R. Malik. (2008). Breed-related disorders of cats. *Journal of Small Animal Practice* 49 (4): 167–68.

Ingham, P. W., and M. Placzek. (2006). Orchestrating ontogenesis: Variations on a theme by sonic hedgehog. *Nature Reviews. Genetics* 7 (11): 841–50.

Johnson, W. E., E. Eizirik, J. Pecon-Slattery, W. J. Murphy, A. Antunes, E. Teeling, and S. J. O'Brien. (2006). The late Miocene radiation of modern Felidae: A genetic assessment. *Science* 311 (5757): 73–77.

Kehler, J. S., V. A. David, A. A. Schäffer, K. Bajema, E. Eizirik, D. K. Ryugo, . . . and M. Menotti-Raymond. (2007). Four independent mutations in the feline fibroblast growth factor 5 gene determine the long-haired phenotype in domestic cats. *Journal of Heredity* 98 (6): 555–66.

Lange, A., H. L. Nemeschkal, and G. B. Müller. (2013). Biased polyphenism in polydactylous cats carrying a single point mutation: The Hemingway model for digit novelty. *Evolutionary Biology* 41: 262–75.

Lettice, L. A., A. E. Hill, P. S. Devenney, and R. E. Hill. (2008). Point mutations in a distant sonic hedgehog *cis*-regulator generate a variable regulatory output responsible for preaxial polydactyly. *Human Molecular Genetics* 17 (7): 978–85.

Linderholm, A., and G. Larson. (2013). The role of humans in facilitating and sustaining coat colour variation in domestic animals. *Seminars in Cell & Developmental Biology* 24 (6–7): 587–93.

Linseele, V., W. Van Neer, and S. Hendricks. (2007). Evidence for early cat taming in Egypt. *Journal of Archaeological Science* 34 (12): 2081–90.

Lynch, M. (1991). The genetic interpretation of inbreeding depression and outbreeding depression. *Evolution* 45 (3): 622–29.

Lyons, L. A., D. L. Imes, H. C. Rah, and R. A. Grahn. (2005). Tyrosinase mutations associated with Siamese and Burmese patterns in the domestic cat (*Felis catus*). *Animal Genetics* 36 (2): 119–26.

Macdonald, D. W. (1983). The ecology of carnivore social behaviour. *Nature* 301 (5899): 379–84.

Málek, J. (1997). *The Cat in Ancient Egypt*. Philadelphia: University of Pennsylvania Press.

Mendl, M., and R. Harcourt. (2000). "Individuality in the Domestic Cat: Origins, Development and Stability." In *The Domestic Cat: The Biology of Its Behaviour*, 2nd ed., edited by D. C. Turner and P. Bateson, 47–64. Cambridge: Cambridge University Press.

Menotti-Raymond, M., V. A. David, L. A. Lyons, A. A. Schäffer, J. F. Tomlin, M. K. Hutton, and S. J. O'Brien. (1999). A genetic linkage map of microsatellites in the domestic cat (*Felis catus*). *Genomics* 57 (1): 9–23.

Menotti-Raymond, M., V. A. David, S. Pflueger, M. E. Roelke, J. Kehler, S. J. O'Brien, and

K. Narfström. (2010). Widespread retinal degenerative disease mutation (rdAc) discovered among a large number of popular cat breeds. *Veterinary Journal* 186 (1): 32–38.

Menotti-Raymond, M., V. A. David, A. A. Schäffer, R. Stephens, D. Wells, R. Kumar-Singh, . . . and K. Narfström. (2007). Mutation in *CEP290* discovered for cat model of human retinal degeneration. *Journal of Heredity* 98 (3): 211–20.

Menotti-Raymond, M., V. A. David, A. A. Schäffer, J. F. Tomlin, E. Eizirik, C. Phillip, . . . and S. J. O'Brien. (2009). An autosomal genetic linkage map of the domestic cat, *Felis silvestris catus. Genomics* 93 (4): 305–13.

Morris, D. (1999). *Cat Breeds of the World.* New York: Viking.

Natoli, E., and E. de Vito. (1991). Agonistic behaviour, dominance rank and copulatory success in a large multi-male feral cat, *Felis catus* L., colony in central Rome. *Animal Behaviour* 42 (2): 227–41.

Nicastro, N. (2004). Perceptual and acoustic evidence for species-level differences in meow vocalizations by domestic cats (*Felis catus*) and African wild cats (*Felis silvestris lybica*). *Journal of Comparative Psychology* 118 (3): 287.

O'Brien, S. J., and W. E. Johnson. (2007). The evolution of cats. *Scientific American* 297 (1): 68–75.

O'Brien, S. J., W. Johnson, C. Driscoll, J. Pontius, J. Pecon-Slattery, and M. Menotti-Raymond. (2008). State of cat genomics. *Trends in Genetics* 24 (6): 268–79.

O'Brien, S. J., M. Menotti-Raymond, W. J. Murphy, W. G. Nash, J. Wienberg, R. Stanyon, . . . and J. A. M. Graves. (1999). The promise of comparative genomics in mammals. *Science* 286 (5439): 458–81.

O'Brien, S. J., and W. G. Nash. (1982). Genetic mapping in mammals: Chromosome map of domestic cat. *Science* 216 (4543): 257–65.

Pontius, J. U., J. C. Mullikin, D. R. Smith, K. Lindblad-Toh, S. Gnerre, M. Clamp, . . . and K. McKernan. (2007). Initial sequence and comparative analysis of the cat genome. *Genome Research* 17 (11): 1675–89.

Schmidt-Küntzel, A., E. Eizirik, S. J. O'Brien, and M. Menotti-Raymond. (2005). Tyrosinase and tyrosinase related protein 1 alleles specify domestic cat coat color phenotypes of the albino and brown loci. *Journal of Heredity* 96 (4): 289–301.

Serpell, J. A. (2000). "Domestication and History of the Cat." In *The Domestic Cat: The Biology of Its Behaviour,* 2nd ed., edited by D. C. Turner and P. Bateson, 179. Cambridge: Cambridge University Press.

Stephens, G. (2001). *Legacy of the Cat: The Ultimate Illustrated Guide,* 2nd ed. San Francisco: Chronicle Books.

Taylor, D. (1989). *The Ultimate Cat Book.* New York: Simon and Schuster.

Todd, N. B. (1977). Cats and commerce. *Scientific American* 237 (5): 100–107.

Van Valkenburgh, B., and C. B. Ruff. (1987). Canine tooth strength and killing behaviour in large carnivores. *Journal of Zoology* 212 (3): 379–97.

Vigne, J. D., J. Guilaine, K. Debue, L. Haye, and P. Gérard. (2004). Early taming of the cat in Cyprus. *Science* 304 (5668): 259.

Wastlhuber, J. (1991). "History of Domestic Cat and Cat Breeds." In *Feline Husbandry: Diseases and Management in the Multiple-Cat Environment,* edited by N. C. Pedersen, 1–59. Goleta, CA: American Veterinary Publications.

Zatta, P., and A. Frank. (2007). Copper deficiency and neurological disorders in man and animals. *Brain Research Reviews* 54 (1): 19–33.

Zeuner, F. E. (1958). Dog and cat in the Neolithic of Jericho. *Palestine Exploration Quarterly* 90 (1): 52–55.

Zeuner, F. E. (1963). *A History of Domesticated Animals*. London: Hutchinson.

Zhigachev, A. I., and M. V. Vladimirova. (2002). Analysis of the inheritance of taillessness in the Baikuzino population of cats from Udmurtia. *Russian Journal of Genetics* 38 (9): 1051–53.

Chapter 4: OTHER PREDATORS

NOTES

1. Gehrt and Fritzell 1998.
2. Hauver et al. 2013.
3. Zeveloff 2002.
4. Prange, Gehrt, and Wiggers 2003.
5. MacClintock 1981.
6. Ikeda et al. 2004.
7. Michler and Hohmann 2005.
8. West-Eberhard 1989; Pigliucci 2001; Stearns 1989; Hunt et al. 2011.
9. Other fierce predators in this family include wolverines, American badgers, and ratels.
10. Harris and Yalden 2008, 485–87; Sato et al. 2003; Hernádi et al. 2012.
11. Blandford 1987; Heptner and Sludskii 2002.
12. Harris and Yalden 2008.
13. Lodé 1996, 2008.
14. Sato et al. 2003; Hemmer 1990.
15. Hemmer 1990.
16. A. P. Thomson 1951.
17. S. Brown, "History of the Ferret," Small Mammal Health Series, VeterinaryPartner.com, http://www.veterinarypartner.com/Content.plx?P=A&A=496.
18. A. P. Thomson 1951.
19. Brown, "History of the Ferret"; Plummer 2001.
20. Davison et al. 1999.
21. Brown, "History of the Ferret"
22. G. M. Thomson 2011.
23. Feral ferrets tend to take hold where there are no similar-sized predators.
24. Poole 1972.
25. Ibid.
26. Hernádi et al. 2012.
27. Ibid.
28. Driscoll, Macdonald, and O'Brien 2009.
29. Lindeberg 2008.
30. O'Regan and Kitchener 2005.
31. Ibid.

32. Hansen and Damgaard 2009.
33. Kruska 1996.
34. Bowman et al. 2007.
35. See, for example, McGinnity et al. 2003 and Fraser et al. 2008.
36. We owe the term "Baldwin effect" to George Gaylord Simpson (Simpson 1953), who was critical of it. The original formulation of the idea was Baldwin 1896.
37. For example, Morgan 1896; Osborn 1896. See also Baldwin et al. 1902.
38. Much of this discussion is ultimately derived from Crispo 2007.
39. See, for example, Turney, Whitely, and Anderson 1996; various contributors in Weber and Depew 2003; and Dennett 2003. Baldwin was a psychologist, which may be one reason for this narrow reading.
40. See Crispo 2007 for a particularly good discussion.
41. Waddington 1953.
42. Waddington 1953, 1956, 1961.
43. West-Eberhard's (2005) concept of phenotypic accommodation would subsume both genetic assimilation and the Baldwin effect.

REFERENCES

Baldwin, J. M. (1896). A new factor in evolution. *American Naturalist* 30: 441–51, 536–53.

Baldwin, J. M., H. F. Osborn, C. L. Morgan, E. B. Poulton, F. W. Headley, and H. W. Conn. (1902). *Development and Evolution: Including Psychophysical Evolution, Evolution by Orthoplasy, and the Theory of Genetic Modes.* New York: Macmillan.

Blandford, P. R. S. (1987). Biology of the polecat *Mustela putorius*: A literature review. *Mammal Review* 17 (4): 155–98.

Bowman, J., A. G. Kidd, R. M. Gorman, and A. I. Schulte-Hostedde. (2007). Assessing the potential for impacts by feral mink on wild mink in Canada. *Biological Conservation* 139 (1): 12–18.

Crispo, E. (2007). The Baldwin effect and genetic assimilation: Revisiting two mechanisms of evolutionary change mediated by phenotypic plasticity. *Evolution* 61 (11): 2469–79.

Davison, A., J. D. S. Birks, H. I. Griffiths, A. C. Kitchener, D. Biggins, and R. K. Butlin. (1999). Hybridization and the phylogenetic relationship between polecats and domestic ferrets in Britain. *Biological Conservation* 87 (2): 155–61.

Dennett, D. (2003). "The Baldwin Effect: A Crane, Not a Skyhook." In *Evolution and Learning: The Baldwin Effect Reconsidered*, edited by B. H. Weber and D. J. Depew, 66–79. Cambridge, MA: MIT Press.

Driscoll, C. A., J. Clutton-Brock, A. C. Kitchener, and S. J. O'Brien. (2009). The taming of the cat. *Scientific American* 300 (6): 68–75.

Driscoll, C. A., D. W. Macdonald, and S. J. O'Brien. (2009). From wild animals to domestic pets, an evolutionary view of domestication. *Proceedings of the National Academy of Sciences USA* 106 (suppl. 1): 9971–78.

Fraser, D. J., A. M. Cook, J. D. Eddington, P. Bentzen, and J. A. Hutchings. (2008). Mixed evidence for reduced local adaptation in wild salmon resulting from interbreeding with escaped farmed salmon: Complexities in hybrid fitness. *Evolutionary Applications* 1 (3): 501–12.

Gehrt, S. D., and E. K. Fritzell. (1998). Resource distribution, female home range dispersion and male spatial interactions: Group structure in a solitary carnivore. *Animal Behaviour* 55 (5): 1211–27.

Hansen, S. W., and B. M. Damgaard. (2009). Running in a running wheel substitutes for stereotypies in mink (*Mustela vison*) but does it improve their welfare? *Applied Animal Behaviour Science* 118 (1): 76–83.

Harris, S., and D. W. Yalden. (2008). *Mammals of the British Isles: Handbook.* Southampton, UK: Mammal Society.

Hauver, S., B. T. Hirsch, S. Prange, J. Dubach, and S. D. Gehrt. (2013). Age, but not sex or genetic relatedness, shapes raccoon dominance patterns. *Ethology* 119 (9): 769–78.

Hemmer, H. (1990). *Domestication: The Decline of Environmental Appreciation*, 2nd ed. Translated by Neil Beckhaus. Cambridge: Cambridge University Press.

Heptner, V., and A. Sludskii. (2002). *Mammals of the Soviet Union*, vol. 2, part 1b, *Carnivores (Mustelidae and Procyonidae).* Washington, DC: Smithsonian Institution Libraries and National Science Foundation.

Hernádi, A., A. Kis, B. Turcsán, and J. Topál. (2012). Man's underground best friend: Domestic ferrets, unlike the wild forms, show evidence of dog-like social-cognitive skills. *PLoS One* 7 (8): e43267.

Hunt, B. G., L. Ometto, Y. Wurm, D. Shoemaker, V. Y. Soojin, L. Keller, and M. A. Goodisman. (2011). Relaxed selection is a precursor to the evolution of phenotypic plasticity. *Proceedings of the National Academy of Sciences USA* 108 (38): 15936–41.

Ikeda, T., M. Asano, Y. Matoba, and G. Abe. (2004). Present status of invasive alien raccoon and its impact in Japan. *Global Environmental Research* 8 (2): 125–31.

Kruska, D. (1996). The effect of domestication on brain size and composition in the mink (*Mustela vison*). *Journal of Zoology* 239 (4): 645–61.

Lindeberg, H. (2008). Reproduction of the female ferret (*Mustela putorius furo*). *Reproduction in Domestic Animals* 43: 150–56.

Lodé, T. (1996). Conspecific tolerance and sexual segregation in the use of space and habitats in the European polecat. *Acta Theriologica* 41 (2): 171–76.

Lodé, T. (2008). Kin recognition versus familiarity in a solitary mustelid, the European polecat (*Mustela pistorius*). Comptes Rendus Biologies 331 (3) 248–54.

MacClintock, D. (1981). *Natural History of Raccoons.* New York: Scribner.

McGinnity, P., P. Prodöhl, A. Ferguson, R. Hynes, N. O. Maoiléidigh, N. Baker, . . . and G. Rogan. (2003). Fitness reduction and potential extinction of wild populations of Atlantic salmon, *Salmo salar*, as a result of interactions with escaped farm salmon. *Proceedings of the Royal Society. Series B: Biological Sciences* 270 (1532): 2443–50.

Michler, F., and U. Hohmann. (2005). "Investigations on the Ethological Adaptations of the Raccoon (*Procyon lotor* L., 1758) in the Urban Habitat Using the Example of the City of Kassel, North Hessen (Germany), and the Resulting Conclusions for Conflict Management." Paper presented at the XXVII IUGB-CONGRESS, Hannover, Germany.

Morgan, C. L. (1896). *Habit and Instinct.* London: E. Arnold.

O'Regan, H. J., and A. C. Kitchener. (2005). The effects of captivity on the morphology of captive, domesticated and feral mammals. *Mammal Review* 35 (3–4): 215–30.

Osborn, H. F. (1896). A mode of evolution requiring neither natural selection nor the inheritance of acquired characters. *Transactions of the New York Academy of Sciences* 15: 141–42.

Pigliucci, M. (2001). *Phenotypic Plasticity: Beyond Nature and Nurture.* Baltimore: Johns Hopkins University Press.

Plummer, D. B. (2001). *In Pursuit of Coney.* Machynlleth, Wales: Coch-y-Bonddu Books.

Poole, T. B. (1972). Some behavioural differences between the European polecat, *Mustela putorius,* the ferret, *M. furo,* and their hybrids. *Journal of Zoology* 166 (1): 25–35.

Prange, S., S. D. Gehrt, and E. P. Wiggers. (2003). Demographic factors contributing to high raccoon densities in urban landscapes. *Journal of Wildlife Management* 67 (2): 324–33.

Sato, J. J., T. Hosoda, M. Wolsan, K. Tsuchiya, M. Yamamoto, and H. Suzuki. (2003). Phylogenetic relationships and divergence times among mustelids (Mammalia: Carnivora) based on nucleotide sequences of the nuclear interphotoreceptor retinoid binding protein and mitochondrial cytochrome *b* genes. *Zoological Science* 20 (2): 243–64.

Simpson, G. G. (1953). The Baldwin effect. *Evolution* 7 (2): 110–17.

Stearns, S. C. (1989). The evolutionary significance of phenotypic plasticity. *BioScience* 39 (7): 436–45.

Thomson, A. P. (1951). A history of the ferret. *Journal of the History of Medicine and Allied Sciences* 6 (Autumn): 471–80.

Thomson, G. M. (2011). *The Naturalisation of Animals and Plants in New Zealand.* Cambridge: Cambridge University Press.

Turney, P., D. Whitely, and R. Anderson. (1996). *Evolutionary Computation, Evolution, Learning and Instinct: One Hundred Years of the Baldwin Effect.* Cambridge, MA: MIT Press.

Waddington, C. H. (1953). Genetic assimilation of an acquired character. *Evolution* 7 (2): 118–26.

Waddington, C. H. (1956). Genetic assimilation of the bithorax phenotype. *Evolution* 10 (1): 1–13.

Waddington, C. H. (1961). Genetic assimilation. *Advances in Genetics* 1961 (10): 257–93.

Weber, B. H., and D. J. Depew. (2003). *Evolution and Learning: The Baldwin Effect Reconsidered.* Cambridge, MA: MIT Press.

West-Eberhard, M. J. (1989). Phenotypic plasticity and the origins of diversity. *Annual Review of Ecology and Systematics* 20: 249–78.

West-Eberhard, M. J. (2005). Developmental plasticity and the origin of species differences. *Proceedings of the National Academy of Sciences USA* 102 (suppl. 1): 6543–49.

Zeveloff, S. I. (2002). *Raccoons: A Natural History.* Washington, DC: Smithsonian Institution Press.

Chapter 5: Evolutionary Interlude

NOTES

1. Dawkins 2004.
2. Helfman et al. 2009.
3. Sire and Huysseune 2003; Sire and Akimenko 2004.
4. For example, stonefishes (family Synanceiidae) and anglerfishes (Antennariidae).
5. See, for example, Hall 2003.
6. Shubin, Tabin, and Carroll 2009: Evo-devo (evolutionary developmental) research has

uncovered much more homology in even quite distantly related taxa (e.g., insects and vertebrates) than was previously thought.

7. See Lauder 2012; Hall 2012.

8. Akam 1995; Davidson and Erwin 2006; Heffer and Pick 2013.

9. Alan Wilson was a pioneer in this regard; see, for example, King and Wilson 1975.

10. Some genes code for transcription factors, which, by binding to regulatory elements, influence gene expression.

11. Carroll 2005a, 2005b, 2008.

12. Biémont and Vieira 2006. But some of the junk DNA is truly junky—that is, of no functional significance contra intelligent design.

13. Godfrey-Smith 2009 is particularly good on this point. See also Baum, Smith, and Donovan 2013.

14. Gisolfi and Mora 2000.

REFERENCES

Akam, M. (1995). Hox genes and the evolution of diverse body plans. *Philosophical Transactions of the Royal Society. B: Biological Sciences* 349 (1329): 313–19.

Baum, D. A., S. D. Smith, and S. S. S. Donovan. (2005). The tree-thinking challenge. *Science* 310 (5750): 979–80.

Biémont, C., and C. Vieira. (2006). Genetics: Junk DNA as an evolutionary force. *Nature* 443 (7111): 521–24.

Carroll, S. B. (2005a). *Endless Forms Most Beautiful: The New Science of Evo Devo and the Making of the Animal Kingdom.* New York: Norton.

Carroll, S. B. (2005b). Evolution at two levels: On genes and form. *PLoS Biology* 3 (7): e245.

Carroll, S. B. (2008). Evo-devo and an expanding evolutionary synthesis: A genetic theory of morphological evolution. *Cell* 134 (1): 25–36.

Coyne, J. A. (2005). Switching on evolution. *Nature* 435 (7045): 1029–30.

Craig, L. R. (2009). Defending evo-devo: A response to Hoekstra and Coyne. *Philosophy of Science* 76 (3): 335–44.

Davidson, E. H., and D. H. Erwin. (2006). Gene regulatory networks and the evolution of animal body plans. *Science* 311 (5762): 796–800.

Dawkins, R. (1976). *The Selfish Gene.* Oxford: Oxford University Press.

Dawkins, R. (2004). *The Ancestor's Tale: A Pilgrimage to the Dawn of Evolution.* Boston: Houghton Mifflin.

Dermitzakis, E. T., A. Reymond, and S. E. Antonarakis. (2005). Conserved non-genic sequences—An unexpected feature of mammalian genomes. *Nature Reviews. Genetics* 6 (2): 151–57.

Gisolfi, C. V., and M. T. Mora. (2000). *The Hot Brain: Survival, Temperature, and the Human Body.* Cambridge, MA: MIT Press.

Godfrey-Smith, P. (2009). *Darwinian Populations and Natural Selection.* Oxford: Oxford University Press.

Hall, B. K. (1992). Waddington's legacy in development and evolution. *American Zoologist* 32 (1): 113–22.

Hall, B. K. (2003). Descent with modification: The unity underlying homology and homoplasy as seen through an analysis of development and evolution. *Biological Reviews* 78 (3): 409–33.

Hall, B. K. (2004). In search of evolutionary developmental mechanisms: The 30-year gap between 1944 and 1974. *Journal of Experimental Zoology. Part B: Molecular and Developmental Evolution* 302B (1): 5–18.

Hall, B. K., ed. (2012). *Homology: The Hierarchical Basis of Comparative Biology*. San Diego, CA: Academic Press.

Heffer, A., and L. Pick. (2013). Conservation and variation in Hox genes: How insect models pioneered the evo-devo field. *Annual Review of Entomology* 58: 161–79.

Helfman, G., B. B. Collette, D. E. Facey, and B. W. Bowen. (2009). *The Diversity of Fishes: Biology, Evolution, and Ecology*, 2nd ed. Hoboken, NJ: Wiley-Blackwell.

Hoekstra, H. E., and J. A. Coyne. (2007). The locus of evolution: Evo devo and the genetics of adaptation. *Evolution* 61 (5): 995–1016.

Keller, E. F. (1984). *A Feeling for the Organism: The Life and Work of Barbara McClintock*, 10th anniversary ed. New York: Freeman.

King, M.-C., and A. C. Wilson. (1975). Evolution at two levels in humans and chimpanzees. *Science* 188 (4184): 107–16.

Lauder, G. V. (2012). "Homology, Form, and Function." In *Homology: The Hierarchial Basis of Comparative Biology*, edited by B. K. Hall, 151. San Diego, CA: Academic Press.

Lavoué, S., M. Miya, M. E. Arnegard, P. B. McIntyre, V. Mamonekene, and M. Nishida. (2010). Remarkable morphological stasis in an extant vertebrate despite tens of millions of years of divergence. *Proceedings of the Royal Society. B: Biological Sciences*, published online before print September 29, 2010.

McClintock, B. (1987). *The Collected Papers of Barbara McClintock*. New York: Garland.

Pigliucci, M. (2007). Do we need an extended evolutionary synthesis? *Evolution* 61 (12): 2743–49.

Pigliucci, M., and G. B. Muller. (2010). *Evolution, the Extended Synthesis*. Cambridge, MA: MIT Press.

Shubin, N., C. Tabin, and S. Carroll. (2009). Deep homology and the origins of evolutionary novelty. *Nature* 457 (7231): 818–23.

Sire, J.-Y., and M.-A. Akimenko. (2004). Scale development in fish: A review, with description of sonic hedgehog (*shh*) expression in the zebrafish (*Danio rerio*). *International Journal of Developmental Biology* 48: 233–48.

Sire, J.-Y., and A. N. N. Huysseune. (2003). Formation of dermal skeletal and dental tissues in fish: A comparative and evolutionary approach. *Biological Reviews* 78 (2): 219–49.

Tauber, A. I. (2010). Reframing developmental biology and building evolutionary theory's new synthesis. *Perspectives in Biology and Medicine* 53 (2): 257–70.

Toth, A. L., and G. E. Robinson. (2007). Evo-devo and the evolution of social behavior. *Trends in Genetics* 23 (7): 334–41.

West-Eberhard, M. J. (2003). *Developmental Plasticity and Evolution*. Oxford: Oxford University Press.

West-Eberhard, M. J. (2005). Developmental plasticity and the origin of species differences. *Proceedings of the National Academy of Sciences USA* 102 (suppl. 1): 6543–49.

Williams, G. (1966). *Adaptation and Natural Selection*. Princeton, NJ: Princeton University Press.

Chapter 6: PIGS

NOTES

1. Pre-Viking Swedish warriors, for example, wore helmets with boar crests (Simoons 1994).
2. Kiple and Ornelas, 2000.
3. These distinctive masks are used in initiation rites.
4. See Meggitt 1974 for Mae Enga, a highland tribe. In some highland tribes the symbolic wealth function of pigs virtually precludes their being eaten, despite protein malnutrition (Jelliffe and Maddocks 1964).
5. Harris 1989, 1997, 2001.
6. Lobban 1998.
7. Ibid.
8. Zeuner 1963; Lobban 1998.
9. Zeuner 1963.
10. "And the pig, because it parts the hoof and is cloven-footed but does not chew the cud, is unclean to you" (Leviticus 11:7).
11. *Diacodexis* is one of the earliest artiodactyl fossils known (Rose 1982).
12. Mizelle 2011.
13. Frantz et al. 2013. Mona, Randi, and Tommaseo-Ponzetta 2007, with smaller sample, estimate 5 mya.
14. Larson et al. 2005.
15. Zeder 1982, 2011, 2012.
16. Zeder 1982 (commensal route); Clutton-Brock 1999 (human management route).
17. Kim et al. 1994 (China); Zeder 2008 (Near East); Peters et al. 1999.
18. Groenen et al. 2012; Giuffra et al. 2000; Wu et al. 2007; Larson et al. 2005; Bruford, Bradley, and Luikart 2003; but see Larson and Burger 2013 for a deflationary view.
19. Wu et al. 2007.
20. Ervynk et al. 2001 (molar size changes at Çayönü Tepesi, southeastern Anatolia).
21. Giuffra et al. 2000; Larson et al. 2005; Ottoni et al. 2013.
22. Larson et al. 2005; see Larson and Burger 2013 for why such genetic affinities are weak evidence and probably misleading.
23. Vigne et al. 2009.
24. Oppenheimer 2004; Gray and Jordan 2000; but see Oppenheimer and Richards 2001 for the slow-boat hypothesis.
25. Dobney, Cucchi, and Larson 2008.
26. Larson et al. 2007.
27. Larson et al. 2005.
28. Megens et al. 2008.
29. Larson et al. 2010.
30. Fang et al. 2005.
31. Porter 1993.
32. Fang et al. 2005.
33. Giuffra et al. 2000. Recently the reverse has occurred: European breeds are now imported into China to improve native Chinese breeds (Megens et al. 2008).

34. Jones, Rothschild, and Ruvinsky 1998.
35. Gea-Izquierdo, Cañellas, and Montero 2008.
36. Dohner 2001.
37. Clark and Dzieciolowski 1991.
38. Ekarius 2008.
39. For example, Knipple and Knipple 2011; Wood et al. 2008 (scientific analysis); Gibson 2012 (increased use of lard).
40. Mayer and Brisbin 2008.
41. Kruska and Röhrs 1974; Kruska 1988.
42. Röhrs and Ebinger 1999.
43. Kruska 1970, 1972; Kruska and Stephan 1973; see also Raichlen and Gordon 2011 for the relation between exercise capacity and brain size.
44. Maselli et al. 2014. This is an example of Dollo's law, according to which the loss of complex traits is often irreversible.
45. Pukite 1999.
46. Caro 2005; Cieslak et al. 2011.
47. Cieslak et al. 2011.
48. "Landrace" is a somewhat confusing name for a breed. Most breeds of any domesticated mammal are derived from landraces, but in pigs there are a number of breeds called "landrace"—for example, American Landrace and German Landrace. The first landrace breed was Danish, and it has contributed in varying degrees to the development of all others.
49. Six traditional pig types in China all have a mutation promoting black coloration (Fang et al. 2009).
50. Giuffra et al. 2000.
51. Mayer and Brisbin 2008; De Marinis and Asprea 2006. This conservation of domestic coloration is more pronounced where there is no introgression from wild pigs.
52. Parés-Casanova 2013. Another marker of sexual dimorphism is the size of the third molar (Albarella et al. 2006).
53. The pattern of sexual dimorphism in domestic pig breeds also violates Rensch's rule, according to which the degree of sexual dimorphism should increase with increasing body size (Parés-Casanova 2013).
54. Pigs may have first arrived in the United States during de Soto's 1539–40 expedition (Mayer and Brisbin 2008).
55. It is noteworthy, in this regard, that another sexually dimorphic trait in pigs—the third molar—is also late developing.
56. Fang et al. 2009.
57. Li et al. 2010; Yuan and Flad 2005.
58. Rothschild et al. 1994; Lamberson et al. 1991; Drake, Fraser, and Weary 2008.
59. Jeon et al. 1999; Fan et al. 2011; Rehfeldt et al. 2000; Rehfeldt, Henning, and Fiedler 2008; Chen et al. 2012.
60. Rutherford et al. 2013; Fernandez-Rodriguez et al. 2011; Baxter et al. 2013.
61. Giuffra et al. 2000.
62. Li et al. 2010; see also Fang et al. 2009.
63. Johansson et al. 2005.

64. Lai et al. 2007.
65. Amaral et al. 2011; Cherel et al. 2011.
66. Bidanel et al. 2001.
67. Groenen et al. 2012; Amaral et al. 2011.
68. Liu et al. 2007.
69. Nikitin et al. 2010.
70. Megens et al. 2008.
71. Ibid.
72. Gustafsson et al. 1999 (maternal behavior); Fraser and Thompson 1991 (piglets born with little tusklike teeth).
73. Held, Cooper, and Mendl 2008.

REFERENCES

Albarella, U., A. Tagliacozzo, K. Dobney, and P. Rowley-Conwy. (2006). Pig hunting and husbandry in prehistoric Italy: A contribution to the domestication debate. *Proceedings of the Prehistoric Society* 72: 193–227.

Amaral, A. J., L. Ferretti, H.-J. Megens, R. P. M. A. Crooijmans, H. Nie, S. E. Ramos-Onsins, . . . and M. A. M. Groenen. (2011). Genome-wide footprints of pig domestication and selection revealed through massive parallel sequencing of pooled DNA. *PLoS One* 6 (4): e14782.

Baxter, E. M., K. M. D. Rutherford, R. B. D'Eath, G. Arnott, S. P. Turner, P. Sandøe, . . . and A. B. Lawrence. (2013). The welfare implications of large litter size in the domestic pig II: Management factors. *Animal Welfare* 22: 219–38.

Bidanel, J.-P., D. Milan, N. Iannuccelli, Y. Amigues, M.-Y. Boscher, F. Bourgeois, . . . and H. Lagant. (2001). Detection of quantitative trait loci for growth and fatness in pigs. *Genetics, Selection, Evolution: GSE* 33 (3): 289–310.

Bruford, M. W., D. G. Bradley, and G. Luikart. (2003). DNA markers reveal the complexity of livestock domestication. *Nature Reviews. Genetics* 4 (11): 900–910.

Caro, T. (2005). The adaptive significance of coloration in mammals. *BioScience* 55 (2): 125–36.

Chen, K., R. Hawken, G. H. Flickinger, S. L. Rodriguez-Zas, L. A. Rund, M. B. Wheeler, . . . and L. B. Schook. (2012). Association of the porcine transforming growth factor beta type I receptor (TGFBR1) gene with growth and carcass traits. *Animal Biotechnology* 23 (1): 43–63.

Cherel, P., J. Pires, J. Glenisson, D. Milan, N. Iannuccelli, F. Herault, . . . and P. Le Roy. (2011). Joint analysis of quantitative trait loci and major-effect causative mutations affecting meat quality and carcass composition traits in pigs. *BMC Genetics* 12 (1): 76.

Cieslak, M., M. Reissmann, M. Hofreiter, and A. Ludwig. (2011). Colours of domestication. *Biological Reviews* 86 (4): 885–99.

Clark, C. M. H., and R. M. Dzieciolowski. (1991). Feral pigs in the northern South Island, New Zealand. *Journal of the Royal Society of New Zealand* 21 (3): 237–47.

Clutton-Brock, J. (1999). *A Natural History of Domesticated Mammals*, 2nd ed. Cambridge: Cambridge University Press.

De Marinis, A. M., and A. Asprea. (2006). Hair identification key of wild and domestic ungulates from southern Europe. *Wildlife Biology* 12 (3): 305–20.

Dobney, K., T. Cucchi, and G. Larson. (2008). The pigs of Island Southeast Asia and the

Pacific: New evidence for taxonomic status and human-mediated dispersal. *Asian Perspectives* 47 (1): 59–74.

Dohner, J. V. (2001). *The Encyclopedia of Historic and Endangered Livestock and Poultry Breeds*. New Haven, CT: Yale University Press.

Drake, A., D. Fraser, and D. M. Weary. (2008). Parent-offspring resource allocation in domestic pigs. *Behavioral Ecology and Sociobiology* 62 (3): 309–19.

Ekarius, C. (2008). *Storey's Illustrated Breed Guide to Sheep, Goats, Cattle and Pigs*. North Adams, MA: Storey.

Ervynck, A., K. Dobney, H. Hongo, and R. Meadow. (2001). Born free? New evidence for the status of *Sus scrofa* at Neolithic Çayönü Tepesi (Southeastern Anatolia, Turkey). *Paléorient* 27 (2): 47–73.

Fan, B., S. K. Onteru, Z.-Q. Du, D. J. Garrick, K. J. Stalder, and M. F. Rothschild. (2011). Genome-wide association study identifies loci for body composition and structural soundness traits in pigs. *PLoS One* 6 (2): e14726.

Fang, M., X. Hu, T. Jiang, M. Braunschweig, L. Hu, Z. Du, … and N. Li. (2005). The phylogeny of Chinese indigenous pig breeds inferred from microsatellite markers. *Animal Genetics* 36 (1): 7–13.

Fang, M., G. Larson, H. S. Ribeiro, N. Li, and L. Andersson. (2009). Contrasting mode of evolution at a coat color locus in wild and domestic pigs. *PLoS Genetics* 5 (1): e 1000341.

Fernandez-Rodriguez, A., M. Munoz, A. Fernandez, R. N. Pena, A. Tomas, J. L. Noguera, … and A. I. Fernandez. (2011). Differential gene expression in ovaries of pregnant pigs with high and low prolificacy levels and identification of candidate genes for litter size. *Biology of Reproduction* 84 (2): 299–307.

Frantz, L. A., J. G. Schraiber, O. Madsen, H.-J. Megens, M. Bosse, Y. Paudel, … and M. A. M. Groenen. (2013). Genome sequencing reveals fine scale diversification and reticulation history during speciation in *Sus. Genome Biology* 14 (9): R107.

Fraser, D., and B. K. Thompson. (1991). Armed sibling rivalry among suckling piglets. *Behavioral Ecology and Sociobiology* 29 (1): 9–15.

Gea-Izquierdo, G., I. Cañellas, and G. Montero. (2008). Acorn production in Spanish holm oak woodlands. *Forest Systems* 15 (3): 339–54.

Gibson, T. (2012). *Lard: The Lost Art of Cooking with Your Grandmother's Secret Ingredient*. Kansas City, MO: Andrews McMeel.

Giuffra, E., J. Kijas, V. Amarger, O. Carlborg, J. Jeon, and L. Andersson. (2000). The origin of the domestic pig: Independent domestication and subsequent introgression. *Genetics* 154: 1785–91.

Gray, R., and F. Jordan. (2000). Language trees support the express-train sequence of Austronesian expansian. *Nature* 405: 1052–54.

Groenen, M. A., A. L. Archibald, H. Uenishi, C. K. Tuggle, Y. Takeuchi, M. F. Rothschild, … and H.-J. Megens. (2012). Analyses of pig genomes provide insight into porcine demography and evolution. *Nature* 491 (7424): 393–98.

Gustafsson, M., P. Jensen, F. H. de Jonge, and T. Schuurman. (1999). Domestication effects on foraging strategies in pigs (*Sus scrofa*). *Applied Animal Behaviour Science* 62 (4): 305–17.

Harris, M. (1989). *Cows, Pigs, Wars, & Witches: The Riddles of Culture*. New York: Random House.

Harris, M. (1997). "The Abominable Pig." In *Food and Culture: A Reader*, edited by C. Counihan and P. Van Esterik, 67–79. New York: Routledge.

Harris, M. (2001). *Cultural Materialism: The Struggle for a Science of Culture*, updated ed. Walnut Creek, CA: AltaMira.

Held, S., J. J. Cooper, and M. T. Mendl. (2008). "Advances in the Study of Cognition, Behavioural Priorities and Emotions." In *The Welfare of Pigs*, edited by J. N. Marchant-Forde, 47–94. New York: Springer.

Jelliffe, D. B., and I. Maddocks. (1964). Notes on ecologic malnutrition in the New Guinea Highlands. *Clinical Pediatrics* 3 (7): 432–38.

Jeon, J.-T., Ö. Carlborg, A. Törnsten, E. Giuffra, V. Amarger, P. Chardon, . . . and K. Lundström. (1999). A paternally expressed QTL affecting skeletal and cardiac muscle mass in pigs maps to the IGF2 locus. *Nature Genetics* 21 (2): 157–58.

Johansson, A., G. Pielberg, L. Andersson, and I. Edfors-Lilja. (2005). Polymorphism at the porcine dominant white/KIT locus influence coat colour and peripheral blood cell measures. *Animal Genetics* 36 (4): 288–96.

Jones, G., M. Rothschild, and A. Ruvinsky. (1998). "Genetic Aspects of Domestication, Common Breeds and Their Origin." In *The Genetics of the Pig*, edited by M. F. Rothschild and A. Ruvinsky, 17–50. Wallingford, UK: CAB International.

Kim, S.-O., C. M. Antonaccio, Y. K. Lee, S. M. Nelson, C. Pardoe, J. Quilter, and A. Rosman. (1994). Burials, pigs, and political prestige in Neolithic China [and comments and reply]. *Current Anthropology* 35 (2): 119–41.

Kiple, K. F., and K. C. Ornelas, eds. (2000). *Cambridge World History of Food*. Cambridge: Cambridge University Press.

Knipple, A., and P. Knipple. (2011). "Barbecue as Slow Food." In *The Slaw and the Slow Cooked: Culture and Barbecue in the Mid-south*, edited by J. R. Veteto and E. M. Maclin, 151. Nashville, TN: Vanderbilt University Press.

Kruska, D. (1970). Comparative cytoarchitectonic investigations in brains of wild and domestic pigs. *Vergleichend cytoarchitektonische Untersuchungen an Gehirnen von Wild- und Hausschweinen* 131 (4): 291–324.

Kruska, D. (1972). A volumetric comparison of some visual centers in the brains of wild boars and domestic pigs. *Volumenvergleich optischer Hirnzentren bei Wild- und Hausschweinen* 138 (3): 265–82.

Kruska, D. (1988). "Mammalian Domestication and Its Effect on Brain Structure and Behavior." In *Intelligence and Evolutionary Biology*, edited by H. J. Jerison and I. Jerison, 211–50. Berlin: Springer.

Kruska, D., and M. Röhrs. (1974). Comparative-quantitative investigations on brains of feral pigs from the Galapagos Islands and of European domestic pigs. *Anatomy and Embryology* 144 (1): 61–73.

Kruska, D., and H. Stephan. (1973). Volumetric comparisons in allocortical brain centers of wild and domestic pigs. *Volumenvergleich allokortikaler Hirnzentren bei Wild- und Hausschweinen* 84 (3): 387–415.

Lai, F., J. Ren, H. Ai, N. Ding, J. Ma, D. Zeng, . . . and L. Huang. (2007). Chinese white Rongchang pig does not have the dominant white allele of KIT but has the dominant black allele of MC1R. *Journal of Heredity* 98 (1): 84–87.

Lamberson, W., R. Johnson, D. R. Zimmerman, and T. Long. (1991). Direct responses to selection for increased litter size, decreased age at puberty, or random selection following selection for ovulation rate in swine. *Journal of Animal Science* 69 (8): 3129–43.

Larson, G., and J. Burger. (2013). A population genetics view of animal domestication. *Trends in Genetics* 29 (4): 197–205.

Larson, G., T. Cucchi, M. Fujita, E. Matisoo-Smith, J. Robins, A. Anderson, . . . and K. Dobney. (2007). Phylogeny and ancient DNA of *Sus* provides insights into Neolithic expansion in Island Southeast Asia and Oceania. *Proceedings of the National Academy of Sciences USA* 104 (12): 4834–39.

Larson, G., K. Dobney, U. Albarella, M. Fang, E. Matisoo-Smith, J. Robins, . . . and E. Willerslev. (2005). Worldwide phylogeography of wild boar reveals multiple centers of pig domestication. *Science* 307 (5715): 1618–21.

Larson, G., R. Liu, X. Zhao, J. Yuan, and N. Li. (2010). Patterns of East Asian pig domestication, migration, and turnover revealed by modern and ancient DNA. *Proceedings of the National Academy of Sciences USA* 107 (17): 7686–91.

Li, J., H. Yang, J. R. Li, H. P. Li, T. Ning, X. R. Pan, . . . and Y. P. Zhang. (2010). Artificial selection of the melanocortin receptor 1 gene in Chinese domestic pigs during domestication. *Heredity* 105 (3): 274–81.

Liu, G., D. G. Jennen, E. Tholen, H. Juengst, T. Kleinwächter, M. Hölker, . . . and K. Wimmers. (2007). A genome scan reveals QTL for growth, fatness, leanness and meat quality in a Duroc-Pietrain resource population. *Animal Genetics* 38 (3): 241–52.

Lobban, R. A. (1998). Pigs in ancient Egypt. *MASCA Research Papers in Science and Archaeology* 15: 137–47.

Maselli, V., G. Polese, G. Larson, P. Raia, N. Forte, D. Rippa, . . . and D. Fulgione. (2014). A dysfunctional sense of smell: The irreversibility of olfactory evolution in free-living pigs. *Evolutionary Biology* 41: 229–39.

Mayer, J. J., and I. L. Brisbin. (2008). *Wild Pigs in the United States: Their History, Comparative Morphology, and Current Status.* Athens, GA: University of Georgia Press.

Megens, H.-J., R. Crooijmans, M. San Cristobal, X. Hui, N. Li, and M. Groenen. (2008). Biodiversity of pig breeds from China and Europe estimated from pooled DNA samples: Differences in microsatellite variation between two areas of domestication. *Genetics, Selection, Evolution: GSE* 40 (1): 103–28.

Meggitt, M. J. (1974). "Pigs are our hearts!" The Te exchange cycle among the Mae Enga of New Guinea. *Oceania* 44 (3): 165–203.

Mizelle, B. (2011). *Pig.* London: Reaktion Books.

Mona, S., E. Randi, and M. Tommaseo-Ponzetta. (2007). Evolutionary history of the genus *Sus* inferred from cytochrome b sequences. *Molecular Phylogenetics and Evolution* 45 (2): 757–62.

Nikitin, S. V., N. S. Yudin, S. P. Knyazev, R. B. Aitnazarov, V. A. Bekenev, V. S. Deeva, . . . and V. I. Ermolaev. (2010). Differentiation of wild boar and domestic pig populations based on the frequency of chromosomes carrying endogenous retroviruses. *Natural Science* 2 (6): 527–34.

Oppenheimer, S. (2004). The "Express Train from Taiwan to Polynesia": On the congruence of proxy lines of evidence. *World Archaeology* 36 (4): 591–600.

Oppenheimer, S., and M. Richards. (2001). Fast trains, slow boats, and the ancestry of the Polynesian Islanders. *Science Progress* 84 (3): 157–81.

Ottoni, C., L. Girdland Flink, A. Evin, C. Geörg, B. De Cupere, W. Van Neer, . . . and G. Larson. (2013). Pig domestication and human-mediated dispersal in western Eurasia revealed through ancient DNA and geometric morphometrics. *Molecular Biology and Evolution* 30 (4): 824–32.

Parés-Casanova, P. M. (2013). Sexual size dimorphism in swine denies Rensch's rule. *Asian Journal of Agriculture and Food Sciences* 1 (4): 112–18.

Peters, J., A. von den Driesch, D. Helmer, and M. Saña Segui. (1999). Early animal husbandry in the Northern Levant. *Paléorient* 25 (2): 27–48.

Porter, V. (1993). *Pigs: A Handbook to the Breeds of the World*. Ithaca, NY: Comstock.

Pukite, J. (1999). *A Field Guide to Pigs*. Helena, MT: Falcon.

Raichlen, D. A., and A. D. Gordon. (2011). Relationship between exercise capacity and brain size in mammals. *PLoS One* 6 (6): e20601.

Rehfeldt, C., I. Fiedler, G. Dietl, and K. Ender. (2000). Myogenesis and postnatal skeletal muscle cell growth as influenced by selection. *Livestock Production Science* 66 (2): 177–88.

Rehfeldt, C., M. Henning, and I. Fiedler. (2008). Consequences of pig domestication for skeletal muscle growth and cellularity. *Livestock Science* 116 (1–3): 30–41.

Röhrs, M., and P. Ebinger. (1999). Verwildert ist nicht gleich wild: Die Hirngewichte verwilderter Haussäugetiere [Feral animals are not really wild: The brain weights of wild domestic mammals]. *Berliner und Münchener tierärztliche Wochenschrift* 112 (6–7): 234–38.

Rose, K. D. (1982). Skeleton of *Diacodexis*, oldest known artiodactyl. *Science* 216 (4546): 621–23.

Rothschild, M., C. Jacobson, D. Vaske, C. Tuggle, T. Short, S. Sasaki, . . . and D. McLaren. (1994). "A Major Gene for Litter Size in Pigs." Paper presented at the Proceedings of the 5th World Congress on Genetics Applied to Livestock Production.

Rutherford, K. M. D., E. M. Baxter, R. B. D'Eath, S. P. Turner, G. Arnott, R. Roehe, . . . and A. B. Lawrence. (2013). The welfare implications of large litter size in the domestic pig I: biological factors. *Animal Welfare* 22 (2): 199–218.

Simoons, F. J. (1994). *Eat Not This Flesh: Food Avoidances from Prehistory to the Present*, 2nd ed. Madison: University of Wisconsin Press.

Vigne, J.-D., A. Zazzo, J.-F. Saliège, F. Poplin, J. Guilaine, and A. Simmons. (2009). Pre-Neolithic wild boar management and introduction to Cyprus more than 11,400 years ago. *Proceedings of the National Academy of Sciences* 106 (38): 16135–38.

Wood, J. D., M. Enser, A. V. Fisher, G. R. Nute, P. R. Sheard, R. I. Richardson, . . . and F. M. Whittington. (2008). Fat deposition, fatty acid composition and meat quality: A review. *Meat Science* 78 (4): 343–58.

Wu, G.-S., Y.-G. Yao, K.-X. Qu, Z.-L. Ding, H. Li, M. G. Palanichamy, . . . and Y.-P. Zhang. (2007). Population phylogenomic analysis of mitochondrial DNA in wild boars and domestic pigs revealed multiple domestication events in East Asia. *Genome Biology* 8 (11): R245.

Yuan, J., and R. Flad. (2005). New zooarchaeological evidence for changes in Shang Dynasty animal sacrifice. *Journal of Anthropological Archaeology* 24 (3): 252–70.

Zeder, M. A. (1982). The domestication of animals. *Reviews in Anthropology* 9 (4): 321–27.

Zeder, M. A. (2008). Domestication and early agriculture in the Mediterranean Basin: Origins, diffusion, and impact. *Proceedings of the National Academy of Sciences USA* 105 (33): 11597–604.

Zeder, M. A. (2011). The origins of agriculture in the Near East. *Current Anthropology* 52 (S4): S221–35.

Zeder, M. A. (2012). "Pathways to Animal Domestication." In *Biodiversity in Agriculture: Domestication, Evolution, and Sustainability*, edited by P. Gepts, T. R. Famula, R. L. Bettinger, S. B. Brush, A. B. Damania, P. E. McGuire, and C. O. Qualset, 227. Cambridge: Cambridge University Press.

Zeuner, F. (1963). *A History of Domesticated Animals*. London: Hutchinson.

Chapter 7: CATTLE

NOTES

1. Burke 1998, originally published as "On the Sublime and Beautiful" (1798).
2. Guthrie 2005 (chap. 5) discusses various views on motivation for cave art involving large mammals generally, from aesthetic motivations to magical thinking to shamanistic religious rituals.
3. About 10,500 BP: Helmer et al. 2005; Hongo et al. 2009; Twiss and Russell 2009 (bull heads).
4. Numbers 24:8 (Revised Standard Version; "the horns of a wild ox"), cited in Richard L. Atkins, "The Worship of Yahweh as a Bull," *Atkins Light Quest* (blog), http://www.atkinslightquest.com/Documents/Religion/Hebrew-Myths/Worship-of-Yahweh-as-a-Bull.htm.
5. Evans 1921.
6. Vivekanandan and Alagumalai 2013.
7. This form of bullfighting is called *course Landaise* in the south of France.
8. See, for example, Wolin 1979.
9. Strömberg 2011.
10. Janis 2007.
11. Bibi 2013: 19.3–16.6 mya; Hassanin et al. 2012: 27.6–22.4 mya; Meredith et al. 2011: 19.1–16.4 mya.
12. Solounias et al. 1995.
13. Bibi and Vrba 2010.
14. For example, chamois, American mountain goats, serows, and gorals.
15. Hassanin et al. 2013.
16. Cis van Vuure 2005.
17. "*Bos primigenius*," IUCN Red List of Threatened Species, http://www.iucnredlist.org/details/136721/0. The wild ancestor of the zebu is sometimes referred to as *Bos primigenius nomadicus*.
18. Julius Caesar, *The Gallic Wars* 6.28; italics added.
19. Loftus et al. 1994; Helmer et al. 2005.
20. Bradley et al. 1996, 1998; Hanotte et al. 2002; Pérez-Pardal et al. 2010.
21. Achilli et al. 2008, but with some admixture of native African aurochs (Decker et al. 2014).
22. Zeder 2011; Bollongino et al. 2012.
23. Pinhasi, Fort, and Ammerman 2005; Edwards et al. 2007.

24. Teasdale and Bradley 2012.
25. Bramanti et al. 2009.
26. Cymbron et al. 2005; Beja-Pereira et al. 2006; Negrini et al. 2007. There is some dispute as to how much interbreeding occurred with local aurochs during the European expansion (Edwards et al. 2007; Bonfiglio et al. 2010; Pérez-Pardal et al. 2010; Götherström et al. 2005; Bollongino et al. 2008). From what I can distill, there was relatively little genetic introgression from local European aurochs, and it came primarily by way of bulls and occurred mostly in southern Europe, where domestic cattle ranged more freely (see, for example, Teasdale and Bradley 2012).
27. Murray et al. 2010.
28. Fuller 2006.
29. Chen et al. 2010.
30. Baig et al. 2005.
31. Around 3500 BP, where they met taurine breeds that had arrived about a thousand years earlier (Zhang et al. 2013).
32. Hanotte et al. 2002.
33. Ikram 1995.
34. Freeman et al. 2004; Hanotte et al. 2002.
35. Epstein and Mason 1984.
36. Ikram 1995; Kendall 1998.
37. Epstein 1971.
38. Sanga × zebu breeds are called Zenga; they occur mostly in East Africa (J. E. O. Rege 1999).
39. E. Rege 2003.
40. Vigne, Peters, and Helmer 2005.
41. Sherratt 1981, 1983.
42. For example, see Vigne and Helmer 2007; for a general review of the hypothesis, see Greenfield 2010.
43. Near East aurochs may have been much smaller (more "gracile") than European aurochs (Edwards et al. 2007) and hence easier to control for milking purposes. Nonetheless, it would have taken multiple persons to tether the animal so that it could do no harm.
44. Evershed et al. 2008.
45. For evidence of dairying in the Carpathian Basin, see Craig et al. 2005.
46. Leonardi et al. 2012.
47. For example, Edwards et al. 2007; Achilli et al. 2008; Bollongino et al. 2008.
48. Helmer et al. 2005; Polák and Frynta 2010 found intersexual convergence for shoulder height but not body mass.
49. Ingram et al. 2009; Itan et al. 2009.
50. Gerbault et al. 2011.
51. Check 2006; Holden and Mace 2002; Reilly 2012.
52. Harvey et al. 1998.
53. Romero et al. 2012.
54. Itan et al. (2009) suggest that dairying originated in the northern Balkans; Gerbault et al. (2011) suggest it started in the Near East.
55. Burger et al. 2007.
56. Gerbault et al. 2011.

57. Tandon et al. 1981; Romero et al. 2012.
58. Ingram et al. 2007.
59. Bayoumi et al. 1982.
60. Hijazi et al. 1983.
61. Elsik, Tellam, and Worley 2009; the name of the cow was "L1 Dominette 01449" (Burt 2009).
62. Stothard et al. 2011: Holstein black angus; Lee et al. 2013: Hanwoo (Korean); Canavez 2012: Nellore (zebu); Tsuda et al. 2013: the endangered Japanese breed Mishima-Ushi.
63. These studies were primarily genome-wide association studies (GWASs). Bolormaa et al. (2011) found mutations related to growth; Raven, Cocks, and Hayes (2014) identified mutations related to milk production; and Hayes et al. (2010) identified mutations related to coat color and milk fat.
64. See Gautier and Naves 2011 for footprints of selection in creole cattle.
65. Lewin 2013; Qanbari et al. 2010.
66. Qanbari et al. 2010.
67. Gautier and Naves 2011; Porto-Neto et al. 2013.
68. See, for example, Fadista et al. 2010; Liu et al. 2010; Hou et al. 2011; Zhan et al. 2011.
69. Stothard et al. 2011.
70. Stapely et al. 2010; Liu and Bickhart 2012: There are a number of reasons for this, including CNV-induced changes in gene structure, dosage effects, and the exposure of recessive alleles to selection. CNVs can also affect larger parts of the genome than point mutations.
71. Felius 1995.
72. Gilliam et al. 1936 for beta-carotene levels in Friesian (Holstein) versus Guernsey cattle.
73. Lewin 2013. The father was named Pawnee Farm Arlinda Chief ("Chief"); and his son, Walkway Chief Mark ("Mark").

REFERENCES

Achilli, A., S. Bonfiglio, A. Olivieri, A. Malusa, M. Pala, B. Kashani, . . . and O. Semino. (2009). The multifaceted origin of taurine cattle reflected by the mitochondrial genome. *PLoS One* 4 (6): e5753.

Achilli, A., A. Olivieri, M. Pellecchia, C. Uboldi, L. Colli, N. Al-Zahery, . . . and U. Perego. (2008). Mitochondrial genomes of extinct aurochs survive in domestic cattle. *Current Biology: CB* 18: R157–58.

Ajmone-Marsan, P., J. F. Garcia, and J. A. Lenstra. (2010). On the origin of cattle: How aurochs became cattle and colonized the world. *Evolutionary Anthropology* 19 (4): 148–57.

Baig, M., A. Beja-Pereira, R. Mohammad, K. Kulkarni, S. Farah, and G. Luikart. (2005). Phylogeography and origin of Indian domestic cattle. *Current Science* 89 (1): 38–40.

Bayoumi, R. A. L., S. D. Flatz, W. Kühnau, and G. Flatz. (1982). Beja and Nilotes: Nomadic pastoralist groups in the Sudan with opposite distributions of the adult lactase phenotypes. *American Journal of Physical Anthropology* 58 (2): 173–78.

Beja-Pereira, A., D. Caramelli, C. Lalueza-Fox, C. Vernesi, N. Ferrand, A. Casoli, . . . and M. Lari. (2006). The origin of European cattle: Evidence from modern and ancient DNA. *Proceedings of the National Academy of Sciences USA* 103: 8113–18.

Bibi, F. (2013). A multi-calibrated mitochondrial phylogeny of extant Bovidae (Artiodactyla, Ruminantia) and the importance of the fossil record to systematics. *BMC Evolutionary Biology* 13 (1): 1–15.

Bibi, F., and E. Vrba. (2010). Unraveling bovid phylogeny: Accomplishments and challenges. *BMC Biology* 8 (1): 50.

Blott, S. C., J. L. Williams, and C. S. Haley. (1998). Genetic relationships among European cattle breeds. *Animal Genetics* 29 (4): 273–82.

Bollongino, R., J. Burger, A. Powell, M. Mashkour, J.-D. Vigne, and M. G. Thomas. (2012). Modern taurine cattle descended from small number of Near-Eastern founders. *Molecular Biology and Evolution* 29 (9): 2101–4.

Bollongino, R., J. Elsner, J. Vigne, and J. Burger. (2008). Y-SNPs do not indicate hybridisation between European aurochs and domestic cattle. *PLoS One* 3 (10): e3418.

Bolormaa, S., B. Hayes, K. Savin, R. Hawken, W. Barendse, P. Arthur, . . . and M. Goddard. (2011). Genome-wide association studies for feedlot and growth traits in cattle. *Journal of Animal Science* 89 (6): 1684–97.

Bonfiglio, S., A. Achilli, A. Olivieri, R. Negrini, L. Colli, L. Liotta, . . . and L. Ferretti. (2010). The enigmatic origin of bovine mtDNA haplogroup R: Sporadic interbreeding or an independent event of *Bos primigenius* domestication in Italy? *PLoS One* 5 (12): e15760.

Bradley, D. G., R. T. Loftus, P. Cunningham, and D. E. MacHugh. (1998). Genetics and domestic cattle origins. *Evolutionary Anthropology* 6 (3): 79–86.

Bradley, D. G., D. E. MacHugh, P. Cunningham, and R. T. Loftus. (1996). Mitochondrial diversity and the origin of the African and European cattle. *Proceedings of the National Academy of Sciences USA* 93: 5131–35.

Bramanti, B., M. G. Thomas, W. Haak, M. Unterlaender, P. Jores, K. Tambets, . . . and J. Burger. (2009). Genetic discontinuity between local hunter-gatherers and central Europe's first farmers. *Science* 326 (5949): 137–40.

Burger, J., M. Kirchner, B. Bramanti, W. Haak, and M. G. Thomas. (2007). Absence of the lactase-persistence-associated allele in early Neolithic Europeans. *Proceedings of the National Academy of Sciences USA* 104 (10): 3736–41.

Burke, E. (1998). *A Philosophical Enquiry into the Origin of Our Ideas of the Sublime and Beautiful: And Other Pre-evolutionary Writings*, edited by D. Womersley. London: Penguin.

Burt, D. (2009). The cattle genome reveals its secrets. *Journal of Biology* 8 (4): 36.

Canavez, F. C., D. D. Luche, P. Stothard, K. R. Leite, J. M. Sousa-Canavez, G. Plastow, . . . and S. S. Moore. (2012). Genome sequence and assembly of *Bos indicus. Journal of Heredity* 103 (3): 342–48.

Check, E. (2006). Human evolution: How Africa learned to love the cow. *Nature* 444 (7122): 994–96.

Chen, S., B.-Z. Lin, M. Baig, B. Mitra, R. J. Lopes, A. M. Santos, . . . and A. Beja-Pereira. (2010). Zebu cattle are an exclusive legacy of the South Asia Neolithic. *Molecular Biology and Evolution* 27 (1): 1–6.

Craig, O. E., J. Chapman, C. Heron, L. H. Willis, L. Bartosiewicz, G. Taylor, . . . and M. Collins. (2005). Did the first farmers of central and eastern Europe produce dairy foods? *Antiquity* 79 (306): 882–94.

Cymbron, T., A. R. Freeman, M. I. Malheiro, J. D. Vigne, and D. G. Bradley. (2005). Microsatellite diversity suggests different histories for Mediterranean and Northern European cattle populations. *Proceedings of the Royal Society. B: Biological Sciences* 272 (1574): 1837–43.

Cymbron, T., R. T. Loftus, M. I. Malheiro, and D. G. Bradley. (1999). Mitochondrial sequence variation suggests an African influence in Portuguese cattle. *Proceedings of the Royal Society. B: Biological Sciences* 266 (1419): 597–603.

D'Andrea, M., L. Pariset, D. Matassino, A. Valentini, J. A. Lenstra, G. Maiorano, and F. Pilla. (2011). Genetic characterization and structure of the Italian Podolian cattle breed and its relationship with some major European breeds. *Italian Journal of Animal Science* 10 (4): 237–43.

Decker, J. E., S. D. McKay, M. M. Rolf, J. Kim, A. M. Alcalá, T. S. Sonstegard, . . . and L. Praharani. (2014). Worldwide patterns of ancestry, divergence, and admixture in domesticated cattle. *PLoS Genetics*, March 27.

Del Bo, L., M. Polli, M. Longeri, G. Ceriotti, C. Looft, A. Barre-Dirie, . . . and M. Zanotti. (2001). Genetic diversity among some cattle breeds in the Alpine area. *Journal of Animal Breeding and Genetics* 118 (5): 317–25.

Edwards, C., R. Bollongino, A. Scheu, A. Chamberlain, A. Tresset, J. Vigne, . . . and T. Heupink. (2007). Mitochondrial DNA analysis shows a Near Eastern Neolithic origin for domestic cattle and no indication of domestication of European aurochs. *Proceedings of the Royal Society. B: Biological Sciences* 274: 1377–85.

Elsik, C. G., R. L. Tellam, and K. C. Worley. (2009). The genome sequence of taurine cattle: A window to ruminant biology and evolution. *Science* 324 (5926): 522–28.

Epstein, H. (1971). *The Origin of the Domestic Animals of Africa*, 1:185–455. New York: African Publishing Corporation.

Epstein, H., and I. L. Mason. (1984). "Cattle." In *Evolution of Domesticated Animals*, edited by I. L. Mason, 25. London: Longman.

Evans, A. (1921). On a Minoan bronze group of a galloping bull and acrobatic figure from Crete, with glyptic comparisons and a note on the Oxford relief showing the taurokathapsia. *Journal of Hellenic Studies* 41 (2): 247–59.

Evershed, R. P., S. Payne, A. G. Sherratt, M. S. Copley, J. Coolidge, D. Urem-Kotsu . . . and M. M. Burton. (2008). Earliest date for milk use in the Near East and southeastern Europe linked to cattle herding. *Nature* 455 (7212): 528–31.

Fadista, J., B. Thomsen, L. Holm, and C. Bendixen. (2010). Copy number variation in the bovine genome. *BMC Genomics* 11 (1): 284.

Felius, M. (1995). *Cattle Breeds: An Encyclopedia*. London: Trafalgar Square Books.

Felius, M., P. A. Koolmees, B. Theunissen, and J. A. Lenstra. (2011). On the breeds of cattle—Historic and current classifications. *Diversity* 3 (4): 660–92.

Fernandez, M. H., and E. S. Vrba. (2005). A complete estimate of the phylogenetic relationships in Ruminantia: A dated species-level supertree of the extant ruminants. *Biological Reviews* 80 (2): 269–302.

Freeman, A., C. Meghen, D. Machugh, R. Loftus, M. Achukwi, A. Bado, . . . and D. Bradley. (2004). Admixture and diversity in West African cattle populations. *Molecular Ecology* 13 (11): 3477–87.

Fuller, D. Q. (2006). Agricultural origins and frontiers in South Asia: A working synthesis. *Journal of World Prehistory* 20 (1): 1–86.

Gautier, M., D. Laloë, and K. Moazami-Goudarzi. (2010). Insights into the genetic history of French cattle from dense SNP data on 47 worldwide breeds. *PLoS One* 5 (9): e13038.

Gautier, M., and M. Naves. (2011). Footprints of selection in the ancestral admixture of a New World Creole cattle breed. *Molecular Ecology* 20 (15): 3128–43.

Gerbault, P., A. Liebert, Y. Itan, A. Powell, M. Currat, J. Burger, . . . and M. G. Thomas. (2011). Evolution of lactase persistence: An example of human niche construction. *Philosophical Transactions of the Royal Society. B: Biological Sciences* 366 (1566): 863–77.

Gilliam, A. E., I. M. Heilbron, W. S. Ferguson, and S. J. Watson. (1936). Variations in the carotene and vitamin A values of the milk fat of cattle of typical English breeds. *Biochemical Journal* 30 (9): 1728–34.

Ginja, C., M. Penedo, L. Melucci, J. Quiroz, O. Martinez Lopez, M. Revidatti, . . . and L. Gama. (2010). Origins and genetic diversity of New World Creole cattle: Inferences from mitochondrial and Y chromosome polymorphisms. *Animal Genetics* 41 (2): 128–41.

Götherström, A., C. Anderung, L. Hellborg, R. Elburg, C. Smith, D. G. Bradley, and H. Ellegren. (2005). Cattle domestication in the Near East was followed by hybridization with aurochs bulls in Europe. *Proceedings of the Royal Society. B: Biological Sciences* 272 (1579): 2345–51.

Greenfield, H. J. (2010). The secondary products revolution: The past, the present and the future. *World Archaeology* 42 (1): 29–54.

Guthrie, R. D. (2005). *The Nature of Paleolithic Art*. Chicago: University of Chicago Press.

Hanotte, O., D. G. Bradley, J. W. Ochieng, Y. Verjee, E. W. Hill, and J. E. O. Rege. (2002). African pastoralism: Genetic imprints of origins and migrations. *Science* 296 (5566): 336–39.

Harris, M. (1992). The cultural ecology of India's sacred cattle. *Current Anthropology* 33 (1): 261–76.

Harvey, C. B., E. J. Hollox, M. Poulter, Y. Wang, M. Rossi, S. Auricchio, . . . and D. M. Swallow. (1998). Lactase haplotype frequencies in Caucasians: Association with the lactase persistence/non-persistence polymorphism. *Annals of Human Genetics* 62 (3): 215–23.

Hassanin, A., J. An, A. Ropiquet, T. T. Nguyen, and A. Couloux. (2013). Combining multiple autosomal introns for studying shallow phylogeny and taxonomy of Laurasiatherian mammals: Application to the tribe Bovini (Cetartiodactyla, Bovidae). *Molecular Phylogenetics and Evolution* 66 (3): 766–75.

Hassanin, A., F. Delsuc, A. Ropiquet, C. Hammer, B. Jansen Van Vuuren, C. Matthee, . . . and A. Couloux. (2012). Pattern and timing of diversification of Cetartiodactyla (Mammalia, Laurasiatheria), as revealed by a comprehensive analysis of mitochondrial genomes. *Comptes Rendus Biologies* 335 (1): 32–50.

Hayes, B. J., J. Pryce, A. J. Chamberlain, P. J. Bowman, and M. E. Goddard. (2010). Genetic architecture of complex traits and accuracy of genomic prediction: Coat colour, milk-fat percentage, and type in Holstein cattle as contrasting model traits. *PLoS Genetics* 6 (9): e1001139.

Helmer, D., L. Gourichon, H. Monchot, J. Peters, and M. Sana Segui. (2005). "Identifying Early Domestic Cattle from Pre-Pottery Neolithic Sites on the Middle Euphrates Using Sexual Dimorphism." In *The First Steps of Animal Domestication: New Archaeozoological Approaches*, edited by J. D. Vigne, J. Peters, and D. Helmer, 86–95. Oxford: Oxbow.

Hijazi, S., A. Abulaban, Z. Ammarin, and G. Flatz. (1983). Distribution of adult lactase phenotypes in Bedouins and in urban and agricultural populations of Jordan. *Tropical and Geographical Medicine* 35 (2): 157–61.

Holden, C., and R. Mace. (2002). "Pastoralism and the Evolution of Lactase Persistence." *The Human Biology of Pastoral Populations*, edited by W. R. Leonard and M. H. Crawford,

280–307. Cambridge Studies in Biological and Evolutionary Anthropology 30. Cambridge: Cambridge University Press.

Hongo, H., J. Pearson, B. Öksüz, and G. Ilgezdi. (2009). The process of ungulate domestication at Cayönü, southeastern Turkey: A multidisciplinary approach focusing on *Bos* sp. and *Cervus elaphus. Anthropozoologica* 44 (1): 63–78.

Hou, Y., G. Liu, D. Bickhart, M. Cardone, K. Wang, E.-S. Kim, . . . and C. Van Tassell. (2011). Genomic characteristics of cattle copy number variations. *BMC Genomics* 12 (1): 127.

Ikram, S. (1995). *Choice Cuts: Meat Production in Ancient Egypt.* Orientalia Lovaniensia Analecta 69. Leuven, Belgium: Peeters.

Ingram, C. J., M. F. Elamin, C. A. Mulcare, M. E. Weale, A. Tarekegn, T. O. Raga, . . . and D. M. Swallow. (2007). A novel polymorphism associated with lactose tolerance in Africa: Multiple causes for lactase persistence? *Human Genetics* 120 (6): 779–88.

Ingram, C. J., C. A. Mulcare, Y. Itan, M. G. Thomas, and D. M. Swallow. (2009). Lactose digestion and the evolutionary genetics of lactase persistence. *Human Genetics* 124 (6): 579–91.

Itan, Y., A. Powell, M. A. Beaumont, J. Burger, and M. G. Thomas. (2009). The origins of lactase persistence in Europe. *PLoS Computational Biology* 5 (8): e1000491.

Janis, C. (2007). "Artiodactyla Paleoecology and Evolutionary Trends." In *The Evolution of Artiodactyls*, edited by D. R. Prothero and S. E. Foss, 292–315. Baltimore: Johns Hopkins University Press.

Joshi, B., A. Singh, and R. Gandhi, R. (2001). Performance evaluation, conservation and improvement of Sahiwal cattle in India. *Animal Genetic Resources Information* 31: 43–54.

Joshi, N. R., and R. W. Phillips. (1953). *Zebu Cattle of India and Pakistan.* New York: FAO.

Kantanen, J., C. Edwards, D. Bradley, H. Viinalass, S. Thessler, Z. Ivanova, . . . and S. Stojanović. (2009). Maternal and paternal genealogy of Eurasian taurine cattle (*Bos taurus*). *Heredity* 103 (5): 404–15.

Kendall, T. (1998). *Proceedings of the Ninth Conference of the International Society of Nubian Studies.* Boston: Northeastern University.

Kidd, K., and L. Cavalli-Sforza. (1974). The role of genetic drift in the differentiation of Icelandic and Norwegian cattle. *Evolution* 28 (3): 381.

Lee, K. T., W. H. Chung, S. Y. Lee, J. W. Choi, J. Kim, D. Lim, . . . and T. H. Kim. (2013). Whole-genome resequencing of Hanwoo (Korean cattle) and insight into regions of homozygosity. *BMC Genomics* 14 (1): 519.

Leonardi, M., P. Gerbault, M. G. Thomas, and J. Burger. (2012). The evolution of lactase persistence in Europe. A synthesis of archaeological and genetic evidence. *International Dairy Journal* 22 (2): 88–97.

Lewin, H. (2013). "Genomic Footprints of Selection after 50 Years of Dairy Cattle Breeding." Paper presented at the Plant and Animal Genome XXI Conference.

Liu, G. E., and D. M. Bickhart. (2012). Copy number variation in the cattle genome. *Functional & Integrative Genomics* 12 (4): 609–24.

Liu, G. E., Y. Hou, B. Zhu, M. F. Cardone, L. Jiang, A. Cellamare, . . . and J. W. Keele. (2010). Analysis of copy number variations among diverse cattle breeds. *Genome Research* 39: 693–703.

Loftus, R., D. MacHugh, D. Bradley, P. Sharp, and P. Cunningham. (1994). Evidence for two independent domestications of cattle. *Proceedings of the National Academy of Sciences USA* 91: 2757–61.

MacHugh, D. E., M. D. Shriver, R. T. Loftus, P. Cunningham, and D. G. Bradley. (1997). Microsatellite DNA variation and the evolution, domestication and phylogeography of taurine and zebu cattle (*Bos taurus* and *Bos indicus*). *Genetics* 146 (3): 1071–86.

Magee, D., C. Meghen, S. Harrison, C. Troy, T. Cymbron, C. Gaillard, . . . and D. Bradley. (2002). A partial African ancestry for the Creole cattle populations of the Caribbean. *Journal of Heredity* 93 (6): 429–32.

Manwell, C., and C. Baker. (1980). Chemical classification of cattle. 2. Phylogenetic tree and specific status of the Zebu. *Animal Blood Groups and Biochemical Genetics* 11 (2): 151–62.

Maretto, F., J. Ramljak, F. Sbarra, M. Penasa, R. Mantovani, A. Ivanković, and G. Bittante. (2012). Genetic relationships among Italian and Croatian Podolian cattle breeds assessed by microsatellite markers. *Livestock Science* 150 (1–3): 256–64.

Maudet, C., G. Luikart, and P. Taberlet. (2002). Genetic diversity and assignment tests among seven French cattle breeds based on microsatellite DNA analysis. *Journal of Animal Science* 80 (4): 942–50.

McDowell, R. (1985). Crossbreeding in tropical areas with emphasis on milk, health, and fitness. *Journal of Dairy Science* 68 (9): 2418–35.

Meredith, R., J. Janecka, J. Gatesy, O. Ryder, C. Fisher, E. Teeling, . . . and W. Murphy. (2011). Impacts of the Cretaceous Terrestrial Revolution and KPg extinction on mammal diversification. *Science* 334 (6055): 521–24.

Miretti, M., S. Dunner, M. Naves, E. Contel, and J. Ferro. (2004). Predominant African-derived mtDNA in Caribbean and Brazilian Creole cattle is also found in Spanish cattle (*Bos taurus*). *Journal of Heredity* 95 (5): 450–53.

Mirol, P., G. Giovambattista, J. Lirón, and F. Dulout. (2003). African and European mitochondrial haplotypes in South American Creole cattle. *Heredity* 91 (3): 248–54.

Mukesh, M., M. Sodhi, S. Bhatia, and B. Mishra. (2004). Genetic diversity of Indian native cattle breeds as analysed with 20 microsatellite loci. *Journal of Animal Breeding and Genetics* 121 (6): 416–24.

Murray, C., E. Huerta-Sanchez, F. Casey, and D. G. Bradley. (2010). Cattle demographic history modelled from autosomal sequence variation. *Philosophical Transactions of the Royal Society. B: Biological Sciences* 365 (1552): 2531–39.

Negrini, R., I. Nijman, E. Milanesi, K. Moazami-Goudarzi, J. Williams, G. Erhardt, . . . and D. Bradley. (2007). Differentiation of European cattle by AFLP fingerprinting. *Animal Genetics* 38 (1): 60–66.

Pariset, L., M. Mariotti, A. Nardone, M. I. Soysal, E. Ozkan, J. L. Williams, . . . and A. Valentini. (2010). Relationships between Podolic cattle breeds assessed by single nucleotide polymorphisms (SNPs) genotyping. *Journal of Animal Breeding and Genetics* 127 (6): 481–88.

Pellecchia, M., R. Negrini, L. Colli, M. Patrini, E. Milanesi, A. Achilli, . . . and P. Ajmone-Marsan. (2007). The mystery of Etruscan origins: Novel clues from *Bos taurus* mitochondrial DNA. *Proceedings of the Royal Society. B: Biological Sciences* 274 (1614): 1175–79.

Pérez-Pardal, L., L. J. Royo, A. Beja-Pereira, S. Chen, R. J. Cantet, A. Traoré, . . . and I. Fernández. (2010). Multiple paternal origins of domestic cattle revealed by Y-specific interspersed multilocus microsatellites. *Heredity* 105 (6): 511–19.

Pérez-Pardal, L., L. J. Royo, A. Beja-Pereira, I. Curik, A. Traoré, I. Fernández, . . . and F. Goyache. (2010). Y-specific microsatellites reveal an African subfamily in taurine (*Bos taurus*) cattle. *Animal Genetics* 41 (3): 232–41.

Pinhasi, R., J. Fort, and A. J. Ammerman. (2005). Tracing the origin and spread of agriculture in Europe. *PLoS Biology* 3 (12): e410.

Polák, J., and D. Frynta. (2010). Patterns of sexual size dimorphism in cattle breeds support Rensch's rule. *Evolutionary Ecology* 24 (5): 1255–66.

Porto-Neto, L. R., T. S. Sonstegard, G. E. Liu, D. M. Bickhart, M. V. Da Silva, M. A. Machado, . . . and C. P. Van Tassell. (2013). Genomic divergence of zebu and taurine cattle identified through high-density SNP genotyping. *BMC Genomics* 14 (1): 876.

Qanbari, S., E. Pimentel, J. Tetens, G. Thaller, P. Lichtner, A. Sharifi, and H. Simianer. (2010). A genome-wide scan for signatures of recent selection in Holstein cattle. *Animal Genetics* 41 (4): 377–89.

Raven, L.-A., B. G. Cocks, and B. J. Hayes. (2014). Multibreed genome wide association can improve precision of mapping causative variants underlying milk production in dairy cattle. *BMC Genomics* 15 (1): 62.

Rege, E. (2003). "Defining Livestock Breeds in the Context of Community-Based Management of Farm Animal Genetic Resources." In *Community-Based Management of Animal Genetic Resources: Proceedings of the Workshop Held in Mbabane, Swaziland, 7–11 May 2001*, 27–36. [New York: Food and Agricultural Organization of the United Nations.]

Rege, J. E. O. (1999). The state of African cattle genetic resources I. Classification framework and identification of threatened and extinct breeds. *Animal Genetic Resources Information* 25: 1–25.

Reilly, B. J. (2012). Revisiting Bedouin desert adaptations: Lactase persistence as a factor in Arabian Peninsula history. *Journal of Arabian Studies* 2 (2): 93–107.

Romero, I. G., C. B. Mallick, A. Liebert, F. Crivellaro, G. Chaubey, Y. Itan, . . . and R. Villems. (2012). Herders of Indian and European cattle share their predominant allele for lactase persistence. *Molecular Biology and Evolution* 29 (1): 249–60.

Sherratt, A. (1981). *Plough and Pastoralism: Aspects of the Secondary Products Revolution*. New York: Cambridge University Press.

Sherratt, A. (1983). The secondary exploitation of animals in the Old World. *World Archaeology* 15 (1): 90–104.

Solounias, N., J. C. Barry, R. L. Bernor, E. H. Lindsay, and S. M. Raza. (1995). The oldest bovid from the Siwaliks, Pakistan. *Journal of Vertebrate Paleontology* 15 (4): 806–14.

Stapely, J., J. Reger, P. G. D. Feulner, C. Smajda, J. Galindo, R. Ekblom, . . . and J. Slate. (2010). Adaptation genomics: The next generation. *Trends in Ecology and Evolution* 25 (12): 705–12.

Stothard, P., J.-W. Choi, U. Basu, J. Sumner-Thomson, Y. Meng, X. Liao, and S. Moore. (2011). Whole genome resequencing of black Angus and Holstein cattle for SNP and CNV discovery. *BMC Genomics* 12 (1): 559.

Strömberg, C. A. E. (2011). Evolution of grasses and grassland ecosystems. *Annual Review of Earth and Planetary Sciences* 39 (1): 517–44.

Tandon, R. K., Y. K. Joshi, D. S. Singh, M. Narendranathan, V. Balakrishnan, and K. Lal. (1981). Lactose intolerance in North and South Indians. *American Journal of Clinical Nutrition* 34 (5): 943–46.

Teasdale, M. D., and D. G. Bradley. (2012). "The Origins of Cattle." In *Bovine Genomics*, edited by J. Womack, 1–10 Ames, IA: Wiley-Blackwell.

Tsuda, K., R. Kawahara-Miki, S. Sano, M. Imai, T. Noguchi, Y. Inayoshi, and T. Kono. (2013).

Abundant sequence divergence in the native Japanese cattle *Mishima-Ushi* (*Bos taurus*) detected using whole-genome sequencing. *Genomics* 102 (4): 372–78.

Twiss, K. C., and N. Russell. (2009). Taking the bull by the horns: Ideology, masculinity, and cattle horns at Çatalhöyük (Turkey). *Paléorient* 35 (2): 19–32.

Van Vuure, C. (2005). *Retracing the Aurochs: History, Morphology and Ecology of an Extinct Wild Ox*. Sofia-Moscow: Pensoft.

Vigne, J.-D., and D. Helmer. (2007). Was milk a secondary product in the Old World Neolithisation process? Its role in the domestication of cattle, sheep and goats. *Anthropozoologica* 42 (2): 9–40.

Vigne, J.-D., J. Peters, and D. Helmer, eds. (2005). *The First Steps of Animal Domestication: New Archaeozoological Approaches*. Oxford: Oxbow.

Vivekanandan, P., and V. Alagumalai. (2013). *Community Conservation of Local Livestock Breeds*. Tamilnadu, India: SEVA.

Wolin, M. J. (1979). "The Rumen Fermentation: A Model for Microbial Interactions in Anaerobic Ecosystems." In *Advances in Microbial Ecology*, edited by M. Alexander, 49–77. New York: Springer.

Wurzinger, M., D. Ndumu, R. Baumung, A. Drucker, A. Okeyo, D. Semambo, . . . and J. Sölkner. (2006). Comparison of production systems and selection criteria of Ankole cattle by breeders in Burundi, Rwanda, Tanzania and Uganda. *Tropical Animal Health and Production* 38 (7–8): 571–81.

Zeder, M. A. (2011). The origins of agriculture in the Near East. *Current Anthropology* 52 (S4): S221–35.

Zhan, B., J. Fadista, B. Thomsen, J. Hedegaard, F. Panitz, and C. Bendixen. (2011). Global assessment of genomic variation in cattle by genome resequencing and high-throughput genotyping. *BMC Genomics* 12 (1): 557.

Zhang, H., J. L. Paijmans, F. Chang, X. Wu, G. Chen, C. Lei, . . . and L. Orlando. (2013). Morphological and genetic evidence for early Holocene cattle management in northeastern China. *Nature Communications* 4: 2755.

Chapter 8: SHEEP AND GOATS

NOTES

1. The scapegoats were always male and selected by lot (Leviticus 16:5–10; Feinberg 1958).

2. The goat was released 10 miles from the community to make its return less likely (David Guzik, "Leviticus 16—The Day of Atonement," *Enduring Word Media*, 2004, http://www.enduringword.com/commentaries/0316.htm).

3. Matthew 25:31–33.

4. Perkins and Roselli 2007. Rams have become a model system for the study of male homosexuality.

5. Morton and Avanzo 2011.

6. Kaminski et al. 2005.

7. Slijper 1942.

8. Bibi et al. 2009; Bibi and Vrba 2010; Ropiquet and Hassanin 2005; Mathee and Davis 2001.

9. Pirastru et al. 2009.

10. The so-called mountain goats of North America are not true goats, or even members of the Caprini tribe; they belong to the tribe Rupicaprini—along with the European chamois, serows, and gorals—and hence are more distantly related to true goats than are sheep.

11. Rezeai et al. 2010.

12. Peters et al. 1999.

13. Meadows et al. 2007; Meadows, Hiendleder, and Kijas 2011.

14. Vigne et al. 2011; Zeder 2012.

15. Arbuckle 2008.

16. Zeder 2008, 2011.

17. Zeder 2008. For the first arrival of domestic sheep in India, see Singh et al. 2013; for their first arrival in Africa, see Muigai and Hanotte 2013.

18. Zeder 2008; Chessa et al. 2009; Tresset and Vigne 2011.

19. Chessa et al. 2009.

20. Ibid.

21. Ryder 1983; Sherratt 1981.

22. Chessa et al. 2009.

23. Cai et al. 2011; Muigai and Hanotte 2013; Chessa et al. 2009.

24. Chessa et al. 2009.

25. Chessa et al. 2009; Bollvåg 2010; Tapio et al. 2010.

26. Ryder 1964; Chessa et al. 2009.

27. Chessa et al. 2009.

28. Zeder 1999, 2006; Zeder and Hesse 2000.

29. Zeder 2008, 1999, including northern and central Zagros Mountains.

30. Naderi et al. 2007, 2008.

31. Fernández et al. 2006; Vigne 2011.

32. Zeder, Smith, and Bradley 2006; Luikart et al. 2001. For China, see Chen et al. 2005.

33. Horwitz and Bar-Gal 2006; Zeder 2008; Hatziminaoglou and Boyazoglu 2004.

34. For evidence of Neanderthal predation on wild goats, see Speth and Tchernov 2001; for predation by anatomically modern humans, see Marean 1998; Shea 1998; Otte et al. 2007. Munro 2004 describes predation on wild goats by Natufians.

35. Hecker 1982; Zeder 1999; Zeder and Hesse 2000; but see Arbuckle 2008 and Arbuckle and Atici 2013 for later dates of selective culling.

36. The general term for this sort of mating system is polygyny; see Zeder 2006 and Vigne et al. 2011.

37. Fernández et al. 2006; Luikart et al. 2006.

38. Smith and Horwitz 1984; Zeder 2006; Zohary, Tchernov, and Horwitz 1998; but see Haber and Dayan 2004 for later dates. Part of the explanation for some of these phenotypic alterations is phenotypic plasticity, especially as it relates to fodder provisions (Makarewicz and Tuross 2012). See Polák and Frynta 2009 for estimates of sexual size dimorphism in modern breeds.

39. For example, Coltman et al. 1999; Milner, Elston, and Albon 1999; Coulson et al. 2001; Clutton-Brock and Sheldon 2010; Catchpole et al. 2000.

40. Clutton-Brock 2009; Robinson and Kruuk 2007.

41. Ryder 1981; Catchpole et al. 2000.
42. Pemberton et al. 1996; Robinson and Kruuk 2007; Johnston et al. 2013.
43. Campbell and Donlan 2005.
44. Dubeuf and Boyazoglu 2009.
45. Pieters et al. 2009.
46. International Sheep Genomics Consortium et al. 2010; Chessa et al. 2009; but not in the United States (Blackburn et al. 2011).
47. The phylogeographic signal for sheep breeds is low (Meadows et al. 2005); for goats, however, the phylogeographic signal is strong—stronger even than that of cattle, which are much less portable. In part this difference reflects the fact that goat breeds are more recently derived from local landraces.
48. Sanchez Belda and Trujillano 1979; Diez-Tascón et al. 2000.
49. Chessa et al. 2009.
50. Bruford, Bradley, and Luikart 2003. Kijas et al. 2012; Muigai and Hanotte 2013.
51. Handley et al. 2007.
52. Chessa et al. 2009.
53. Handley et al. 2007.
54. As with cattle, there were two separate routes by which domestic sheep reached Europe: the Danubian and the Mediterranean. See Pereira et al. 2006 for Iberian breeds, and Peter et al. 2007 for Alpine breeds.
55. Chessa et al. 2009.
56. Peter et al. 2007. See Fontanesi et al. 2011 for sheep, and Fontanesi et al. 2012 for goats.
57. Ryder 1981; Ferencakovic et al. 2013; Drăgănescu 2007; Kusza et al. 2008.
58. Kijas et al. 2012; Moradi et al. 2012.
59. International Sheep Genomics Consortium et al. 2010. Kijas et al. 2013 conducted a SNP search on goats.
60. Dong et al. 2013 (pituitary); International Sheep Genomics Consortium et al. 2010.
61. Fontanesi et al. 2010, 2011. See also the analysis of CNVs in goats in Fontanesi et al. 2009.
62. Fontanesi et al. 2011, 2012.
63. Fontanesi et al. 2009 (goats); Fontanesi et al. 2010 (sheep).
64. Fontanesi et al. 2009.
65. Ibid.
66. Dong et al. 2013.
67. Ibid.
68. Reverse transcription is the construction of DNA from RNA.
69. Chessa et al. 2009.
70. Haenlein 2004.
71. Ceballos et al. 2009.

REFERENCES

Arbuckle, B. S. (2008). Revisiting Neolithic caprine exploitation at Suberde, Turkey. *Journal of Field Archaeology* 33 (2): 219–36.

Arbuckle, B. S., and L. Atici. (2013). Initial diversity in sheep and goat management in Neolithic south-western Asia. *Levant* 45 (2): 219–35.

Bibi, F., M. Bukhsianidze, A. W. Gentry, D. Geraads, D. S. Kostopoulos, and E. S. Vrba. (2009). The fossil record and the evolution of Bovidae: State of the field. *Palaeontologica Electronica* 12 (3): 1–11.

Bibi, F., and E. Vrba. (2010). Unraveling bovin phylogeny: Accomplishments and challenges. *BMC Biology* 8 (1): 50.

Blackburn, H. D., S. R. Paiva, S. Wildeus, W. Getz, D. Waldron, R. Stobart, . . . and M. Brown. (2011). Genetic structure and diversity among sheep breeds in the United States: Identification of the major gene pools. *Journal of Animal Science* 89 (8): 2336–48.

Bollvåg, A. Ø. (2010). "Mitochondrial Ewe—Application of Ancient DNA Typing to the Study of Domestic Sheep (*Ovis aries*) in Mediaeval Norway." Master's thesis, University of Oslo, Norway.

Bruford, M., D. Bradley, and G. Luikart. (2003). DNA markers reveal the complexity of livestock domestication. *Nature Reviews. Genetics* 4: 900–910.

Cai, D., Z. Tang, H. Yu, L. Han, X. Ren, X. Zhao, . . . and H. Zhou. (2011). Early history of Chinese domestic sheep indicated by ancient DNA analysis of Bronze Age individuals. *Journal of Archaeological Science* 38 (4): 896–902.

Campbell, K., and C. Donlan. (2005). Feral goat eradications on islands. *Conservation Biology* 19 (5): 1362–74.

Catchpole, E. A., B. J. T. Morgan, T. N. Coulson, S. N. Freeman, and S. D. Albon. (2000). Factors influencing Soay sheep survival. *Journal of the Royal Statistical Society. Series C, Applied Statistics* 49 (4): 453–72.

Ceballos, L. S., E. R. Morales, G. de la Torre Adarve, J. D. Castro, L. P. Martínez, and M. R. S. Sampelayo. (2009). Composition of goat and cow milk produced under similar conditions and analyzed by identical methodology. *Journal of Food Composition and Analysis* 22 (4): 322–29.

Chen, S.-Y., Y.-H. Su, S.-F. Wu, T. Sha, and Y.-P. Zhang. (2005). Mitochondrial diversity and phylogeographic structure of Chinese domestic goats. *Molecular Phylogenetics and Evolution* 37 (3): 804–14.

Chessa, B., F. Pereira, F. Arnaud, A. Amorim, F. Goyache, I. Mainland, . . . and M. Palmarini. (2009). Revealing the history of sheep domestication using retrovirus integrations. *Science* 324 (5926): 532–36.

Clutton-Brock, T. (2009). Sexual selection in females. *Animal Behaviour* 77 (1): 3–11.

Clutton-Brock, T., and B. C. Sheldon. (2010). Individuals and populations: The role of long-term, individual-based studies of animals in ecology and evolutionary biology. *Trends in Ecology & Evolution* 25 (10): 562–73.

Coltman, D. W., J. G. Pilkington, J. A. Smith, and J. M. Pemberton. (1999). Parasite-mediated selection against inbred Soay sheep in a free-living, island population. *Evolution* 53 (4): 1259–67.

Coulson, T., E. Catchpole, S. Albon, B. Morgan, J. Pemberton, T. Clutton-Brock, . . . and B. Grenfell. (2001). Age, sex, density, winter weather, and population crashes in Soay sheep. *Science* 292 (5521): 1528–31.

Diez-Tascón, C., R. P. Littlejohn, P. A. R. Almeida, and A. M. Crawford. (2000). Genetic variation within the Merino sheep breed: Analysis of closely related populations using microsatellites. *Animal Genetics* 31 (4): 243–51.

Dong, Y., M. Xie, Y. Jiang, N. Xiao, X. Du, W. Zhang, . . . and J. Liang. (2013). Sequencing and automated whole-genome optical mapping of the genome of a domestic goat (*Capra hircus*). *Nature Biotechnology* 31 (2): 135–41.

Drăgănescu, C. (2007). A note on Balkan sheep breeds origin and their taxonomy. *Archiva Zootechnica* 10: 90–101.

Dubeuf, J.-P., and J. Boyazoglu. (2009). An international panorama of goat selection and breeds. *Livestock Science* 120 (3): 225–31.

Feinberg, C. L. (1958). The scapegoat of Leviticus Sixteen. *Bibliotheca Sacra* 115 (460): 320–33.

Ferencakovic, M., I. Curik, L. Pérez-Pardal, L. J. Royo, V. Cubric-Curik, I. Fernandez, . . . and K. Krapinec. (2013). Mitochondrial DNA and Y-chromosome diversity in East Adriatic sheep. *Animal Genetics* 44 (2): 184–92.

Fernández, H., S. Hughes, J.-D. Vigne, D. Helmer, G. Hodgins, C. Miquel, . . . and P. Taberlet. (2006). Divergent mtDNA lineages of goats in an Early Neolithic site, far from the initial domestication areas. *Proceedings of the National Academy of Sciences USA* 103 (42): 15375–79.

Fontanesi, L., F. Beretti, P. Martelli, M. Colombo, S. Dall'Olio, M. Occidente, . . . and V. Russo. (2011). A first comparative map of copy number variations in the sheep genome. *Genomics* 97 (3): 158–65.

Fontanesi, L., F. Beretti, V. Riggio, E. Gómez Gonzáles, S. Dall'Olio, R. Davioli, . . . and B. Portolano. (2009). Copy number variation and missense mutations of the agouti signaling protein (ASIP) gene in goat breeds with different coat colors. *Cytogenetic and Genome Research* 126 (4): 333–47.

Fontanesi, L., P. Martelli, F. Beretti, V. Riggio, S. Dall'Olio, M. Colombo, . . . and B. Portolano. (2010). An initial comparative map of copy number variations in the goat (*Capra hircus*) genome. *BMC Genomics* 11: 639.

Fontanesi, L., A. Rustempašić, M. Brka, and V. Russo. (2012). Analysis of polymorphisms in the agouti signalling protein (*ASIP*) and melanocortin 1 receptor (*MC1R*) genes and association with coat colours in two Pramenka sheep types. *Small Ruminant Research* 105 (1): 89–96.

Haber, A., and T. Dayan. (2004). Analyzing the process of domestication: Hagoshrim as a case study. *Journal of Archaeological Science* 31 (11): 1587–1601.

Haenlein, G. F. W. (2004). Goat milk in human nutrition. *Small Ruminant Research* 51 (2): 155–63.

Handley, L. L., K. Byrne, F. Santucci, S. Townsend, M. Taylor, M. Bruford, and G. Hewitt. (2007). Genetic structure of European sheep breeds. *Heredity* 99 (6): 620–31.

Hatziminaoglou, Y., and J. Boyazoglu. (2004). The goat in ancient civilisations: From the Fertile Crescent to the Aegean Sea. *Small Ruminant Research* 51 (2): 123–29.

Hecker, H. M. (1982). Domestication revisited: Its implications for faunal analysis. *Journal of Field Archaeology* 9 (2): 217–36.

Horwitz, L. K., and G. K. Bar-Gal. (2006). The origin and genetic status of insular caprines in the eastern Mediterranean: A case study of free-ranging goats (*Capra aegagrus cretica*) on Crete. *Human Evolution* 21 (2): 123–38.

International Sheep Genomics Consortium, A. L. Archibald, N. E. Cockett, B. P. Dalrymple, T. Faraut, J. W. Kijas, . . . and X. Xun. (2010). The sheep genome reference sequence: A work in progress. *Animal Genetics* 41 (5): 449–53.

Johnston, S. E., J. Gratten, C. Berenos, J. G. Pilkington, T. H. Clutton-Brock, J. M. Pemberton, and J. Slate. (2013). Life history trade-offs at a single locus maintain sexually selected genetic variation. *Nature* 502: 93–95.

Kaminski, J., J. Riedel, J. Call, and M. Tomasello. (2005). Domestic goats, *Capra hircus*, follow gaze direction and use social cues in an object choice task. *Animal Behaviour* 69 (1): 11–18.

Kijas, J. W., J. A. Lenstra, B. Hayes, S. Boitard, L. R. Porto Neto, M. San Cristobal, . . . and other members of the International Sheep Genomics Consortium. (2012). Genome-wide analysis of the world's sheep breeds reveals high levels of historic mixture and strong recent selection. *PLoS Biology* 10 (2): e1001258.

Kijas, J. W., J. S. Ortiz, R. McCulloch, A. James, B. Brice, B. Swain, . . . and the International Goat Genome Consortium. (2013). Genetic diversity and investigation of polledness in divergent goat populations using 52 088 SNPs. *Animal Genetics* 44 (3): 325–35.

Kusza, S., I. Nagy, Z. Sasvári, A. Stágel, T. Németh, A. Molnár, . . . and S. Kukovics. (2008). Genetic diversity and population structure of Tsigai and Zackel type of sheep breeds in the Central-, Eastern- and Southern-European regions. *Small Ruminant Research* 78 (1): 13–23.

Luikart, G., H. Fernandez, M. Mashkour, P. R. England, and P. Taberlet. (2006). Origins and diffusion of domestic goats inferred from DNA markers. In: *Documenting Domestication: New Genetic and Archaeological Paradigms*, edited by M. A. Zeder, D. G. Bradley, E. Emshwiller, and B. D. Smith, 294–305. Berkeley: University of California Press.

Luikart, G., L. Giellly, L. Excoffier, J. Vigne, J. Bouuvet, and P. Taberlet. (2001). Multiple maternal origins and weak phylogeographic structure in domestic goats. *Proceedings of the National Academy of Sciences USA* 98: 5927–32.

Makarewicz, C., and N. Tuross. (2012). Finding fodder and tracking transhumance: Isotopic detection of goat domestication processes in the Near East. *Current Anthropology* 53 (4): 495–505.

Marean, C. W. (1998). A critique of the evidence for scavenging by Neandertals and early modern humans: New data from Kobeh Cave (Zagros Mountains, Iran) and Die Kelders Cave 1 Layer 10 (South Africa). *Journal of Human Evolution* 35 (2): 111–36.

Matthee, C. A., and S. K. Davis. (2001). Molecular insights into the evolution of the family Bovidae: A nuclear DNA perspective. *Molecular Biology and Evolution* 18 (7): 1220–30.

Meadows, J. R. S., I. Cemal, O. Karaca, E. Gootwine, and J. W. Kijas. (2007). Five ovine mitochondrial lineages identified from sheep breeds of the Near East. *Genetics* 175 (3): 1371–79.

Meadows, J. R. S., S. Hiendleder, and J. W. Kijas. (2011). Haplogroup relationships between domestic and wild sheep resolved using a mitogenome panel. *Heredity* 106 (4): 700–706.

Meadows, J. R., K. Li, J. Kantanen, M. Tapio, W. Sipos, V. Pardeshi, . . . and J. W. Kijas. (2005). Mitochondrial sequence reveals high levels of gene flow between breeds of domestic sheep from Asia and Europe. *Journal of Heredity* 96 (5): 494–501.

Milner, J., D. Elston, and S. Albon. (1999). Estimating the contributions of population density and climatic fluctuations to interannual variation in survival of Soay sheep. *Journal of Animal Ecology* 68 (6): 1235–47.

Moradi, M. H., A. Nejati-Javaremi, M. Moradi-Shahrbabak, K. G. Dodds, and J. C. McEwan. (2012). Genomic scan of selective sweeps in thin and fat tail sheep breeds for identifying of candidate regions associated with fat deposition. *BMC Genetics* 13 (1): 10.

Morton, A. J., and L. Avanzo. (2011). Executive decision-making in the domestic sheep. *PLoS One* 6 (1): e15752.

Muigai, A. W. T., and O. Hanotte. (2013). The origin of African sheep: Archaeological and genetic perspectives. *African Archaeological Review* 30 (1): 39–50.

Munro, N. (2004). Zooarchaeological measures of hunting pressure and occupation intensity in the Natufian. *Current Anthropology* 45 (S4): S5–34.

Naderi, S., H.-R. Rezaei, F. Pompanon, M. G. B. Blum, R. Negrini, H.-R. Naghash, . . . and P. Taberlet. (2008). The goat domestication process inferred from large-scale mitochondrial DNA analysis of wild and domestic individuals. *Proceedings of the National Academy of Sciences USA* 105 (46): 17659–64.

Naderi, S., H.-R. Rezaei, P. Taberlet, S. Zundel, S.-A. Rafat, H.-R. Naghash, . . . and for the Econogene Consortium. (2007). Large-scale mitochondrial DNA analysis of the domestic goat reveals six haplogroups with high diversity. *PLoS One* 2 (10): e1012.

Otte, M., F. Biglari, D. Flas, S. Shidrang, N. Zwyns, M. Mashkour, . . . and V. Radu. (2007). The Aurignacian in the Zagros region: New research at Yafteh Cave, Lorestan, Iran. *Antiquity* 81 (311): 82–96.

Pemberton, J. M., J. A. Smith, T. N. Coulson, T. C. Marshall, J. Slate, S. Paterson, . . . and P. Sneath. (1996). The maintenance of genetic polymorphism in small island populations: Large mammals in the Hebrides [and discussion]. *Philosophical Transactions of the Royal Society. B: Biological Sciences* 351 (1341): 745–52.

Pereira, F., S. J. Davis, L. Pereira, B. McEvoy, and A. Amorin. (2006). Genetic signatures of a Mediterranean influence in Iberian Peninsula sheep husbandry. *Molecular Biology and Evolution* 23 (7): 1420–26.

Perkins, A., and C. E. Roselli. (2007). The ram as a model for behavioral neuroendocrinology. *Hormones and Behavior* 52 (1): 70–77.

Peter, C., M. Bruford, T. Perez, S. Dalamitra, G. Hewitt, G. Erhardt, and the Econogene Consortium. (2007). Genetic diversity and subdivision of 57 European and Middle-Eastern sheep breeds. *Animal Genetics* 38 (1): 37–44.

Peters, J., A. von den Driesch, D. Helmer, and M. Saña Segui. (1999). Early animal husbandry in the Northern Levant. *Paléorient* 25 (2): 27–48.

Pieters, A., E. van Marle-Köster, C. Visser, and A. Kotze. (2009). South African developed meat type goats: A forgotten animal genetic resource? *Animal Genetic Resources Information* 44: 33–43.

Pirastru, M., C. Multineddu, P. Mereu, M. Sannai, E. S. el Sherbini, E. Hadjisterkotis, . . . and B. Masala. (2009). The sequence and phylogenesis of the α-globin genes of Barbary sheep (*Ammotragus lervia*), goat (*Capra hircus*), European mouflon (*Ovis aries musimon*) and Cyprus mouflon (*Ovis aries ophion*). *Comparative Biochemistry and Physiology. D, Genomics and Proteomics* 4 (3): 168–73.

Polák, J., and D. Frynta. (2009). Sexual size dimorphism in domestic goats, sheep, and their wild relatives. *Biological Journal of the Linnean Society* 98 (4): 872–83.

Rezaei, H. R., S. Naderi, I. C. Chintauan-Marquier, P. Taberlet, A. T. Virk, H. R. Naghash, . . . and F. Pompanon. (2010). Evolution and taxonomy of the wild species of the genus *Ovis* (Mammalia, Artiodactyla, Bovidae). *Molecular Phylogenetics and Evolution* 54 (2): 315–26.

Robinson, M. R., and L. E. Kruuk. (2007). Function of weaponry in females: The use of horns in intrasexual competition for resources in female Soay sheep. *Biology Letters* 3 (6): 651–54.

Ropiquet, A., and A. Hassanin. (2005). Molecular phylogeny of caprines (Bovidae, Antilopinae): The question of their origin and diversification during the Miocene. *Journal of Zoological Systematics and Evolutionary Research* 43 (1): 49–60.

Ryder, M. L. (1964). The history of sheep breeds in Britain (continued). *Agricultural History Review* 12 (2): 65–82.

Ryder, M. L. (1981). A survey of European primitive breeds of sheep. *Annales de Génétique et de Sélection Animale* 13: 381–418.

Ryder, M. L. (1983). A re-assessment of Bronze-Age wool. *Journal of Archaeological Science* 10 (4): 327–31.

Sanchez Belda, A., and S. Trujillano. (1979). *Spanish Breeds of Sheep.* Madrid: Publicaciones de Extension Agraria.

Shea, J. J. (1998). Neandertal and early modern human behavioral variability: A regional-scale approach to lithic evidence for hunting in the Levantine Mousterian. *Current Anthropology* 39 (S1): S45–78.

Sherratt, A. (1981). *Plough and Pastoralism: Aspects of the Secondary Products Revolution.* New York: Cambridge University Press.

Singh, S., S. Kumar Jr., A. P. Kolte, and S. Kumar. (2013). Extensive variation and sub-structuring in lineage A mtDNA in Indian sheep: Genetic evidence for domestication of sheep in India. *PLoS One* 8 (11): e77858.

Slijper, E. (1942). Biologic-anatomical investigations on the bipedal gait and upright posture in mammals, with special reference to a little goat, born without forelegs. *Proceedings of the Koninklijke Nederlandsche Akademie van Wetenschappen* 45: 288–95.

Smith, P., and L. K. Horwitz. (1984). Radiographic evidence for changing patterns of animal exploitation in the Southern Levant. *Journal of Archaeological Science* 11 (6): 467–75.

Speth, J. D., and E. Tchernov. (2001). "Neandertal Hunting and Meat-Processing in the Near East." In *Meat-Eating and Human Evolution,* edited by C. B. Stanford and H. T. Bunn, 52–72. Oxford: Oxford University Press.

Tapio, M., M. Ozerov, I. Tapio, M. A. Toro, N. Marzanov, M. Ćinkulov, . . . and J. Kantanen. (2010). Microsatellite-based genetic diversity and population structure of domestic sheep in northern Eurasia. *BMC Genetics* 11: 76.

Tresset, A., and J.-D. Vigne. (2011). Last hunter-gatherers and first farmers of Europe. *Comptes Rendus Biologies* 334 (3): 182–89.

Vigne, J. D. (2011). The origins of animal domestication and husbandry: A major change in the history of humanity and the biosphere. *Comptes Rendus Biologies* 334 (3) 171–81.

Vigne, J. D., I. Carrère, F. Briois, and J. Guilaine. (2011). The early process of mammal domestication in the Near East: New evidence from the Pre-Neolithic and Pre-Pottery Neolithic in Cyprus. *Current Anthropology* 52 (suppl. 4): S255–71.

West-Eberhard, M. J. (2003). *Developmental Plasticity and Evolution.* New York: Oxford University Press.

Zeder, M. A. (1999). Animal domestication in the Zagros: A review of past and current research. *Paléorient* 25 (2): 11–25.

Zeder, M. A. (2006). "Archaeological Approaches to Documenting Animal Domestication." In *Documenting Domestication: New Genetic and Archaeological Paradigms,* edited by M. A. Zeder, D. Bradley, E. Emshwiller, and B. D. Smith, 171–81. Berkeley: University of California Press.

Zeder, M. A. (2008). Domestication and early agriculture in the Mediterranean Basin: Origins, diffusion, and impact. *Proceedings of the National Academy of Sciences USA* 105: 11597–604.

Zeder, M. A. (2011). The origins of agriculture in the Near East. *Current Anthropology* 52 (S4): S221–35.

Zeder, M. A. (2012). "Pathways to Animal Domestication." In *Biodiversity in Agriculture: Domestication, Evolution, and Sustainability*, edited by P. Gepts, T. R. Famula, R. L. Bettinger, S. B. Brush, A. B. Damania, P. E. McGuire, and C. O. Qualset, 227. Cambridge: Cambridge University Press.

Zeder, M. A., and B. Hesse. (2000). The initial domestication of goats (*Capra hircus*) in the Zagros Mountains 10,000 years ago. *Science* 287: 2254–57.

Zeder, M. A., B. D. Smith, and D. G. Bradley. (2006). Documenting domestication: The intersection of genetics and archaeology. *Trends in Genetics* 22 (3): 139–55.

Zohary, D., E. Tchernov, and L. Horwitz. (1998). The role of unconscious selection in the domestication of sheep and goats. *Journal of Zoology* 245: 129–35.

Chapter 9: REINDEER

NOTES

1. Burrows and Wallace 1999.
2. Irving and de Bonniville 1884.
3. Moore 1934.
4. Allen 1996, reviewed in Dugan 2008.
5. Yamin-Pasternak 2010.
6. Fitzhugh 2009; see DePriest and Beaubien 2011 for a review.
7. Vitebsky 2005.
8. See Vitebsky 2005 for a good discussion of megaliths.
9. This early deer had antlers (Pitra et al. 2004).
10. For example, Markusson and Folstad 1997.
11. Zhang and Zhang 2012.
12. Meiri et al. 2014; Guthrie 1968.
13. Lichens are generally toxic to humans; though not toxic for reindeer, they are not very nutritious and generally are consumed in lieu of better forage (Heggberget, Gaare, and Ball 2010; Airaksinen et al. 1986).
14. This practice may go back to the Neanderthals; Buck and Stringer (forthcoming).
15. Rue 2004.
16. Timisjärvi, Nieminen, and Sippola 1984.
17. Hogg et al. 2011; but according to Douglas and Jeffery 2014, UV perception is not that rare in mammals.
18. Hogg et al. 2011. For seasonal changes in retinal sensitivity to UV, see Stokkan et al. 2013.
19. Hogg et al. 2011.
20. Bahn 1977.

21. For the role of reindeer meat in the Neanderthal diet, see Enloe 2003; for reindeer in the diet of humans (Magdalenian culture), see Patou-Mathis 2000.

22. Straus 1996. See Bricker, Mellars, and Peterkin 1993 for atlatl use on reindeer hunts.

23. Hayden et al. 1981.

24. Straus 1987.

25. Kuntz and Costamagno 2011.

26. Baskin 2010.

27. Flagstad and Røed 2003.

28. Some describe reindeer as "semidomesticated"—that is, not fully domesticated.

29. Mirov 1945; Aronsson 1991.

30. Røed et al. 2008.

31. Ingold 1980; Grøn 2011.

32. Grøn 2011.

33. Mirov 1945; Vitebsky 2005.

34. Mirov 1945.

35. Mirov 1945; Laufer 1917.

36. Cronin et al. 1995.

37. Brännlund and Axelsson 2011.

38. Forbes et al. 2009; Forbes and Kumpula 2009; Krupnik 2000.

39. Ingold 1986; Müller-Willie et al. 2006.

40. Baskin 2000, 2010.

41. Vitebsky 2005.

42. Mirov 1945.

43. Laufer 1917; see also Mirov 1945 and Gordon 2003.

44. Reviewed in Mirov 1945; Storli 1996.

45. Røed et al. 2008.

46. Ibid.

47. Røed et al. 2011.

48. Mazullo 2010.

49. Puputti and Niskanen 2009.

50. Rødven et al. 2009; Baskin 2010.

51. Andrén et al. 2006; Tveraa et al. 2003.

52. Folstad and Karter 1992.

53. Oksanen et al. 1992.

54. Rødven et al. 2009.

55. Ibid.

56. R. Harris, "The Deer That Reigns," *Cultural Survival Quarterly* 31.3 (Fall 2007), http://www.culturalsurvival.org/publications/cultural-survival-quarterly/finland/deer-reigns.

57. For general discussion of the relationship between sexual selection and antler size, see Clutton-Brock 1982 and Caro et al. 2003; for sexual selection in reindeer, see Røed et al. 2002.

58. It would be interesting, in this regard, to compare the reindeer of the Evenki, which are among the most domesticated, to those of the Sami.

59. Ashman 2003; Chenoweth, Rundle, and Blows 2008; Francis 2004.

60. Berglund 2013; Stankowich and Caro 2009; Packer 1983; Estes 1991.
61. Lincoln 1992. Alternatively, female antlers have been hypothesized to function in their competition with young males over sparse winter forage (Espmark 1964; Holand et al. 2004.).
62. Jacobsen, Colman, and Reimers (2011) suggest that antlerless female populations evolved through genetic drift. According to Weladji et al. 2005, antlerlessness is a function of poor environmental conditions. See also Cronin, Haskell, and Ballard 2010; Reimers 2011.
63. Reimers, Røed, and Colman 2012.
64. Vitebsky 2005.

REFERENCES

Airaksinen, M. M., P. Peura, L. Ala-Fossi-Salokangas, S. Antere, J. Lukkarinen, M. Saikkonen, and F. Stenbäck. (1986). Toxicity of plant material used as emergency food during famines in Finland. *Journal of Ethnopharmacology* 18 (3): 273–96.

Allen, W. (1996). "Shamanic Manipulation of Conspecifics: An Analysis of the Prehistory and Ethnohistory of Hallucinogens and Psychological Legerdemain." In *Foods of the Gods: Eating and the Eaten in Fantasy and Science Fiction*, edited by G. Westfahl, G. Slusser, and E. S. Rabkin, 39. Athens: University of Georgia Press.

Andrén, H., J. D. Linnell, O. Liberg, R. Andersen, A. Danell, J. Karlsson, . . . and T. Kvam. (2006). Survival rates and causes of mortality in Eurasian lynx (*Lynx lynx*) in multi-use landscapes. *Biological Conservation* 131 (1): 23–32.

Aronsson, K.-Å. (1991). *Forest Reindeer Herding AD 1–1800*. Archaeology and Environment 10. Umeå, Sweden: Umeå University.

Ashman, T. L. (2003). Constraints on the evolution of males and sexual dimorphism: Field estimates of genetic architecture of reproductive traits in three populations of gynodioecious *Fragaria virginiana*. *Evolution* 57 (9): 2012–25.

Bahn, P. G. (1977). Seasonal migration in south-west France during the late glacial period. *Journal of Archaeological Science* 4 (3): 245–57.

Baskin, L. M. (2000). Reindeer husbandry/hunting in Russia in the past, present and future. *Polar Research* 19 (1): 23–29.

Baskin, L. M. (2010). Differences in the ecology and behaviour of reindeer populations in the USSR. *Rangifer* 6 (2): 333–40.

Berglund, A. (2013). Why are sexually selected weapons almost absent in females? *Current Zoology* 59 (4): 564–68.

Brännlund, I., and P. Axelsson. (2011). Reindeer management during the colonization of Sami lands: A long-term perspective of vulnerability and adaptation strategies. *Global Environmental Change* 21 (3): 1095–105.

Bricker, H. M., P. Mellars, and G. L. Peterkin. (1993). Introduction: The study of Palaeolithic and Mesolithic hunting. *Archeological Papers of the American Anthropological Association* 4 (1): 1–9.

Buck, L. T., and C. B. Stringer. Forthcoming. Having the stomach for it: A contribution to Neanderthal diets? *Quaternary Science Reviews*.

Burrows, E. G., and M. Wallace. (1999). *Gotham: A History of New York City to 1898*. New York: Oxford University Press.

Caro, T., C. Graham, C. Stoner, and M. Flores. (2003). Correlates of horn and antler shape in bovids and cervids. *Behavioral Ecology and Sociobiology* 55 (1): 32–41.

Chenoweth, S. F., H. D. Rundle, and M. W. Blows. (2008). Genetic constraints and the evolution of display trait sexual dimorphism by natural and sexual selection. *American Naturalist* 171 (1): 22–34.

Clutton-Brock, T. (1982). The functions of antlers. *Behavior* 79: 108–25.

Cronin, M. A., S. P. Haskell, and W. B. Ballard. (2010). The frequency of antlerless female caribou and reindeer in Alaska. *Rangifer* 23 (2): 67–70.

Cronin, M. A., L. Renecker, B. J. Pierson, and J. C. Patton. (1995). Genetic variation in domestic reindeer and wild caribou in Alaska. *Animal Genetics* 26 (6): 427–34.

DePriest, P. T., and H. F. Beaubien. (2011). "Case Study: Deer Stones of Mongolia after Three Millennia." In *Biocolonization of Stone: Control and Preventive Methods: Proceedings from the MCI Workshop Series*, edited by A. E. Charola, C. McNamara, and R. J. Koestler, 103–8. Smithsonian Contributions to Museum Conservation 2. Washington, DC: Smithsonian Institution Scholarly Press.

Douglas, R., and G. Jeffery. (2014). The spectral transmission of ocular media suggests ultraviolet sensitivity is widespread among mammals. *Proceedings of the Royal Society. B: Biological Sciences* 281 (1780): 20132995.

Dugan, F. M. (2008). Fungi, folkways and fairy tales: Mushrooms and mildews in stories, remedies and rituals, from Oberon to the Internet. *North American Fungi* 3: 23–72.

Enloe, J. G. (2003). Acquisition and processing of reindeer in the Paris Basin. *BAR International Series* 1144: 23–32.

Espmark, Y. (1964). Studies in dominance-subordination relationship in a group of semi-domestic reindeer (*Rangifer tarandus* L.). *Animal Behaviour* 12 (4): 420–26.

Estes, R. D. (1991). The significance of horns and other male secondary sexual characters in female bovids. *Applied Animal Behaviour Science* 29 (1): 403–51.

Fitzhugh, W. W. (2009). Stone shamans and flying deer of northern Mongolia: Deer goddess of Siberia or chimera of the steppe? *Arctic Anthropology* 46 (1–2): 72–88.

Flagstad, Ø., and K. H. Røed. (2003). Refugial origins of reindeer (*Rangifer tarandus* L.) inferred from mitochondrial DNA sequences. *Evolution* 57 (3): 658–70.

Folstad, I., and A. Karter. (1992). Parasites, bright males, and the immunocompetence handicap. *American Naturalist* 139: 603–22.

Forbes, B. C., and T. Kumpula. (2009). The ecological role and geography of reindeer (*Rangifer tarandus*) in northern Eurasia. *Geography Compass* 3 (4): 1356–80.

Forbes, B. C., F. Stammler, T. Kumpula, N. Meschtyb, A. Pajunen, and E. Kaarlejärvi. (2009). High resilience in the Yamal-Nenets social-ecological system, West Siberian Arctic, Russia. *Proceedings of the National Academy of Sciences USA* 106 (52): 22041–48.

Francis, R. C. (2004). *Why Men Won't Ask for Directions: The Seductions of Sociobiology*. Princeton, NJ: Princeton University Press.

Gordon, B. (2003). Rangifer and man: An ancient relationship. *Rangifer* 23 (special issue no. 14).

Grøn, O. (2011). Reindeer antler trimming in modern large-scale reindeer pastoralism and parallels in an early type of hunter-gatherer reindeer herding system: Evenk ethnoarchaeology in Siberia. *Quaternary International* 238 (1–2): 76–82.

Guthrie, R. D. (1968). Paleoecology of the large-mammal community in interior Alaska during the late Pleistocene. *American Midland Naturalist* 79 (2): 346–63.

Hassanin, A., and E. J. Douzery. (2003). Molecular and morphological phylogenies of Ruminantia and the alternative position of the Moschidae. *Systematic Biology* 52 (2): 206–28.

Hayden, B., S. Bowdler, K. W. Butzer, M. N. Cohen, M. Druss, R. C. Dunnell, . . . and J. Kamminga. (1981). Research and development in the Stone Age: Technological transitions among hunter-gatherers [and comments and reply]. *Current Anthropology* 22 (5): 519–48.

Heggberget, T. M., E. Gaare, and J. P. Ball. (2010). Reindeer (*Rangifer tarandus*) and climate change: Importance of winter forage. *Rangifer* 22 (1): 13–31.

Hogg, C., M. Neveu, K.-A. Stokkan, L. Folkow, P. Cottrill, R. Douglas, . . . and G. Jeffery. (2011). Arctic reindeer extend their visual range into the ultraviolet. *Journal of Experimental Biology* 214 (12): 2014–19.

Holand, Ø., H. Gjøstein, A. Losvar, J. Kumpula, M. Smith, K. Røed, . . . and R. Weladji. (2004). Social rank in female reindeer (*Rangifer tarandus*): Effects of body mass, antler size and age. *Journal of Zoology* 263 (4): 365–72.

Ingold, T. (1980). *Hunters, Pastoralists, and Ranchers: Reindeer Economies and Their Transformation.* Cambridge: Cambridge University Press.

Ingold, T. (1986). Reindeer economies: And the origins of pastoralism. *Anthropology Today* 2 (4): 5–10.

Irving, W., and W. L. E. de Bonniville. (1884). *Knickerbocker History of New York.* New York: Lovell.

Jacobsen, B. W., J. E. Colman, and E. Reimers. (2011). The frequency of antlerless females among Svalbard reindeer. *Rangifer* 18 (2): 81–84.

Krupnik, I. (2000). Reindeer pastoralism in modern Siberia: Research and survival during the time of crash. *Polar Research* 19 (1): 49–56.

Kuntz, D., and S. Costamagno. (2011). Relationships between reindeer and man in southwestern France during the Magdalenian. *Quaternary International* 238 (1–2): 12–24.

Laufer, B. (1917). *The Reindeer and Its Domestication.* Lancaster, PA: Corinthian Press.

Lincoln, G. A. (1992). Biology of antlers. *Journal of Zoology* 226 (3): 517–28.

Markusson, E., and I. Folstad. (1997). Reindeer antlers: Visual indicators of individual quality? *Oecologia* 110 (4): 501–7.

Mazzullo, N. (2010). "More than Meat on the Hoof? Social Significance of Reindeer among Finnish Saami in a Rationalized Pastoralist Economy." In *Good to Eat, Good to Live With: Nomads and Animals in Northern Eurasia and Africa,* edited by F. Stammler and H. Takakura, 101–22. Northeast Asian Study Series 11. Sendai, Japan: Center for Northeast Asia Studies, Tohoku University.

Meiri, M., A. M. Lister, M. J. Collins, N. Tuross, T. Goebel, S. Blockley, . . . and I. Barnes. (2014). Faunal record identifies Bering isthmus conditions as constraint to end-Pleistocene migration to the New World. *Proceedings of the Royal Society. B: Biological Sciences* 281 (1776): 20132067.

Mirov, N. (1945). Notes on the domestication of reindeer. *American Anthropologist* 47 (3): 393–408.

Moore, C. C. (1934). *The Night before Christmas: (A Visit from St. Nicholas).* New York: Courier Dover.

Müller-Wille, L., D. Heinrich, V.-P. Lehtola, P. Aikio, Y. Konstantinov, and V. Vladimirova. (2006). "Dynamics in Human-Reindeer Relations: Reflections on Prehistoric, Historic and Contemporary Practices in Northernmost Europe." In *Reindeer Management in Northernmost Europe*, edited by B. C. Forbes, M. Bölter, L. Müller-White, J. Hukkinen, F. Müller, N. Gunslay, and Y. Konstantinov, 27–45. Ecological Studies 184. Berlin: Springer.

Oksanen, A., M. Nieminen, T. Soveri, and K. Kumpula. (1992). Oral and parenteral administration of ivermectin to reindeer. *Veterinary Parasitology* 41 (3): 241–47.

Packer, C. (1983). Sexual dimorphism: The horns of African antelopes. *Science* 221 (4616): 1191–93.

Patou-Mathis, M. (2000). Neanderthal subsistence behaviours in Europe. *International Journal of Osteoarchaeology* 10 (5): 379–95.

Pitra, C., J. Fickel, E. Meijaard, and C. Groves. (2004). Evolution and phylogeny of Old World deer. *Molecular Phylogenetics and Evolution* 33 (3): 880–95.

Puputti, A.-K., and M. Niskanen. (2009). Identification of semi-domesticated reindeer (*Rangifer tarandus tarandus*, Linnaeus 1758) and wild forest reindeer (*Rt fennicus*, Lönnberg 1909) from postcranial skeletal measurements. *Mammalian Biology—Zeitschrift für Säugetierkunde* 74 (1): 49–58.

Reimers, E. (2011). Antlerless females among reindeer and caribou. *Canadian Journal of Zoology* 71 (7): 1319–25.

Reimers, E., K. H. Røed, and J. E. Colman. (2012). Persistence of vigilance and flight response behaviour in wild reindeer with varying domestic ancestry. *Journal of Evolutionary Biology* 25 (8): 1543–54.

Rødven, R., I. Männikkö, R. A. Ims, N. G. Yoccoz, and I. Folstad. (2009). Parasite intensity and fur coloration in reindeer calves—Contrasting artificial and natural selection. *Journal of Animal Ecology* 78 (3): 600–607.

Røed, K. H., Ø. Flagstad, G. Bjørnstad, and A. K. Hufthammer. (2011). Elucidating the ancestry of domestic reindeer from ancient DNA approaches. *Quaternary International* 238 (1–2): 83–88.

Røed, K. H., Ø. Flagstad, M. Nieminen, Ø. Holand, M. J. Dwyer, N. Røv, and C. Vilà, C. (2008). Genetic analyses reveal independent domestication origins of Eurasian reindeer. *Proceedings of the Royal Society. B: Biological Sciences* 275 (1645): 1849–55.

Røed, K. H., O. Holand, M. E. Smith, H. Gjøstein, J. Kumpula, and M. Niemenen. (2002). Reproductive success in reindeer males in a herd with varying sex ratio. *Molecular Ecology* 11 (7): 1239–43.

Rue, L. L. 2004. *The Encyclopedia of Deer: Your Guide to the World's Deer Species Including Whitetails, Mule Deer, Caribou, Elk, Moose and More*. Minneapolis, MN: Voyageur Press.

Stankowich, T., and T. Caro. (2009). Evolution of weaponry in female bovids. *Proceedings of the Royal Society. B: Biological Sciences* 276 (1677): 4329–34.

Stokkan, K.-A., L. Folkow, J. Dukes, M. Neveu, C. Hogg, S. Siefken, . . . and G. Jeffery. (2013). Shifting mirrors: Adaptive changes in retinal reflections to winter darkness in Arctic reindeer. *Proceedings of the Royal Society. B: Biological Sciences* 280 (1773): 20132451.

Storli, I. (1996). On the historiography of Sami reindeer pastoralism. *Acta Borealia* 13 (1): 81–115.

Straus, L. G. (1987). "Hunting in Late Upper Paleolithic Western Europe." In *The Evolution of Human Hunting*, edited by M. H. Nitecki and D. V. Nitecki, 147–76. New York: Plenum.

Straus, L. G. (1996). "The Archaeology of the Pleistocene–Holocene Transition in Southwest Europe." In *Humans at the End of the Ice Age: The Archaeology of the Pleistocene–Holocene Transition*, edited by L. G. Strauss, B. V. Eriksen, J. M. Erlandson, and D. R. Yesner, 83–99. New York: Plenum.

Timisjärvi, J., M. Nieminen, and A.-L. Sippola. (1984). The structure and insulation properties of the reindeer fur. *Comparative Biochemistry and Physiology. A, Physiology* 79 (4): 601–9.

Tveraa, T., P. Fauchald, C. Henaug, and N. G. Yoccoz. (2003). An examination of a compensatory relationship between food limitation and predation in semi-domestic reindeer. *Oecologia* 137 (3): 370–76.

Vitebsky, P. (2005). *The Reindeer People: Living with Animals and Spirits in Siberia*. Boston: Houghton Mifflin.

Weladji, R., Ø. Holand, G. Steinheim, J. Colman, H. Gjøstein, and A. Kosmo. (2005). Sexual dimorphism and intercohort variation in reindeer calf antler length is associated with density and weather. *Oecologia* 145 (4): 549–55.

Yamin-Pasternak, S. (2010). Shroom: A cultural history of the magic mushroom. *Ethnobiology Letters* 1: 26–27.

Zhang, W. Q., and M. H. Zhang. (2012). Phylogeny and evolution of Cervidae based on complete mitochondrial genomes. *Genetics and Molecular Research: GMR* 11 (1): 628–35.

Chapter 10: CAMELS

NOTES

1. Bulliet 1990.
2. Prothero 2009.
3. Janis, Theodor, and Boisvert 2002.
4. Webb 1977.
5. But see Cui et al. 2007.
6. Ibid.
7. For early dates of camel domestication, see Frifelt 1990, 1991. For later dates, see M. Uerpmann and Uerpmann 2012; H.-P. Uerpmann 1999.
8. M. Uerpmann and Uerpmann 2012 argue the latter; see also Bulliet 1990 for milk-based theory of camel domestication.
9. See Sapir-Hen and Ben-Yosef 2013 for the role of camels in hauling copper; Grigson 2012; Ben-Yosef et al. 2012.
10. Farrokh 2007.
11. Nabhan 2008.
12. Drucker, Edwards, and Saalfeld 2010; see also Edwards, Saalfeld, and Clifford 2005.
13. Khalaf 1999, 2000.
14. Wani et al. 2010.
15. See Bener et al. 2005 for camel jockey injuries; see also Caine and Caine 2005; Tinson et al. 2007.

16. Bulliet 1990.
17. Sharp 2012; see also Nagy, Skidmore, and Juhasz 2013 for camel milk.
18. Raziq and Younas 2006.
19. Al-Swailem et al. 2007; white camels are also bred in the United States ("White Camel Breeding Program," Lost World Ranch, http://lostworldranch.com/white-camel-breeding-program.php).
20. Abdallah and Faye 2012.
21. Peters 1997; Gauthier-Pilters and Dagg 1981.
22. See Uerpmann and Uerpmann 2012 for a nice series from Timna, Israel; see also Grigson 2012; Ben-Yosef et al. 2012.
23. See, for example, Wardeh 2004.
24. Al-Swailem et al. 2007.
25. Leese 1927.
26. Köhler-Rollefson 1993a.
27. Wardeh 2004.
28. Al-Swailem et al. 2007.
29. Kuz'mina 2008.
30. Ibid.
31. Schmidt-Nielsen 1964.
32. Al-Ali, Husayni, and Power 1988.
33. Aristotle, *Historia Animalium* 2.1.498b9.
34. Cui et al. 2007; but see Trinks et al. 2012 for a more Western origin; Schaller 1998.
35. Potts 2004.
36. Ibid.
37. See Peters and von den Driesch 1997.
38. Potts 2004.
39. There are reportedly 400–500 Bactrian camels left in China (Hare 1997).
40. Nowak et al. 1997. There is an important caveat in comparing domestic Bactrians to the surviving wild Bactrians. None of the widely scattered wild Bactrian populations evidence a recent ancestral relationship to domestic Bactrians. There is a rather large genomic difference between wild and domestic Bactrians (about 3 percent), which indicates that the wild ancestors of domestic camels may have been of a different subspecies than extant wild Bactrians. Hence, the wild ancestors of domestic Bactrians may have more closely resembled the domestic Bactrians living today than do the wild Bactrians of today.
41. Irwin 2010
42. Faye 2013.

REFERENCES

Abdallah, H., and B. Faye. (2012). Phenotypic classification of Saudi Arabian camel (*Camelus dromedarius*) by their body measurements. *Emirates Journal of Food and Agriculture* 24 (3): 272–80.

Al-Ali, A., H. Husayni, and D. Power. (1988). A comprehensive biochemical analysis of the blood of the camel (*Camelus dromedarius*). *Comparative Biochemistry and Physiology. B, Comparative Biochemistry* 89 (1): 35–37.

Al-Swailem, A. M., K. A. Al-Busadah, M. M. Shehata, I. O. Al-Anazi, and E. Askari. (2007). Classification of Saudi Arabian camel (*Camelus dromedarius*) subtypes based on RAPD technique. *Journal of Food, Agriculture & Environment* 5 (1): 143.

Andersson, L. S., M. Larhammar, F. Memic, H. Wootz, D. Schwochow, C.-J. Rubin, . . . and G. Hjälm. (2012). Mutations in DMRT3 affect locomotion in horses and spinal circuit function in mice. *Nature* 488 (7413): 642–46.

Becker, A.-C., K. Stock, and O. Distl. (2011). Genetic correlations between free movement and movement under rider in performance tests of German Warmblood horses. *Livestock Science* 142 (1): 245–52.

Ben-Yosef, E., R. Shaar, L. Tauxe, and H. Ron. (2012). A new chronological framework for Iron Age copper production at Timna (Israel). *Bulletin of the American Schools of Oriental Research* 367: 31–71.

Bener, A., F. H. Al-Mulla, S. M. Al-Humoud, and A. Azhar. (2005). Camel racing injuries among children. *Clinical Journal of Sport Medicine* 15 (5): 290–93.

Bulliet, R. W. (1990). *The Camel and the Wheel*. New York: Columbia University Press.

Caine, D., and C. Caine. (2005). Child camel jockeys: A present-day tragedy involving children and sport. *Clinical Journal of Sport Medicine* 15 (5): 287–89.

Cui, P., R. Ji, F. Ding, D. Qi, H. Gao, H. Meng, . . . and H. Zhang. (2007). A complete mitochondrial genome sequence of the wild two-humped camel (*Camelus bactrianus ferus*): An evolutionary history of Camelidae. *BMC Genomics* 8: 241.

Dagg, A. I. (1974). The locomotion of the camel (*Camelus dromedarius*). *Journal of Zoology* 174 (1): 67–78.

Drucker, A. G., G. P. Edwards, and W. K. Saalfeld. (2010). Economics of camel control in central Australia. *Rangeland Journal* 32 (1): 117–27.

Edwards, G. P., K. Saalfeld, and B. Clifford. (2005). Population trend of feral camels in the Northern Territory, Australia. *Wildlife Research* 31 (5): 509–17.

Farrokh, K. (2007). *Shadows in the Desert: Ancient Persia at War*. Oxford: Osprey.

Faye, B. (2013). "Classification, History and Distribution of the Camel." In *Camel Meat and Meat Products*, edited by I. T. Kadim, O. Mahgoub, B. Faye, and M. M. Farouk, 1–6. Oxford: CAB International.

Frifelt, K. (1990). A third millennium kiln from the Oman Peninsula. *Arabian Archaeology and Epigraphy* 1 (1): 4–15.

Frifelt, K. (1991). The Island of Umm an-Nar, vol. 1, *Third Millenium Graves*. Jutland Archaeological Society Publications. Århus, Denmark: Århus University Press.

Gauthier-Pilters, H., and A. I. Dagg. (1981). *The Camel, Its Evolution, Ecology, Behavior, and Relationship to Man*. Chicago: University of Chicago Press.

Grigson, C. (2012). Camels, copper and donkeys in the early iron age of the Southern Levant: Timna revisited. *Levant* 44 (1): 82–100.

Hare, J. (1997). The wild Bactrian camel *Camelus bactrianus ferus* in China: The need for urgent action. *Oryx* 31 (1): 45–48.

Irwin, R. (2010). *Camel*. London: Reaktion.

Janis, C. M., J. M. Theodor, and B. Boisvert. (2002). Locomotor evolution in camels revisited: A quantitative analysis of pedal anatomy and the acquisition of the pacing gait. *Journal of Vertebrate Paleontology* 22 (1): 110–21.

Khalaf, S. (1999). Camel racing in the Gulf: Notes on the evolution of a traditional cultural sport. *Anthropos* 94 (1–3): 85–106.

Khalaf, S. (2000). Poetics and politics and newly invented traditions in the Gulf: Camel racing in the United Arab Emirates. *Ethnology* 39 (3): 243–61.

Köhler-Rollefson, I. (1993a). About camel breeds: A reevaluation of current classification systems. *Journal of Animal Breeding and Genetics* 110 (1–6): 66–73.

Kuz'mina, E. E. (2008). *The Prehistory of the Silk Road.* Philadelphia: University of Pennsylvania Press.

Leese, S. L. (1927). *A Treatise on the One-Humped Camel: In Health and Disease.* Stamford, England: Haynes and Sons.

Nabhan, G. P. (2008). Camel whisperers: Desert nomads crossing paths. *Journal of Arizona History* 49 (2): 95–118.

Nagy, P., J. Skidmore, and J. Juhasz. (2013). Use of assisted reproduction for the improvement of milk production in dairy camels (*Camelus dromedarius*). *Animal Reproduction Science* 136 (3): 205–10.

Nowak, M. A., M. C. Boerlijst, J. Cooke, and J. M. Smith. (1997). Evolution of genetic redundancy. *Nature* 387: 167–71.

Peters, J. (1997). The dromedary: Ancestry, history of domestication and medical treatment in early historic times. *Tierärztliche Praxis. Ausgabe G, Grosstiere/Nutztiere* 25 (6): 559–65.

Peters, J., and A. von den Driesch. (1997). The two-humped camel (*Camelus bactrianus*): New light on its distribution, management and medical treatment in the past. *Journal of Zoology* 242 (4): 651–79.

Pfau, T., E. Hinton, C. Whitehead, A. Wiktorowicz-Conroy, and J. R. Hutchinson. (2011). Temporal gait parameters in the alpaca and the evolution of pacing and trotting locomotion in the Camelidae. *Journal of Zoology* 283 (3): 193–202.

Potts, D. (2004). Camel hybridization and the role of *Camelus bactrianus* in the ancient Near East. *Journal of the Economic and Social History of the Orient* 47 (2): 143–65.

Promerová, M., L. S. Andersson, R. Juras, M. C. T. Penedo, M. Reissmann, T. Tozaki, . . . and L. Andersson. (2014). Worldwide frequency distribution of the "*Gait keeper*" mutation in the *DMRT3* gene. *Animal Genetics* 45 (2): 274–82.

Prothero, D. R. (2009). Evolutionary transitions in the fossil record of terrestrial hoofed mammals. *Evolution: Education and Outreach* 2 (2): 289–302.

Raziq, A., and M. Younas. (2006). White camels of Balochistan. *Science International (Lahore)* 18 (1): 47.

Sapir-Hen, L., and E. Ben-Yosef. (2013). The introduction of domestic camels to the Southern Levant: Evidence from the Aravah Valley. *Tel Aviv* 40 (2): 277–85.

Schaller, G. B. (1998). *Wildlife of the Tibetan Steppe.* Chicago: University of Chicago Press.

Schmidt-Nielsen, K. (1964). *Desert Animals: Physiological Problems of Heat and Water.* Oxford: Clarendon.

Sharp, N. C. (2012). Animal athletes: A performance review. *Veterinary Record* 171 (4): 87–94.

Tinson, A., K. Kuhad, R. Sambyal, A. Rehman, and J. Al-Masri. (2007). "Evolution of Camel Racing in the United Arab Emirates, 1987–2007." In *Proceedings of the International Camel Conference "Recent Trends in Camelids Research and Future Strategies for Saving Camels,"* Rajasthan, India, 16–17 February 2007, edited by T. Gahlot, 144–50.

Trinks, A., P. Burger, N. Benecke, and J. Burger. (2012). Bactrian camels (*Camelus bactrianus*): Ancient DNA reveals domestication process: The case of the two-humped camel. In *Camels in Asia and North Africa*, edited by E.-M. Knoll and P. Burger, 79–86. Veröffentlichungen zur Sozialanthropologie 18. Vienna: Austrian Academy of Sciences.

Uerpmann, H.-P. (1999). Camel and horse skeletons from protohistoric graves at Mleiha in the Emirate of Sharjah (UAE). *Arabian Archaeology and Epigraphy* 10 (1): 102–18.

Uerpmann, M., and H.-P. Uerpmann. (2012). "Archeozoology of Camels in South-Eastern Arabia." In *Camels in Asia and North Africa*, edited by E.-M. Knoll and P. Burger, 109–22. Veröffentlichungen zur Sozialanthropologie 18. Vienna: Austrian Academy of Sciences Press.

Wani, N. A., U. Wernery, F. Hassan, R. Wernery, and J. Skidmore. (2010). Production of the first cloned camel by somatic cell nuclear transfer. *Biology of Reproduction* 82 (2): 373–79.

Wardeh, M. (2004). Classification of the dromedary camels. *Journal of Camel Science* 1: 1–7.

Webb, S. D. (1972). Locomotor evolution in camels. *Forma et Function* 5: 99–112.

Webb, S. D. (1977). A history of savanna vertebrates in the New World. Part I: North America. *Annual Review of Ecology and Systematics* 8 (1): 355–80.

Zwart, M. (2012). The gene that gates horse gaits. *Journal of Experimental Biology* 215 (23): v–vi.

Chapter 11: HORSES

NOTES

1. The horse was an irascible black Quarter Horse stallion named Prince, which, for her amusement, she insisted I ride bareback, with predictable results.
2. Anthony 2009.
3. Levine 1999; Olsen 2006; Outram et al. 2009.
4. Thieme 2005.
5. Valladas et al. 2001; Pruvost et al. 2011.
6. Anthony 2009.
7. Levine 1998.
8. The term "tarpan" is derived from a Turkic word for wild horse.
9. Froehlich 1999; Gingerich 1991; Prothero 1993.
10. Steiner and Ryder 2011.
11. Shorrocks 2007. The ability of horses to survive on grass is also greatly aided by more predigestive processing in the form of chewing, which increases the digestibility of grasses. To this end, their large, high-crowned molars and powerful jaws are important adaptations.
12. See MacFadden et al. 2012 for a discussion of traditional museum presentations of horse evolution.
13. Weinstock et al. 2005.
14. Shoemaker and Clauset 2014; Kurtén 1980.
15. As in the extinction of other North American megafauna, there is a debate about the relative role of human predation and climate change (Grund, Surovell, and Lyons 2012; Prescott et al. 2012a; Lorenzen et al. 2011; and Koch and Barnosky 2006).

16. Mihlbachler et al. 2011.

17. Orlando et al. 2009; Weinstock et al. 2005.

18. Azzaroli 1983.

19. Clutton-Brock (1992) views the two types of wild horses as subspecies; Groves 1994 considers them two distinct species. The Mongolian and Eurasian wild horses diverged over 700,000 years ago.

20. Benecke et al. 2006; Sommer et al. 2011; Bendrey 2012. Warmuth et al. (2011) provides evidence of another refugium in Iberia.

21. These steppe peoples were collectively referred to as "Kurgan," a name that derives from the earthen mounds that marked their grave sites (Gimbutas 1990). But this term has fallen out of favor with the increasing recognition of their cultural and ethnic diversity.

22. Levine 1998; Olsen 2006; Anthony 2009.

23. Anthony 2009; Dergachev 1989.

24. Anthony 2009; see also Levine 2005; Bendrey 2011.

25. Bökönyi 1978 (Ukraine); Anthony 2009 (Pontic steppe); Levine 1999, 2005; Olsen 2006; see Bendrey 2012 for an overview.

26. Levine 2005; Outram et al. 2009.

27. Another candidate area for early horse domestication is the region just north of the Caspian and Black Seas, known as the Pontic steppe. Anthony (2009) believes horse management could have commenced there as early as 7000 BP, as evidenced by their sacrifice and burial—along with sheep and cattle—in human graves. Warmuth proposes an independent domestication in Iberia because of the high genetic diversity of horse populations there (Warmuth et al. 2011). Genetic diversity, though, can result from a number of processes, including recent hybridization of distantly related stocks; it is therefore not a reliable indicator of a point of origin for domestication

28. Outram et al. 2009.

29. Brown and Anthony 1998; see also Levine 2005 for horse hunting.

30. By this time horse milk had become an important staple for many steppe cultures.

31. Anthony and Brown 2011.

32. Renfrew 1998; Anthony 2009; N. Kim 2011.

33. Sherratt 1983; Anthony 2009.

34. See Moorey 1986 for early use of chariots by the Assyrians; and Shaughnessy 1988 for evidence that the first domestic horses to arrive in China were attached to chariots.

35. For chariot burials in Scotland, see Carter and Hunter 2003; for burials in eastern Europe, see Kuznetsov 2006; for Chinese burials, see Shaughnessy 1988; and for steppe burials, see Outram et al. 2011.

36. Kelekna 2009; Anthony 2009; Clutton-Brock 1992.

37. McMiken 1990; Archer 2010; Anthony 2009.

38. Warmuth et al. 2012.

39. The deliberate introduction of wild mares into domestic herds helped to maintain hardiness in domestic horses for the harsh steppe environment.

40. Cieslak et al. 2010. This was one way to deal with a trade-off between tameness and hardiness.

41. Lindgren 2004; Wade et al. 2009; Kavar and Dovč 2008.

42. Sherratt 1983. See Zeder 2006 for some qualifications as to the utility of this criterion.

43. Anthony and Brown 2011.

44. Bökönyi 1978, 1991.

45. Anthony 2009; Bökönyi 1987. See also Grigson 1993, for evidence of the first domestic horse in the Levant.

46. Anthony 2013. The famous Buhen horse provides some of the earliest evidence for domestic horses in Egypt (Clutton-Brock 1974; Raulwing and Clutton-Brock 2009).

47. Cai et al. 2009.

48. This would have been just prior to the Shang dynasty.

49. Lei et al. 2009; Warmuth et al. 2012.

50. Groves 1994; Olsen 2006.

51. Pruvost et al. 2011.

52. Hofreiter and Schöneberg 2010.

53. Bökönyi 1987; Groves 1994.

54. Hediger 1981.

55. Krueger et al. 2011; Proops, Walton, and McComb 2010.

56. Ludwig 2009.

57. Cieslak et al. 2011.

58. Bellone 2010.

59. Ultimately, this package deal derives from deep—that is arising early—developmental connections. For example, melanocytes derive from embryonic neural crest cells, which, in addition, differentiate into bone, cartilage, and neurons (Le Douarin and Kalcheim 1999). So any alteration in neural crest development will have a high probability of affecting not only skin coloration but the development of these other tissues as well.

60. It seems more likely to me that the brushy manes function in intermale competition (intrasexual selection) given the frequency of neck bites in these contests and the large male canines. If so, the long manes of domestic horses may reflect, in part, a reduction in sexual selection.

61. Hendricks 2007.

62. Olsen 2006.

63. Skorkowski 1970.

64. P. H. J. Maas, "Recreating Extinct Animals by Selective Breeding," The Sixth Extinction, last modified October 2, 2011, http://extinct.petermaas.nl/extinct/articles/selectivebreeding.htm.

65. Heck 1952.

66. Olsen 2006.

67. Tchernov and Horwitz 1991.

68. See Brooks et al. 2010 for an attempt to quantify size variation among horse breeds.

69. Petersen et al. 2013.

70. Goodwin, Levine, and McGreevy 2008; Mostard et al. 2009.

71. See W. G. Hill and Bunger 2004 for implications of epistasis and other nonadditive genetic effects with regard to long-term selection.

72. Horse people are quick to claim that Thoroughbreds are bred for racing ability, not just speed. But speed is obviously a key component of racing ability.

73. Cigar, considered by many the second-best racehorse ever, was infertile.

74. Morris and Allen 2002. Laing and Leech 1975 (mares).

75. Conversely, some of the most successful studs weren't particularly outstanding racers. This gets to the issue of stud fees and foal prices—in essence, the way the marketplace works in horse racing—which are completely out of whack. That is, the correlation between foal prices and lifetime winnings is extremely low—an indication of a high degree of genetic randomness that the horse market has failed to take on board (Wilson and Rambaut 2008).

76. Wisher, Allen, and Wood 2006; Jeffcott et al. 1982.

77. Langlois 1980; Williamson and Beilharz 1998.

78. Pigliucci and Preston 1984; Denny 2008.

79. Barbaro and Ruffian were two of the most famous victims.

80. Wade et al. 2009.

81. E. W. Hill et al. 2012.

82. H. Kim et al. 2013.

83. E. W. Hill et al. 2012.

84. Binns, Boehler, and Lambert 2010; E. W. Hill et al. 2012; McGivney et al. 2012.

85. Binns, Boehler, and Lambert 2010; E. W. Hill et al. 2012.

86. Eynon et al. 2011.

87. Bower et al. 2012.

88. Ibid.

89. Wade et al. 2009; Petersen et al. 2013.

90. Doan, Cohen, Sawyer, et al. 2012.

91. Doan, Cohen, Harrington, et al. 2012.

REFERENCES

Anthony, D. W. (2009). *The Horse, the Wheel, and Language: How Bronze-Age Riders from the Eurasian Steppes Shaped the Modern World*. Princeton, NJ: Princeton University Press.

Anthony, D. W. (2013). "Horses, Ancient Near East and Pharaonic Egypt." In *The Encyclopedia of Ancient History*, edited by R. S. Bagnall, K. Brodersen, C. B. Champion, A. Erskine, and S. R. Huebner, 3311–14. Malden, MA: Wiley-Blackwell.

Anthony, D. W., and D. R. Brown. (2011). The secondary products revolution, horse-riding, and mounted warfare. *Journal of World Prehistory* 24 (2): 131–60.

Archer, R. (2010). "Chariotry to Cavalry: Developments in the Early First Millennium." In *New Perspectives on Ancient Warfare*, edited by G. G. Fagan and M. Trundle, 57. History of Warfare 59. Leiden, Netherlands: Brill.

Azzaroli, A. (1983). Quaternary mammals and the "end-Villafranchian" dispersal event—A turning point in the history of Eurasia. *Palaeogeography, Palaeoclimatology, Palaeoecology* 44 (1): 117–39.

Bellone, R. R. (2010). Pleiotropic effects of pigmentation genes in horses. *Animal Genetics* 41: 100–110.

Bendrey, R. (2011). Some like it hot: Environmental determinism and the pastoral economies of the later prehistoric Eurasian steppe. *Pastoralism* 1 (1): 1–16.

Bendrey, R. (2012). From wild horses to domestic horses: A European perspective. *World Archaeology* 44 (1): 135–57.

Benecke, N., S. Olsen, S. Grant, A. Choyke, and L. Bartosiewicz. (2006). Late prehistoric exploitation of horses in central Germany and neighboring areas—the archaeozoological record. *BAR International Series* 1560: 195.

Binns, M. M., D. A. Boehler, and D. H. Lambert. (2010). Identification of the myostatin locus (*MSTN*) as having a major effect on optimum racing distance in the Thoroughbred horse in the USA. *Animal Genetics* 41: 154–58.

Bökönyi, S. (1978). The earliest waves of domestic horses in East Europe. *Journal of Indo-European Studies* 6 (1–2): 17–76.

Bökönyi, S. (1987). History of horse domestication. *Animal Genetic Resources Information* 6 (1): 29–34.

Bökönyi, S. (1991). "Late Chalcolithic Horses in Anatolia." In *Equids of the Ancient World*, vol. 2, edited by R. H. Meadow and H.-P. Uerpmann, 123–31. Beihefte zum Tübinger Atlas des Vorderen Orients. Reihe A, Naturwissenschaften, 19/2. Wiesbaden: Reichert.

Bower, M. A., B. A. McGivney, M. G. Campana, J. Gu, L. S. Andersson, E. Barrett, . . . and E. W. Hill. (2012). The genetic origin and history of speed in the Thoroughbred racehorse. *Nature Communications* 3.

Brooks, S. A., S. Makvandi-Nejad, E. Chu, J. J. Allen, C. Streeter, E. Gu, . . . and N. B. Sutter. (2010). Morphological variation in the horse: Defining complex traits of body size and shape. *Animal Genetics* 41: 159–65.

Brown, D., and D. Anthony. (1998). Bit wear, horseback riding and the Botai site in Kazakhstan. *Journal of Archaeological Science* 25 (4): 331–47.

Cai, D., Z. Tang, L. Han, C. F. Speller, D. Y. Yang, X. Ma, . . . and H. Zhou. (2009). Ancient DNA provides new insights into the origin of the Chinese domestic horse. *Journal of Archaeological Science* 36 (3): 835–42.

Carter, S., and F. Hunter. (2003). An Iron Age chariot burial from Scotland. *Antiquity* 77 (297): 531–35.

Cieslak, M., M. Pruvost, N. Benecke, M. Hofreiter, A. Morales, M. Reissmann, and A. Ludwig. (2010). Origin and history of mitochondrial DNA lineages in domestic horses. *PLoS One* 5 (12): e15311.

Cieslak, M., M. Reissmann, M. Hofreiter, and A. Ludwig. (2011). Colours of domestication. *Biological Reviews* 86 (4): 885–99.

Clutton-Brock, J. (1974). The Buhen horse. *Journal of Archaeological Science* 1 (1): 89–100.

Clutton-Brock, J. (1992). *Horse Power: A History of the Horse and the Donkey in Human Societies*. Cambridge, MA: Harvard University Press.

Denny, M. W. (2008). Limits to running speed in dogs, horses and humans. *Journal of Experimental Biology* 211 (24): 3836–49.

Dergachev, V. (1989). Neolithic and Bronze Age cultural communities of the steppe zone of the USSR. Antiquity 63 (241): 793–802.

Doan, R., N. Cohen, J. Harrington, K. Veazy, R. Juras, G. Cothran, . . . and S. V. Dindot. (2012). Identification of copy number variants in horses. *Genome Research* 22 (5): 899–907.

Doan, R., N. D. Cohen, J. Sawyer, N. Ghaffari, C. D. Johnson, and S. V. Dindot. (2012). Whole-genome sequencing and genetic variant analysis of a Quarter Horse mare. *BMC Genomics* 13 (1): 78.

Eynon, N., J. R. Ruiz, J. Oliveira, J. A. Duarte, R. Birk, and A. Lucia. (2011). Genes and elite athletes: A roadmap for future research. *Journal of Physiology* 589 (13): 3063–70.

Froehlich, D. J. (1999). Phylogenetic systematics of basal perissodactyls. *Journal of Vertebrate Paleontology* 19 (1): 140–59.

Froehlich, D. J. (2002). Quo vadis eohippus? The systematics and taxonomy of the early Eocene equids (Perissodactyla). *Zoological Journal of the Linnean Society* 134 (2): 141–256.

George, M., and O. A. Ryder. (1986). Mitochondrial DNA evolution in the genus *Equus*. *Molecular Biology and Evolution* 3 (6): 535–46.

Gimbutas, M. (1990). "The Collision of Two Ideologies." In *When Worlds Collide: Indo-Europeans and Pre-Indo-Europeans*, edited by J. A. C. Greppin and T. L. Markey, 171–78. Ann Arbor, MI: Karoma.

Gingerich, P. D. (1991). Systematics and evolution of early Eocene Perissodactyla (Mammalia) in the Clarks Fork Basin, Wyoming. *Contributions from the Museum of Paleontology* 28 (8): 181–213.

Goodwin, D., M. Levine, and P. D. McGreevy. (2008). Preliminary investigation of morphological differences between ten breeds of horses suggests selection for paedomorphosis. *Journal of Applied Animal Welfare Science* 11 (3): 204–12.

Grigson, C. (1993). The earliest domestic horses in the Levant?—New finds from the fourth millennium of the Negev. *Journal of Archaeological Science* 20 (6): 645–55.

Groves, C. P. (1994). "Morphology, Habitat, and Taxonomy." In *Przewalski's Horse: The History and Biology of an Endangered Species*, edited by L. Boyd and K. A. Houpt, 39–60. Albany: State University of New York Press.

Grund, B. S., T. A. Surovell, and S. K. Lyons. (2012). Range sizes and shifts of North American Pleistocene mammals are not consistent with a climatic explanation for extinction. *World Archaeology* 44 (1): 43–55.

Heck, H. (1952). The breeding-back of the Tarpan. *Oryx* 1 (7): 338–42.

Hediger, H. K. (1981). The Clever Hans phenomenon from an animal psychologist's point of view. *Annals of the New York Academy of Sciences* 364 (1): 1–17.

Hendricks, B. L. (2007). *International Encyclopedia of Horse Breeds*. Norman: University of Oklahoma Press.

Hill, E. W., R. G. Fonseca, B. A. McGivney, J. Gu, D. E. MacHugh, and L. M. Katz. (2012). *MSTN* genotype (g.66493737C/T) association with speed indices in Thoroughbred racehorses. *Journal of Applied Physiology* 112 (1): 86–90.

Hill, W. G., and L. Bunger. (2004). Inferences on the genetics of quantitative traits from long-term selection in laboratory and domestic animals. *Plant Breeding Reviews* 24 (2): 169–210.

Hofreiter, M., and T. Schöneberg. (2010). The genetic and evolutionary basis of colour variation in vertebrates. *Cellular and Molecular Life Sciences* 67 (15): 2591–603.

Hulbert, R. C., Jr. (1989). "Phylogenetic Interrelationships and Evolution of North American Late Neogene Equinae." In *The Evolution of Perissodactyls*, edited by D. R. Prothero and R. M. Schoch, 176–93. New York: Clarendon.

Janis, C. M., J. Damuth, and J. M. Theodor. (2002). The origins and evolution of the North American grassland biome: The story from the hoofed mammals. *Palaeogeography, Palaeoclimatology, Palaeoecology* 177 (1–2): 183–98.

Jeffcott, L., P. Rossdale, J. Freestone, C. Frank, and P. Towers-Clark. (1982). An assessment of wastage in Thoroughbred racing from conception to 4 years of age. *Equine Veterinary Journal* 14 (3): 185–98.

Kavar, T., and P. Dovč. (2008). Domestication of the horse: Genetic relationships between domestic and wild horses. *Livestock Science* 116 (1–3): 1–14.

Kelekna, P. (2009). *The Horse in Human History*. Cambridge: Cambridge University Press.

Kim, H., T. Lee, W. Park, J. W. Lee, J. Kim, B.-Y. Lee, . . . and H. Kim. (2013). Peeling back the evolutionary layers of molecular mechanisms responsive to exercise-stress in the skeletal muscle of the racing horse. *DNA Research* 20 (3): 287–98.

Kim, N. (2011). "Cultural Attitudes and Horse Technologies: A View on Chariots and Stirrups from the Eastern End of the Eurasian Continent." In *Science between Europe and Asia: Historical Studies on the Transmission, Adoption and Adaptation of Knowledge*, edited by F. Günergun and D. Raina, 57–73. New York: Springer.

Koch, P. L., and A. D. Barnosky. (2006) Late quaternary extinctions: State of the debate. *Annual Review of Ecology, Evolution, and Systematics*, 37: 215–50.

Krueger, K., B. Flauger, K. Farmer, and K. Maros. (2011). Horses (*Equus caballus*) use human local enhancement cues and adjust to human attention. *Animal Cognition* 14 (2): 187–201.

Kurtén, B. (1980). *Pleistocene Mammals of North America*. New York: Columbia University Press.

Kuznetsov, P. (2006). The emergence of Bronze Age chariots in eastern Europe. *Antiquity* 80 (309): 638–45.

Laing, J., and F. Leech. (1975). The frequency of infertility in Thoroughbred mares. *Journal of Reproduction and Fertility. Supplement*, no. 23: 307–10.

Langlois, B. (1980). Heritability of racing ability in Thoroughbreds—A review. *Livestock Production Science* 7 (6): 591–605.

Le Douarin, N., and C. Kalacheim. (1999). *The Neural Crest*. Cambridge: Cambridge University Press.

Lei, C. Z., R. Su, M. A. Bower, C. J. Edwards, X. B. Wang, S. Weining, . . . and H. Chen. (2009). Multiple maternal origins of native modern and ancient horse populations in China. *Animal Genetics* 40 (6): 933–44.

Leroy, G., L. Callède, E. Verrier, J.-C. Mériaux, A. Ricard, C. Danchin-Burge, and X. Rognon. (2009). Genetic diversity of a large set of horse breeds raised in France assessed by microsatellite polymorphism. *Genetics, Selection, Evolution: GSE* 41 (1): 5.

Levine, M. A. (1998). Eating horses: The evolutionary significance of hippophagy. *Antiquity* 72: 90–100.

Levine, M. A. (1999). Botai and the origins of horse domestication. *Journal of Anthropological Archaeology* 18 (1): 29–78.

Levine, M. A. (2005). "Domestication and Early History of the Horse." In *The Domestic Horse. The Evolution, Development and Management of Its Behaviour*, edited by D. S. Mills and S. M. McDonnell, 5–22. Cambridge: Cambridge University Press.

Lindgren, G., N. Backstrom, J. Swinburne, L. Hellborg, A. Einarsson, K. Sandberg, . . . and H. Ellegren. (2004). Limited number of patrilines in horse domestication. *Nature Genetics* 36 (4): 335–36.

Lindsay, E. H., N. D. Opdyke, and N. M. Johnson. (1980). Pliocene dispersal of the horse *Equus* and late Cenozoic mammalian dispersal events. *Nature* 287: 135–38.

Lorenzen, E. D., D. Nogués-Bravo, L. Orlando, J. Weinstock, J. Binladen, K. A. Marske, . . . and R. Nielsen. (2011). Species-specific responses of Late Quaternary megafauna to climate and humans. *Nature* 479 (7373): 359–64.

Ludwig, A., M. Pruvost, M. Reissmann, N. Benecke, G. Brockmann, P. Castaños, . . . and M.

Hofreiter. (2009). Coat color variation at the beginning of horse domestication. *Science* 324 (5926): 485.

MacFadden, B. J. (1976). Cladistic analysis of primitive equids, with notes on other perissodactyls. *Systematic Biology* 25 (1): 1–14.

MacFadden, B. J. (2005). Fossil horses—Evidence for evolution. *Science* 307 (5716): 1728–30.

MacFadden, B. J., and R. C. Hulbert. (1988). Explosive speciation at the base of the adaptive radiation of Miocene grazing horses. *Nature* 336 (6198): 466–68.

MacFadden, B. J., L. H. Oviedo, G. M. Seymour, and S. Ellis. (2012). Fossil horses, orthogenesis, and communicating evolution in museums. *Evolution: Education and Outreach* 5 (1): 29–37.

McGivney, B. A., J. A. Browne, R. G. Fonseca, L. M. Katz, D. E. MacHugh, R. Whiston, and E. W. Hill. (2012). MTSN genotypes in Thoroughbred horses influence skeletal muscle gene expression and racetrack performance. *Animal Genetics* 43 (6): 810–12.

McMiken, D. (1990). Ancient origins of horsemanship. *Equine Veterinary Journal* 22 (2): 73–78.

Mihlbachler, M. C., F. Rivals, N. Solounias, and G. M. Semprebon. (2011). Dietary change and evolution of horses in North America. *Science* 331 (6021): 1178–81.

Moorey, P. R. S. (1986). The emergence of the light, horse-drawn chariot in the Near-East c. 2000–1500 BC. *World Archaeology* 18 (2): 196–215.

Morris, L. H. A., and W. R. Allen. (2002). Reproductive efficiency of intensively managed Thoroughbred mares in Newmarket. *Equine Veterinary Journal* 34 (1): 51–60.

Mostard, K., D. Goodwin, P. McGreevy, M. Levine, and S. Wendelaar Bonga. (2009). "Initial Evidence of Behavioural Paedomorphosis in Horses." In *Universities Federation for Animal Welfare (UFAW) International Symposium 2009: Darwinian Selection, Selective Breeding and the Welfare of Animals, Bristol, UK, 22–23 Jun 2009*. Bristol, UK: UFAW.

Olsen, S. L. (2006). "Early Horse Domestication on the Eurasian Steppe." In *Documenting Domestication: New Genetic and Archaeological Paradigms*, edited by M. A. Zeder, D. Bradley, E. Emshwiller, and B. D. Smith, 245–69. Berkeley: University of California Press.

Orlando, L., J. L. Metcalf, M. T. Alberdi, M. Telles-Antunes, D. Bonjean, M. Otte, . . . and F. Morello. (2009). Revising the recent evolutionary history of equids using ancient DNA. *Proceedings of the National Academy of Sciences USA* 106 (51): 21754–59.

Outram, A. K., N. A. Stear, R. Bendrey, S. Olsen, A. Kasparov, V. Zaibert, . . . and R. P. Evershed. (2009). The earliest horse harnessing and milking. *Science* 323 (5919): 1332–35.

Outram, A. K., N. A. Stear, A. Kasparov, E. Usmanova, and R. P. Evershed. (2011). Horses for the dead: Funerary foodways in Bronze Age Kazakhstan. *Antiquity* 85 (327): 116–28.

Petersen, J. L., J. R. Mickelson, A. K. Rendahl, S. J. Valberg, L. S. Andersson, J. Axelsson, . . . and A. S. Borges. (2013). Genome-wide analysis reveals selection for important traits in domestic horse breeds. *PLoS Genetics* 9 (1): e1003211.

Pigliucci, M., and K. Preston, eds. (2004). *Phenotypic Integration: Studying the Ecology and Evolution of Complex Phenotypes*. Oxford: Oxford University Press.

Prescott, G. W., D. R. Williams, A. Balmford, R. E. Green, and A. Manica. (2012a). Quantitative global analysis of the role of climate and people in explaining late Quaternary megafaunal extinctions. *Proceedings of the National Academy of Sciences USA* 109 (12): 4527–31.

Proops, L., M. Walton, and K. McComb. (2010). The use of human-given cues by domestic horses, *Equus caballus*, during an object choice task. *Animal Behaviour* 79 (6): 1205–9.

Prothero, D. R. (1993). "Ungulate Phylogeny: Molecular vs. Morphological Evidence." In *Mammal Phylogeny*, vol. 2, *Placentals*, edited by F. S. Szalay, M. J. Novacek, and M. C. McKenna, 173–81. New York: Springer.

Prothero, D. R., and N. Shubin. (1989). "The Evolution of Mid-Oligocene Horses." In *The Evolution of Perissodactyls*, edited by D. R. Prothero and R. M. Schoch, 142–75. New York: Clarendon.

Pruvost, M., R. Bellone, N. Benecke, E. Sandoval-Castellanos, M. Cieslak, T. Kuznetsova, . . . and M. Hofreiter. (2011). Genotypes of predomestic horses match phenotypes painted in Paleolithic works of cave art. *Proceedings of the National Academy of Sciences USA* 108 (46): 18626–30.

Raulwing, P., and J. Clutton-Brock. (2009). The Buhen horse: Fifty years after its discovery (1958–2008). *Journal of Egyptian History* 2 (1–2): 1–2.

Renfrew, C. (1998). "All the King's Horses." In *Creativity in Human Evolution and Prehistory*, edited by S. Mithen, 192. London: Routledge.

Shaughnessy, E. L. (1988). Historical perspectives on the introduction of the chariot into China. *Harvard Journal of Asiatic Studies* 48 (1): 189–237.

Sherratt, A. (1983). The secondary exploitation of animals in the Old World. *World Archaeology* 15 (1): 90–104.

Shoemaker, L., and A. Clauset. (2014). Body mass evolution and diversification within horses (family Equidae). *Ecology Letters* 17 (2): 211–20.

Shorrocks, B. (2007). *The Biology of African Savannahs*. Oxford: Oxford University Press.

Simpson, G. (1961). *Horses*. Garden City, NY: Anchor.

Skorkowski, E. (1970). Tarpans. *Wszechswiat*, no. 7/8: 207–8.

Sommer, R. S., N. Benecke, L. Lõugas, O. Nelle, and U. Schmölcke. (2011). Holocene survival of the wild horse in Europe: A matter of open landscape? *Journal of Quaternary Science* 26 (8): 805–12.

Steiner, C. C., A. Mitelberg, R. Tursi, and O. A. Ryder. (2012). Molecular phylogeny of extant equids and effects of ancestral polymorphism in resolving species-level phylogenies. *Molecular Phylogenetics and Evolution* 65 (2): 573–81.

Steiner, C. C., and O. A. Ryder. (2011). Molecular phylogeny and evolution of the Perissodactyla. *Zoological Journal of the Linnean Society* 163 (4): 1289–1303.

Tchernov, E., and L. Horwitz. (1991). Body size diminution under domestication—Unconscious selection in primeval domesticates. *Journal of Anthropological Archaeology* 10: 54–75.

Thieme, H. (2005). "The Lower Palaeolithic Art of Hunting." In *The Hominid Individual in Context: Archaeological Investigations of Lower and Middle Palaeolithic Landscapes, Locales and Artefacts*, edited by C. Gamble and M. Porr, 115–32. London: Routledge.

Valladas, H., J. Clottes, J.-M. Geneste, M. Garcia, M. Arnold, H. Cachier, and N. Tisnérat-Laborde. (2001). Palaeolithic paintings: Evolution of prehistoric cave art. *Nature* 413 (6855): 479.

Wade, C. M., E. Giulotto, S. Sigurdsson, M. Zoli, S. Gnerre, F. Imsland, . . . and K. Lindblad-Toh. (2009). Genome sequence, comparative analysis, and population genetics of the domestic horse. *Science* 326 (5954): 865–67.

Warmuth, V., A. Eriksson, M. A. Bower, G. Barker, E. Barrett, B. K. Hanks, . . . and A. Manica.

(2012). Reconstructing the origin and spread of horse domestication in the Eurasian steppe. *Proceedings of the National Academy of Sciences USA* 109 (21): 8202–6.

Warmuth, V., A. Eriksson, M. A. Bower, J. Cañon, G. Cothran, O. Distl, . . . and A. Manica. (2011). European domestic horses originated in two Holocene refugia. *PLoS One* 6 (3): e18194.

Weinstock, J., E. Willerslev, A. Sher, W. Tong, S. Y. Ho, D. Rubenstein, . . . and C. Bravi. (2005). Evolution, systematics, and phylogeography of Pleistocene horses in the New World: A molecular perspective. *PLoS Biology* 3 (8): e241.

Williamson, S., and R. Beilharz. (1998). The inheritance of speed, stamina and other racing performance characters in the Australian Thoroughbred. *Journal of Animal Breeding and Genetics* 115 (1–6): 1–16.

Wilson, A. J., and A. Rambaut. (2008). Breeding racehorses: What price good genes? *Biology Letters* 4 (2): 173–75.

Wisher, S., W. R. Allen, and J. L. N. Wood. (2006). Factors associated with failure of Thoroughbred horses to train and race. *Equine Veterinary Journal* 38 (2): 113–18.

Zeder, M. A. (2006). "Archaeological Approaches to Documenting Animal Domestication." In *Documenting Domestication: New Genetic and Archaeological Paradigms*, edited by M. A. Zeder, D. Bradley, E. Emshwiller, and B. D. Smith, 171–81. Berkeley: University of California Press.

Chapter 12: RODENTS

NOTES

1. Pigière et al. 2012; Morales 1995.
2. One idea is that the "guinea" comes from the Guinea coast of West Africa, a resupply area for ships crossing the Atlantic to and from South America, from which Europeans mistakenly believed cavies originated. Another is that "guinea" is a corruption of "Guyana," from which Europe-bound ships exited South America.
3. D'Erchia et al. 1996; the evidence for the move came from mitochondrial DNA.
4. Sullivan and Swofford 1997.
5. Catzeflis, Aguilar, and Jaeger 1992; Steppan, Adkins, and Anderson 2004.
6. Adkins et al. 2001. Another factor is that rodents are prone to allopatric speciation because of their fossorial habits and tendency toward restricted distributions (Steinberg, Patton, and Lacey 2000).
7. Huchon et al. 2002; Blanga-Kanfi et al. 2009; Churakov et al. 2010.
8. Alroy 1999.
9. See, for example, Hedges et al. 1996; Springer 1997; Waddell, Okada, and Hasegawa 1999.
10. Huchon et al. 2002; Wu et al. 2012.
11. Huchon et al. 2002; Asher et al. 2005.
12. Catzeflis et al. 1995. In hystricognaths the masseter medialis passes through the infraorbital foramen and connects to bone on other side. This arrangement is unique to the lineage, but its functional significance, if any, is unknown.
13. Antoine et al. 2012; Voloch et al. 2013.

14. Voloch et al. 2013.

15. Both New and Old World porcupines are hystricognaths, but they come from two different sides of the hystricognath tree. Therefore, their spiny quills are a case of convergent
evolution that is rendered more probable by shared features due to their common ancestry.

16. Spotorno et al. 2004.

17. Guichón and Cassini 1998.

18. Spotorno et al. 2006; Dunnam and Salazar-Bravo 2010.

19. Asher, de Oliveira, and Sachser 2004.

20. Wing 1986.

21. Spotorno et al. 2006.

22. Ibid.

23. Spotorno et al. 2007.

24. Suckow, Stevens, and Wilson 2012.

25. "Where the Ridgebacks Roam," TexCavy's Guinea Pig Pages, http://www.texcavy.com.

26. Botchkarev and Sharov 2004. See Hoekstra 2006 for a general treatise on vertebrate pigmentation. See also Pantalacci et al. 2008 .

27. Castle 1954; Wright 1917a, 1917b.

28. Cavies have played a particularly large role in research on sexual development (Arnold
2009) and fetus-placenta interactions (Kumar and Nankervis 1978; Mess 2007). Another
area of medical research in which cavies loom large is neural tube defects. One reason cavies are preferred over mice and rats for this research is that they have long pregnancies, like
humans.

29. Spotorno et al. 2006.

30. Kruska 1988.

31. Lewejohann et al. 2010. Unfortunately, this comparative research assumed that *C. aperea*
is the ancestor of domestic cavies; but it is not. *C. aperea* is a lowland species, the true
ancestor of domestic cavies; *C. tschudii* is a highland species, as you would expect. *C.
aperea* and *C. tschudii* are certainly closely related, but this research should be repeated
with *C. tschudii*.

32. Künzl and Sachser 1999; Künzl et al. 2003.

33. Sachser 1998; Asher, de Oliveira, and Sachser 2004.

34. Laviola and Terranova 1998; Cairns, Gariépy, and Hood 1990.

35. Steppan, Adkins, and Anderson 2004.

36. Thaler 1986.

37. Boursot et al. 1993.

38. Ibid.

39. Sokal, Oden, and Wilson 1991.

40. N. Royer, "The History of Fancy Mice," American Fancy Rat & Mouse Association, last
modified March 5, 2014, http://www.afrma.org/historymse.htm.

41. Nolan et al. 1995.

42. Keane et al. 2011; Reuveni 2011.

43. Wahlsten, Metten, and Crabbe 2003; Goto et al. 2013.

44. Garland et al. 2011.

45. Austad 2002.

46. Ibid.
47. Kruska 1988.
48. Miller et al. 2002.
49. Miller et al. 2002; Harper et al. 2006.
50. Gärtner 1990, 2012.
51. Krinke 2000.
52. Aplin, Chesser, and ten Have 2003.
53. McCormick 2003.
54. John Berkenhout conferred the scientific name in *Synopsis of the Natural History of Great Britain and Ireland* (1795).
55. Langton 2007.
56. Lockard 1968.
57. Clark and Price 1981.
58. R. J. Blanchard, Flannelly, and Blanchard 1986.
59. D. C. Blanchard et al. 1981.
60. Boice 1981.
61. Belyaev and Borodin 1982.
62. Of necessity, the measure of tameness for rats differed somewhat from that used for the foxes. For foxes, if you recall, tameness was simply the capacity to tolerate human proximity without the flight-or-fight response kicking in. For rats, the tameness criteria focused more on aggression of a particular sort, called defensive aggression, as measured by the response to a gloved hand in a cage in which there is no chance to flee (Naumenko et al. 1989; Plyusnina and Oskina 1997). Defensive aggression, as the name implies, occurs when an animal is under threat, in contrast to offensive aggression, which is directed toward securing resources like food, mates, or territory (R. J. Blanchard and Blanchard 1977).
63. Plyusnina, Solov'eva, and Oskina 2011. See D. C. Blanchard et al. 1994 for the state of things in the 35th generation, by way of comparison
64. Plyusnina, Solov'eva, and Oskina 2011.
65. Oskina, Plyusnina, and Sysoletina 2000; Plyusnina and Oskina 1997.
66. Shishkina, Borodin, and Naumenko 1993.
67. Pylusnina and Oskina 1997; Oskina et al. 2010; Albert et al. 2008, 2009.
68. This is from a study by Albert et al. 2008, which was completely independent of those conducted at Novosibirsk
69. Transgenic mice that overexpress the master stress hormone, corticotropin releasing factor, are learning impaired (Heinrichs et al. 1996).
70. Hoekstra 2006.
71. Panksepp 1998.

REFERENCES

Adkins, R., E. Gelke, D. Rowe, and R. Honeycutt. (2001). Molecular phylogeny and divergence time estimates for major rodent groups: Evidence from multiple genes. *Molecular Biology and Evolution* 18: 777–91.

Albert, F. W., Ö. Carlborg, I. Plyusnina, F. Besnier, D. Hedwig, S. Lautenschläger, . . . and S. Pääbo. (2009). Genetic architecture of tameness in a rat model of animal domestication. *Genetics* 182 (2): 541–54.

Albert, F. W., O. Shchepina, C. Winter, H. Römpler, D. Teupser, R. Palme, . . . and S. Pääbo. (2008). Phenotypic differences in behavior, physiology and neurochemistry between rats selected for tameness and for defensive aggression towards humans. *Hormones and Behavior* 53 (3): 413–21.

Alroy, J. (1999). The fossil record of North American mammals: Evidence for a Paleocene evolutionary radiation. *Systematic Biology* 48 (1): 107–18.

Antoine, P.-O., L. Marivaux, D. A. Croft, G. Billet, M. Ganerød, C. Jaramillo, . . . and A. J. Altamirano. (2012). Middle Eocene rodents from Peruvian Amazonia reveal the pattern and timing of caviomorph origins and biogeography. *Proceedings of the Royal Society. B: Biological Sciences* 279 (1732): 1319–26.

Aplin, K. P., T. Chesser, and J. ten Have. (2003). Evolutionary biology of the genus *Rattus*: Profile of an archetypal rodent pest. *ACIAR Monograph Series* 96: 487–98.

Arnold, A. P. (2009). The organizational-activational hypothesis as the foundation for a unified theory of sexual differentiation of all mammalian tissues. *Hormones and Behavior* 55 (5): 570–78.

Asher, M., E. de Oliveira, and N. Sachser. (2004). Social system and spatial organisation of wild guinea pigs (*Cavia aperea*) in a natural population. *Journal of Mammology* 85 (4): 788–96.

Asher, R. J., J. Meng, J. R. Wible, M. C. McKenna, G. W. Rougier, D. Dashzeveg, and M. J. Novacek. (2005). Stem Lagomorpha and the antiquity of Glires. *Science* 307 (5712): 1091–94.

Austad, S. (2002). A mouse's tale. *Natural History*, April.

Belyaev, D., and P. Borodin. (1982). The influence of stress on variation and its role in evolution. *Biologisches Zentralblatt* 101 (6): 705–14.

Berkenhout, J. (1795). *Synopsis of the Natural History of Great Britain and Ireland*. London: P. Elmsly.

Blanchard, D. C., N. K. Popova, I. Z. Plyusnina, I. L. Velichko, D. Campbell, R. J. Blanchard, . . . and E. M. Nikulina. (1994). Defensive reactions of "wild-type" and "domesticated" wild rats to approach and contact by a threat stimulus. *Aggressive Behavior* 20 (5): 387–97.

Blanchard, D. C., G. Williams, E. M. C. Lee, and R. J. Blanchard. (1981). Taming of wild *Rattus norvegicus* by lesions of the mesencephalic central gray. *Physiological Psychology* 9 (2): 157–63.

Blanchard, R. J., and D. C. Blanchard. (1977). Aggressive behavior in the rat. *Behavioral Biology* 21 (2): 197–224.

Blanchard, R. J., K. J. Flannelly, and D. C. Blanchard. (1986). Defensive behaviors of laboratory and wild *Rattus norvegicus*. *Journal of Comparative Psychology* 100 (2): 101.

Blanga-Kanfi, S., H. Miranda, O. Penn, T. Pupko, R. DeBry, and D. Huchon. (2009). Rodent phylogeny revised: Analysis of six nuclear genes from all major rodent clades. *BMC Evolutionary Biology* 9: 71.

Boice, R. (1981). Behavioral comparability of wild and domesticated rats. *Behavior Genetics* 11 (5): 545–53.

Bonduriansky, R., and T. Day. (2008). Nongenetic inheritance and its evolutionary implications. *Annual Review of Ecology, Evolution, and Systematics* 40 (1): 103.

Botchkarev, V. A., and A. A. Sharov. (2004). BMP signaling in the control of skin development and hair follicle growth. *Differentiation* 72 (9–10): 512–26.

Boursot, P., J. Auffray, J. Britton-Davidian, and F. Bonhomme. (1993). The evolution of house mice. *Annual Review of Ecology and Systematics* 24 (1): 119–52.

Cairns, R. B., J.-L. Gariépy, and K. E. Hood. (1990). Development, microevolution, and social behavior. *Psychological Review* 97 (1): 49–65.

Castle, W. (1954). Coat color inheritance in horses and in other mammals. *Genetics* 39 (1): 35.

Catzeflis, F. M., J.-P. Aguilar, and J.-J. Jaeger. (1992). Muroid rodents: Phylogeny and evolution. *Trends in Ecology & Evolution* 7 (4): 122–26.

Catzeflis, F. M., C. Hänni, P. Sourrouille, and E. Douzery. (1995). Molecular systematics of hystricognath rodents: The contribution of sciurognath mitochondrial 12S rRNA sequences. *Molecular Phylogenetics and Evolution* 4: 357–60.

Churakov, G., M. K. Sadasivuni, K. R. Rosenbloom, D. Huchon, J. Brosius, and J. Schmitz. (2010). Rodent evolution: Back to the root. *Molecular Biology and Evolution* 27 (6): 1315–26.

Clark, B. R., and E. O. Price. (1981). Sexual maturation and fecundity of wild and domestic Norway rats (*Rattus norvegicus*). *Journal of Reproduction and Fertility* 63 (1): 215–20.

Coyne, J. A. (2009). Evolution's challenge to genetics. *Nature* 457 (7228): 382–83.

Danchin, É., A. Charmantier, F. A. Champagne, A. Mesoudi, B. Pujol, and S. Blanchet. (2011). Beyond DNA: Integrating inclusive inheritance into an extended theory of evolution. *Nature Reviews. Genetics* 12 (7): 475–86.

D'Erchia, A. M., C. Gissi, G. Pesole, C. Saccone, and U. Arnason. (1996). The guinea-pig is not a rodent. *Nature* 381: 597–600.

Dunnum, J. L., and J. Salazar-Bravo. (2010). Molecular systematics, taxonomy and biogeography of the genus *Cavia* (Rodentia: Caviidae). *Journal of Zoological Systematics and Evolutionary Research* 48 (4): 376–88.

Feinberg, A. P., and R. A. Irizarry. (2010). Stochastic epigenetic variation as a driving force of development, evolutionary adaptation, and disease. *Proceedings of the National Academy of Sciences USA* 107 (suppl. 1): 1757–64.

Francis, R. C. (2011). *Epigenetics: How Environment Shapes Our Genes.* New York: Norton.

Garland, T., H. Schutz, M. A. Chappell, B. K. Keeney, T. H. Meek, L. E. Copes, . . . and G. van Dijk. (2011). The biological control of voluntary exercise, spontaneous physical activity and daily energy expenditure in relation to obesity: Human and rodent perspectives. *Journal of Experimental Biology* 214 (2): 206–29.

Gärtner, K. (1990). A third component causing random variability beside environment and genotype. A reason for the limited success of a 30 year long effort to standardize laboratory animals? *Laboratory Animals* 24 (1): 71–77.

Gärtner, K. (2012). A third component causing random variability beside environment and genotype. A reason for the limited success of a 30 year long effort to standardize laboratory animals? *International Journal of Epidemiology* 41 (2): 335–41.

Goto, T., A. Tanave, K. Moriwaki, T. Shiroishi, and T. Koide. (2013). Selection for reluctance to avoid humans during the domestication of mice. *Genes, Brain and Behavior* 12 (8): 760–70.

Guichón, M., and M. Cassini. (1998). Role of diet selection in the use of habitat by pampas cavies *Cavia aperea pamparum* (Mammalia, Rodentia). *Mammalia* 62 (1): 23–36.

Harper, J. M., S. J. Durkee, R. C. Dysko, S. N. Austad, and R. A. Miller. (2006). Genetic modulation of hormone levels and life span in hybrids between laboratory and wild-derived mice. *Journals of Gerontology. Series A, Biological Sciences and Medical Sciences* 61 (10): 1019–29.

Hedges, S., P. Parker, C. Sibley, and S. Kumar. (1996). Continental breakup and the ordinal diversification of birds and mammals. *Nature* 381: 226–29.

Heinrichs, S., M. Stenzel-Poore, L. Gold, E. Battenberg, F. Bloom, G. Koob, . . . and E. Merlo Pich. (1996). Learning impairment in transgenic mice with central overexpression of corticotropin-releasing factor. *Neuroscience* 74 (2): 303–11.

Hoekstra, H. (2006). Genetics, development and evolution of adaptive pigmentation in vertebrates. *Heredity* 97 (3): 222–34.

Huchon, D., O. Madsen, M. Sibbald, K. Ament, M. Stanhope, F. Catzeflis, . . . and E. Douzery. (2002). Rodent phylogeny and a timescale for the evolution of Glires: Evidence from an extensive taxon sampling using three nuclear genes. *Molecular Biology and Evolution* 19: 1053–65.

Jablonka, E., and M. J. Lamb. (1998). Epigenetic inheritance in evolution. *Journal of Evolutionary Biology* 11 (2): 159–83.

Jablonka, E., and M. J. Lamb. (2005). *Evolution in Four Dimensions: Genetic, Epigenetic, Behavioral, and Symbolic Variation in the History of Life*. Cambridge, MA: MIT Press.

Jablonka, E., and G. Raz. (2009). Transgenerational epigenetic inheritance: Prevalence, mechanisms, and implications for the study of heredity and evolution. *Quarterly Review of Biology* 84 (2): 131–76.

Keane, T. M., L. Goodstadt, P. Danecek, M. A. White, K. Wong, B. Yalcin, . . . and M. Goodson. (2011). Mouse genomic variation and its effect on phenotypes and gene regulation. *Nature* 477 (7364): 289–94.

Krinke, G. J., ed. (2000). *The Laboratory Rat*. San Diego, CA: Academic Press.

Kruska, D. (1988). Effects of domestication on brain structure and behavior in mammals. *Human Evolution* 3 (6): 473–85.

Kumar, M. L., and G. A. Nankervis. (1978). Experimental congenital infection with cytomegalovirus: A guinea pig model. *Journal of Infectious Diseases* 138 (5): 650–54.

Künzl, C., S. Kaiser, E. Meier, and N. Sachser. (2003). Is a wild mammal kept and reared in captivity still a wild animal? *Hormones and Behavior* 43 (1): 187–96.

Künzl, C., and N. Sachser. (1999). The behavioral endocrinology of domestication: A comparison between the domestic guinea pig (*Cavia aperea* f. *porcellus*) and its wild ancestor, the cavy (*Cavia aperea*). *Hormones and Behavior* 35 (1): 28–37.

Langton, J. (2007). *Rat: How the World's Most Notorious Rodent Clawed Its Way to the Top*. New York: St. Martin's Press.

Laviola, G., and M. L. Terranova. (1998). The developmental psychobiology of behavioural plasticity in mice: The role of social experiences in the family unit. *Neuroscience & Biobehavioral Reviews* 23 (2): 197–213.

Lewejohann, L., T. Pickel, N. Sachser, and S. Kaiser. (2010). Wild genius—domestic fool? Spatial learning abilities of wild and domestic guinea pigs. *Frontiers in Zoology* 7 (1): 9.

Lockard, R. B. (1968). The albino rat: A defensible choice or a bad habit? *American Psychologist* 23 (10): 734.

McCormick, M. (2003). Rats, communications, and plague: Toward an ecological history. *Journal of Interdisciplinary History* 34 (1): 1–25.

Mess, A. (2007). The guinea pig placenta: Model of placental growth dynamics. *Placenta* 28 (8): 812–15.

Miller, R. A., J. M. Harper, R. C. Dysko, S. J. Durkee, and S. N. Austad. (2002). Longer life spans and delayed maturation in wild-derived mice. *Experimental Biology and Medicine* 227 (7): 500–508.

Morales, E. (1995). *The Guinea Pig: Healing, Food, and Ritual in the Andes*. Tucson: University of Arizona Press.

Naumenko, E. V., N. K. Popova, E. M. Nikulina, N. N. Dygalo, G. T. Shishkina, P. M. Borodin, and A. L. Markel. (1989). Behavior, adrenocortical activity, and brain monoamines in Norway rats selected for reduced aggressiveness towards man. *Pharmacology Biochemistry and Behavior* 33 (1): 85–91.

Nolan, P. M., P. J. Sollars, B. A. Bohne, W. J. Ewens, G. E. Pickard, and M. Bućan. (1995). Heterozygosity mapping of partially congenic lines: Mapping of a semidominant neurological mutation, *Wheels* (*Whl*), on mouse chromosome 4. *Genetics* 140 (1): 245–54.

Oskina, I., I. Plyusnina, and A. Y. Sysoletina. (2000). Effect of selection for behavior on the pituitary-adrenal function of norway rats *Rattus norvegicus* in postnatal ontogeny. *Journal of Evolutionary Biochemistry and Physiology* 36 (2): 161–69.

Oskina, I., L. Prasolova, I. Plyusnina, and L. Trut. (2010). Role of glucocorticoids in coat depigmentation in animals selected for behavior. *Cytology and Genetics* 44 (5): 286–93.

Panksepp, J. (1998). *Affective Neuroscience: The Foundations of Human and Animal Emotions*. New York: Oxford University Press.

Pantalacci, S., A. Chaumot, G. Benoît, A. Sadier, F. Delsuc, E. J. P. Douzery, and V. Laudet. (2008). Conserved features and evolutionary shifts of the EDA signaling pathway involved in vertebrate skin appendage development. *Molecular Biology and Evolution* 25 (5): 912–28.

Pigière, F., W. Van Neer, C. Ansieau, and M. Denis. (2012). New archaeozoological evidence for the introduction of the guinea pig to Europe. *Journal of Archaeological Science* 39 (4): 1020–24.

Plyusnina, I., and I. Oskina. (1997). Behavioral and adrenocortical responses to open-field test in rats selected for reduced aggressiveness toward humans. *Physiology & Behavior* 61 (3): 381–85.

Plyusnina, I. Z., M. Y. Solov'eva, and I. N. Oskina. (2011). Effect of domestication on aggression in gray Norway rats. *Behavior Genetics* 41 (4): 583–92.

Rakyan, V., and E. Whitelaw. (2003). Transgenerational epigenetic inheritance. *Current Biology* 13: R6.

Reuveni, E. (2011). The genetic background effect on domesticated species: A mouse evolutionary perspective. *Scientific World Journal* 11: 429–36.

Richards, C. L., O. Bossdorf, and M. Pigliucci. (2010). What role does heritable epigenetic variation play in phenotypic evolution? *BioScience* 60 (3): 232–37.

Sachser, N. (1998). Of domestic and wild guinea pigs: Studies in sociophysiology, domestication, and social evolution. *Naturwissenschaften* 85 (7): 307–17.

Shishkina, G. T., P. M. Borodin, and E. V. Naumenko. (1993). Sexual maturation and seasonal changes in plasma levels of sex steroids and fecundity of wild Norway rats selected for reduced aggressiveness toward humans. *Physiology & Behavior* 53 (2): 389–93.

Sokal, R. R., N. L. Oden, and C. Wilson. (1991). Genetic evidence for the spread of agriculture in Europe by demic diffusion. *Nature* 351: 143–45.

Spotorno, A. E., G. Manríquez, A. Fernández, J. C. Marín, F. González, and J. Wheeler. (2007). "Domestication of Guinea Pigs from a Southern Peru–Northern Chile Wild Species and Their Middle Pre-Colombian Mummies." In *The Quintessential Naturalist: Honoring the*

Life and Legacy of Oliver P. Pearson, edited by D. A. Kelt, E. P. Lessa, J. Salazar-Bravo, and J. L. Patton, 367–88. Berkeley: University of California Press.

Spotorno, A. E., J. C. Marín, G. Manriquez, J. P. Valladares, E. Rico, and C. Rivas. (2006). Ancient and modern steps during the domestication of guinea pigs (*Cavia porcellus* L.). *Journal of Zoology* 270 (1): 57–62.

Spotorno, A., J. P. Valladares, J. C. Marín, and H. Zeballos. (2004). Molecular diversity among domestic guinea-pigs (*Cavia porcellus*) and their close phylogenetic relationship with the Andean wild species *Cavia tschudii*. *Revista Chilena de Historia Natural* 77 (2): 243–50.

Springer, M. S. (1997). Molecular clocks and the timing of the placental and marsupial radiations in relation to the Cretaceous-Tertiary boundary. *Journal of Mammalian Evolution* 4 (4): 285–302.

Steinberg, E. K., J. L. Patton, and E. Lacey. (2000). "Genetic Structure and the Geography of Speciation in Subterranean Rodents: Opportunities and Constraints for Evolutionary Diversification." In *Life Underground: The Biology of Subterranean Rodents*, edited by E. A. Lacey, J. L. Patton, and G. N. Cameron, 301–31. Chicago: University of Chicago Press.

Steppan, S., R. Adkins, and J. Anderson. (2004). Phylogeny and divergence-date estimates of rapid radiations in muroid rodents based on multiple nuclear genes. *Systematic Biology* 53: 533–53.

Suckow, M. A., K. A. Stevens, and R. P. Wilson. (2011). *The Laboratory Rabbit, Guinea Pig, Hamster, and Other Rodents*. London: Academic Press.

Sullivan, J., and D. Swofford. (1997). Are guinea pigs rodents? The importance of adequate models in molecular phylogenetics. *Journal of Mammalian Evolution* 4 (2): 77–86.

Thaler, L. (1986). "Origin and Evolution of Mice: An Appraisal of Fossil Evidence and Morphological Traits." In *The Wild Mouse in Immunology*, edited by M. Potter, J. H. Nadeau, and M. P. Cancro, 3–11. Berlin: Springer.

Trut, L., I. Oskina, and A. Kharlamova. (2009). Animal evolution during domestication: The domesticated fox as a model. *BioEssays* 31 (3): 349–60.

Voloch, C. M., J. F. Vilela, L. Loss-Oliveira, and C. G. Schrago. (2013). Phylogeny and chronology of the major lineages of New World hystricognath rodents: Insights on the biogeography of the Eocene/Oligocene arrival of mammals in South America. *BMC Research Notes* 6 (1): 160.

Waddell, P. J., N. Okada, and M. Hasegawa. (1999). Towards resolving the interordinal relationships of placental mammals. *Systematic Biology* (1999): 1–5.

Wahlsten, D., P. Metten, and J. C. Crabbe. (2003). A rating scale for wildness and ease of handling laboratory mice: Results for 21 inbred strains tested in two laboratories. *Genes, Brain and Behavior* 2 (2): 71–79.

Wing, E. S. (1986). "Domestication of Andean Animals." In *High Altitude Tropical Biogeography*, edited by F. Vuilleumier and M. Monasterio, 246–64. New York: Oxford University Press.

Wong, A. H. C., I. I. Gottesman, and A. Petronis, A. (2005). Phenotypic differences in genetically identical organisms: The epigenetic perspective. *Human Molecular Genetics* 14 (suppl. 1): R11–18.

Wright, S. (1917a). Color inheritance in mammals: Results of experimental breeding can be linked up with chemical researches on pigments—Coat colors of all mammals classified as due to variations in action of two enzymes. *Journal of Heredity* 8 (5): 224–35.

Wright, S. (1917b). Color inheritance in mammals: V. The guinea-pig—Great diversity in coat-pattern, due to interaction of many factors in development—Some factors hereditary, others of the nature of accidents in development. *Journal of Heredity* 8 (10): 476–80.

Wu, S., W. Wu, F. Zhang, J. Ye, X. Ni, J. Sun, . . . and C. L. Organ. (2012). Molecular and paleontological evidence for a post-Cretaceous origin of rodents. *PLoS One* 7 (10): e46445.

Chapter 13: HUMANS—PART I: EVOLUTION

NOTES

1. Lyn and Savage-Rumbaugh 2000. It was particularly significant that the bonobos learned from each other with no visual, but only audio, contact. See also Segerdahl, Fields, and Savage Rumbaugh 2005 for more on Kanzi.
2. Boesch and Boesch 1989.
3. And woman the gatherer (Stanford 1999); Lee and DeVore 1969 is something like the ur-text for this take on human nature.
4. Wrangham and Peterson 1996; Wrangham and Glowacki 2012.
5. Parish 1994; De Waal 1995; Hohmann and Fruth 2000.
6. Parish, De Waal, and Haig et al. 2000.
7. Hare, Wobber, and Wrangham 2012.
8. Wobber, Wrangham, and Hare 2010a, 2010b; Hare, Wobber, and Wrangham 2012.
9. Hare, Wobber, and Wrangham 2012.
10. Shea 1986; Wrangham and Pilbeam 2001.
11. Shea 1983.
12. Penin, Berge, and Baylac 2002; Shea 1989. Several studies found no evidence of heterochrony, much less paedomorphosis, much less neoteny (Williams, Godfrey, and Sutherland 2003; Mitteroecker, Gunz, and Bookstein 2005). Others found some evidence for paedomorphosis in the skull, but not globally. For example, Lieberman et al. (2007) found a moderate degree of paedomorphosis in the neurocranium (braincase), and a smaller degree in the face; others, though, found the opposite pattern (Cobb and O'Higgins 2004; Alba 2002).
13. Lieberman et al. 2007. This form of heterochrony is called postformation or postdisplacement.
14. Wrangham and Pilbeam 2001.
15. Hare, Wobber, and Wrangham 2012.
16. Leigh and Shea 1995; McIntyre et al. 2009.
17. Leigh and Shea 1995. There is also some diminution of sexual dimorphism in the skulls of bonobos relative to those of chimps.
18. Hare, Wobber, and Wrangham 2012.
19. Brosnan 2010; Rosati and Hare 2012.
20. Wobber, Wrangham, and Hare 2010a, 2010b.
21. Gruber, Clay, and Zuberbühler 2010.
22. Hare, Wobber, and Wrangham 2012.
23. Hare 2007; Palagi and Cordoni 2012; Wobber, Wrangham, and Hare 2010a, 2010b.
24. Shea 1989; De Waal and Lanting 1997; Wrangham and Pilbeam 2001 make the case for

the derived nature of bonobos solely by using gorillas as the outgroup. But by the standards typically used in phylogenetic inferences of this sort, such a three-way comparison is woefully inadequate—lacking what phylogeneticists call robustness.

25. Naef 1926; Bolk 1926.

26. Gould 1977; see also Montagu 1955.

27. Gould 1992, chap. 7.

28. *Plesiadapis* and its descendants are sometimes referred to as "primates of modern aspect" by way of distinction from archaic primates of the Paleocene, such as *Carpolestes*; Bloch and Boyer 2002.

29. Perelman et al. 2011. The bases for these dates are estimates of divergence time for living primate lineages.

30. Finlay and Darlington 1995.

31. Jerison 1973.

32. Charnov and Berrigan 1993.

33. Dunbar 1998, 2003; Reader and Layland 2002.

34. Bloch and Boyer 2002: primates of modern aspect. See Perelman et al. 2011 for earlier date based on molecular (DNA) evidence.

35. Schrago and Russo 2003.

36. Kay, Ross, and Williams 1997.

37. Simons et al. 2007.

38. Ward, Walker, and Teaford 1991.

39. Begun, Ward, and Rose 1997. Begun 2005 estimates 15–16 mya for the first Asian primates. See Stewart and Disotell 1998 for earlier date from DNA evidence.

40. Mayr 1950. For a good discussion of the negative impact on anthropology of Mayr's uninformed pronouncements, see Tattersall and Schwartz 2009.

41. Brunet et al. 2005.

42. Richmond and Jungers 2008.

43. Haile-Selassie 2001; Haile-Selassie and WoldeGabriel 2009; *ramidus* has received the most attention and is the species for which the epithet "Ardi" is reserved (e.g., White et al. 2009). In 2009 an entire issue of the journal *Science* was devoted to Ardi.

44. See Lovejoy et al. 2009 for evidence of bipedalism in Ardi. The skull of Ardi had both apelike and hominin features (Suwa, Asfaw, et al. 2009; Suwa, Kono, et al. 2009; Coppens 2006).

45. M. G. Leakey et al. 1998; Macho et al. 2005. See Grine, Ungar, and Teaford 2006 for inferences regarding *A. anamensis* diet.

46. The fossil Lucy is so named because the discoverer was listening to the Beatles song *Lucy in the Sky with Diamonds* when he found her (Johanson and White 1979).

47. McPherron et al. 2010.

48. Berger et al. 2010.

49. Kivell et al. 2011; see Carlson et al. 2011 for the brain.

50. Pickering et al. 2011.

51. L. S. Leakey, Tobias, and Napier 1964. The taxonomic position of *H. habilis* is disputed; many don't consider it a proper member of the genus *Homo* (e.g., Wood and Collard 1999; Tattersall and Schwartz 2009).

52. Wynn and McGrew 1989.

53. Tobias 1987.

54. Tattersall and Schwartz 2009. Some consider *H. ergaster* and *H. erectus* to be a single species (e.g., Bogin and Smith 1996). On this view, *H. ergaster* is an African subspecies of *H. erectus*. Here I am following Tattersall (e.g., Tattersall 2007) in according species status to both.

55. Bramble and Lieberman 2004; see Carrier 1984 for sweat and heat dissipation.

56. For sexual dimorphism in australopithecines, see Collard 2002; Gordon, Green, and Richmond 2008. For reduced sexual dimorphism in early *Homo*, see Frayer and Wolpoff 1985.

57. For example, Ambrose 2001.

58. Vekua et al. 2002.

59. The discoverers of Java Man—Dubois, Trap, and Stechert (1894)—originally assigned it to a new genus, *Pithecanthropus*, rather than to *Homo*.

60. Weidenreich 1935.

61. Brown 2004. The discovery of the "hobbit" elicited quite a brouhaha in the anthropological community regarding its taxonomic status. Some claimed it was a microcephalic modern human. This view is no longer tenable.

62. Carbonell et al. 2008.

63. Rightmire 2004. *H. heidelbergensis* was also the first cosmopolitan (Eurasia and Africa) species (Tattersall and Schwartz 2009).

64. Finlayson 2005. According to Rightmire 1998, humans and Neanderthals speciated almost simultaneously from *H. heidelbergensis*.

65. Ponce de León et al. 2008. The larger adult size of Neanderthal brains came by way of more rapid growth (Rozzi and de Castro 2004; see also Gunz et al. 2010).

66. See, for example, Hall 2003.

67. Wallis 1997. The gestation period for chimpanzees averages 225 days.

68. Zollikofer and Ponce de León 2010.

69. Bogin 1997.

70. Walker et al. 2006.

71. Bogin and Smith 1996; Zolliker and Ponce de León 2010. The growth rate of Turkana Boy was also closer to that of the australopithecines (Tardieu 1998). But see Clegg and Aiello 1999 for growth rate in another *H. ergaster* fossil, called Nariokotome *Homo*.

72. For example, the human foot is hypermorphic relative to that of apes (Lockley and Jackson 2008).

73. For example, there is no evidence that Ardi was a knuckle walker (Lovejoy et al. 2009).

74. Wang et al. 2004.

75. Berge 1998.

76. Berge 1998.

77. Minugh-Purvis and McNamara 2002.

78. Leigh 2004. See Neubauer and Hublin 2012 for the rate of human brain growth compared to other hominids.

79. Rosenberg and Trevethan 1995.

80. Rosenberg and Trevethan 2002.

81. Ponce de León et al. 2008.

82. Miller et al. 2012.

83. Somel et al. 2009; see also Giger et al. 2010 for primate-wide comparisons.

84. King and Wilson 1975. This was based largely on the realization that coding sequences alone could not account for human-chimpanzee divergence (Wilson, Carlson, and White 1977).

85. For genomic evidence of the important role of noncoding DNA sequences in the human-chimp divergence, see Smith, Webster, and Ellegren 2002. See Carroll 2008 for noncoding tissue-specific enhancers of *FOXP2* expression. See Carroll 2005 for a theoretical argument as to the necessity of noncoding cis-regulatory elements in particular. Coyne, unsurprisingly, dissents from this evo-devo view (e.g., Coyne 2005; Hoekstra and Coyne 2007). See Craig 2009 for a rebuttal to Hoekstra and Coyne 2007.

86. These coding gene products are a type of transcription factor, so called because they affect the rate of transcription of DNA to mRNA by binding to noncoding regulatory elements.

87. McHenry 1994; Plavcan and Van Schaik 1997. According to Reno et al. 2003, this reduction commenced earlier, with *Australopithecus*.

88. Plavcan and Van Schaik 1997.

REFERENCES

Alba, D. (2002). "Shape and Stage in Heterochronic Models." In *Human Evolution through Developmental Change*, edited by N. Minugh-Purvis and K. J. McNamara, 28–50. Baltimore: Johns Hopkins University Press.

Ambrose, S. H. (2001). Paleolithic technology and human evolution. *Science* 291 (5509): 1748–53.

Begun, D. R. (2005). *Sivapithecus* is east and *Dryopithecus* is west, and never the twain shall meet. *Anthropological Science* 113 (1): 53.

Begun, D. R., C. V. Ward, and M. D. Rose. (1997). "Events in Hominoid Evolution." In *Function, Phylogeny, and Fossils: Miocene Hominoid Evolution and Adaptations*, edited by D. R. Begun, C. V. Ward, and M. D. Rose, 389–415. New York: Plenum.

Berge, C. (1998). Heterochronic processes in human evolution: An ontogenetic analysis of the hominid pelvis. *American Journal of Physical Anthropology* 105 (4): 441–59.

Berger, L. R., D. J. de Ruiter, S. E. Churchill, P. Schmid, K. J. Carlson, P. H. Dirks, and J. M. Kibii. (2010). *Australopithecus sediba*: A new species of *Homo*-like australopith from South Africa. *Science* 328 (5975): 195–204.

Bloch, J. I., and D. M. Boyer, D. M. (2002). Grasping primate origins. *Science* 298 (5598): 1606–10.

Boesch, C., and H. Boesch. (1989). Hunting behavior of wild chimpanzees in the Tai National Park. *American Journal of Physical Anthropology* 78 (4): 547–73.

Bogin, B. A. (1997). Evolutionary hypotheses for human childhood. *American Journal of Physical Anthropology* 104 (S25): 63–69.

Bogin, B. A., and B. H. Smith. (1996). Evolution of the human life cycle. *American Journal of Human Biology* 8: 703–16.

Bolk, L. (1926). *Das Problem der Menschwerdung* [The problem of human development]: *Vortrag gehalten am 15. April 1926 auf der XXV. Versammlung der Anatomischen Gesellschaft zu Freiburg*. Freiburg, Germany: Fischer.

Bramble, D. M., and D. E. Lieberman. (2004). Endurance running and the evolution of *Homo*. *Nature* 432 (7015): 345–52.

Brosnan, S. F. (2010). Behavioral development: Timing is everything. *Current Biology* 20 (3): R98–100.

Brown, P., T. Sutikna, M. Morwood, R. Soejono, Jatmiko, E. Saptomo, and R. Due. (2004). A new small-bodied hominin from the Late Pleistocene of Flores, Indonesia. *Nature* 431: 1055–61.

Brunet, M., F. Guy, D. Pilbeam, D. E. Lieberman, A. Likius, H. T. Mackaye, . . . and P. Vignaud. (2005). New material of the earliest hominid from the Upper Miocene of Chad. *Nature* 434 (7034): 752–55.

Carbonell, E., J. M. B. de Castro, J. M. Pares, A. Perez-Gonzalez, and J. L. Arsuaga. (2008). The first hominin of Europe. *Nature* 452 (7186): 465–69.

Carlson, K. J., D. Stout, T. Jashashvili, D. J. de Ruiter, P. Tafforeau, K. Carlson, and L. R. Berger. (2011). The endocast of MH1, *Australopithecus sediba*. *Science* 333 (6048): 1402–7.

Carrier, D. R. (1984). The energetic paradox of human running and hominid evolution [and comments and reply]. *Current Anthropology* 25 (4): 483–95.

Carroll, S. B. (2005). Evolution at two levels: On genes and form. *PLoS Biology* 3 (7): e245.

Carroll, S. B. (2008). Evo-devo and an expanding evolutionary synthesis: A genetic theory of morphological evolution. *Cell* 134 (1): 25–36.

Charnov, E. L., and D. Berrigan. (1993). Why do female primates have such long lifespans and so few babies? Or life in the slow lane. *Evolutionary Anthropology: Issues, News, and Reviews* 1 (6): 191–94.

Clegg, M., and L. C. Aiello. (1999). A comparison of the Nariokotome *Homo erectus* with juveniles from a modern human population. *American Journal of Physical Anthropology* 110 (1): 81–93.

Cobb, S., and P. O'Higgins. (2004). Hominins do not share a common postnatal facial ontogenetic shape trajectory. *Journal of Experimental Zoology. Part B: Molecular and Developmental Evolution* 302 (3): 302–21.

Collard, M. (2002). "Grades and Transitions in Human Evolution." In *The Speciation of Modern Homo Sapiens*, edited by T. Crow, 61–102. Oxford: Oxford University Press.

Coppens, Y. (2006). The bunch of ancestors. *Transactions of the Royal Society of South Africa* 61 (1): 1–3.

Coyne, J. A. (2005). Switching on evolution. *Nature* 435 (7045): 1029–30.

Craig, L. R. (2009). Defending evo-devo: A response to Hoekstra and Coyne. *Philosophy of Science* 76 (3): 335–44.

De Waal, F. (1995). Bonobo sex and society. *Scientific American* 272 (3): 82–88.

De Waal, F., and F. Lanting. (1997). *Bonobo: The Forgotten Ape*. Berkeley: University of California Press.

Dubois, E., P. W. M. Trap, and G. Stechert. (1894). *Pithecanthropus erectus: Eine menschenaehnliche Uebergangsform aus Java*. Batavia, Netherlands: Landesdruckerei.

Dunbar, R. I. M. (1998). The social brain hypothesis. *Evolutionary Anthropology* 6: 178–90.

Dunbar, R. I. M. (2003). The social brain: Mind, language, and society in evolutionary perspective. *Annual Review of Anthropology* 32 (3): 163–81.

Finlay, B., and R. Darlington. (1995). Linked regularities in the development and evolution of mammalian brains. *Science* 268 (5217): 1578–84.

Finlayson, C. (2005). Biogeography and evolution of the genus *Homo*. *Trends in Ecology & Evolution* 20 (8): 457–63.

Frayer, D. W., and M. H. Wolpoff. (1985). Sexual dimorphism. *Annual Review of Anthropology* 14 (1): 429–73.

Giger, T., P. Khaitovich, M. Somel, A. Lorenc, E. Lizano, L. W. Harris, … and S. Pääbo. (2010). Evolution of neuronal and endothelial transcriptomes in primates. *Genome Biology and Evolution* 2: 284–92.

Gordon, A. D., D. J. Green, and B. G. Richmond. (2008). Strong postcranial size dimorphism in *Australopithecus afarensis*: Results from two new resampling methods for multivariate data sets with missing data. *American Journal of Physical Anthropology* 135 (3): 311–28.

Gould, S. J. (1977). *Ontogeny and Phylogeny.* Cambridge, MA: Harvard University Press.

Gould, S. J. (1992). *Ever since Darwin: Reflections in Natural History.* New York: Norton.

Grine, F. E., P. S. Ungar, and M. F. Teaford. (2006). Was the Early Pliocene hominin *"Australopithecus" anamensis* a hard object feeder? *South African Journal of Science* 102 (7 & 8): 301–10.

Gruber, T., Z. Clay, and K. Zuberbühler. (2010). A comparison of bonobo and chimpanzee tool use: Evidence for a female bias in the *Pan* lineage. *Animal Behaviour* 80 (6): 1023–33.

Gunz, P., S. Neubauer, B. Maureille, and J.-J. Hublin. (2010). Brain development after birth differs between Neanderthals and modern humans. *Current Biology* 20 (21): R921–22.

Haile-Selassie, Y. (2001). Late Miocene hominids from the middle Awash, Ethiopia. *Nature* 412 (6843): 178–81.

Haile-Selassie, Y., and G. WoldeGabriel. (2009). *Ardipithecus kadabba: Late Miocene Evidence from the Middle Awash, Ethiopia.* Berkeley: University of California Press.

Hall, B. K. (2003). Evo-devo: Evolutionary developmental mechanisms. *International Journal of Developmental Biology* 47 (7/8): 491–96.

Hare, B. (2007). From nonhuman to human mind: What changed and why? *Current Directions in Psychological Science* 16 (2): 60–64.

Hare, B., V. Wobber, and R. Wrangham. (2012). The self-domestication hypothesis: Evolution of bonobo psychology is due to selection against aggression. *Animal Behaviour* 83 (3): 573–85.

Hoekstra, H. E., and J. A. Coyne. (2007). The locus of evolution: Evo devo and the genetics of adaptation. *Evolution* 61 (5): 995–1016.

Hohmann, G., and B. Fruth. (2000). Use and function of genital contacts among female bonobos. *Animal Behaviour* 60 (1): 107–20.

Jerison, H. (1973). *Evolution of the Brain and Intelligence.* New York: Academic Press.

Johanson, D., and T. White. (1979). A systematic assessment of early African hominids. *Science* 203 (4378): 321–30.

Kay, R. F., C. Ross, and B. A. Williams. (1997). Anthropoid origins. *Science* 275 (5301): 797–804.

King, M.-C., and A. C. Wilson. (1975). Evolution at two levels in humans and chimpanzees. *Science* 188 (4184): 107–16.

Kivell, T. L., J. M. Kibii, S. E. Churchill, P. Schmid, and L. R. Berger. (2011). *Australopithecus sediba* hand demonstrates mosaic evolution of locomotor and manipulative abilities. *Science* 333 (6048): 1411–17.

Leakey, L. S., P. V. Tobias, and J. R. Napier. (1964). A new species of the genus *Homo* from Olduvai Gorge. *Nature* 202 (4927): 7–9.

Leakey, M. G., C. S. Feibel, I. McDougall, C. Ward, and A. Walker. (1998). New specimens and confirmation of an early age for *Australopithecus anamensis*. *Nature* 393 (6680): 62–66.

Lee, R. B., and I. DeVore, eds. (1969). *Man the Hunter*. Chicago: Aldine.

Leigh, S. R. (2004). Brain growth, life history, and cognition in primate and human evolution. *American Journal of Primatology* 62 (3): 139–64.

Leigh, S. R., and B. T. Shea. (1995). Ontogeny and the evolution of adult body size dimorphism in apes. *American Journal of Primatology* 36 (1): 37–60.

Lieberman, D. E., J. Carlo, M. Ponce de León, and C. P. E. Zollikofer. (2007). A geometric morphometric analysis of heterochrony in the cranium of chimpanzees and bonobos. *Journal of Human Evolution* 52 (6): 647–62.

Lockley, M., and P. Jackson. (2008). Morphodynamic perspectives on convergence between the feet and limbs of sauropods and humans: Two cases of hypermorphosis. *Ichnos* 15 (3–4): 140–57.

Lovejoy, C. O., G. Suwa, L. Spurlock, B. Asfaw, and T. D. White. (2009). The pelvis and femur of *Ardipithecus ramidus*: The emergence of upright walking. *Science* 326 (5949): 71–71e6.

Lyn, H., and E. S. Savage-Rumbaugh. (2000). Observational word learning in two bonobos (*Pan paniscus*): Ostensive and non-ostensive contexts. *Language & Communication* 20 (3): 255–73.

Macho, G. A., D. Shimizu, Y. Jiang, and I. R. Spears. (2005). *Australopithecus anamensis*: A finite-element approach to studying the functional adaptations of extinct hominins. *Anatomical Record. Part A, Discoveries in Molecular, Cellular, and Evolutionary Biology* 283 (2): 310–18.

Mayr, E. (1950). Taxonomic categories in fossil hominids. *Cold Spring Harbor Symposia on Quantitative Biology* 15: 109–18.

McHenry, H. M. (1994). Behavioral ecological implications of early hominid body size. *Journal of Human Evolution* 27 (1): 77–87.

McIntyre, M. H., E. Herrmann, V. Wobber, M. Halbwax, C. Mohamba, N. de Sousa, . . . and B. Hare. (2009). Bonobos have a more human-like second-to-fourth finger length ratio (2D:4D) than chimpanzees: A hypothesized indication of lower prenatal androgens. *Journal of Human Evolution* 56 (4): 361–65.

McPherron, S. P., Z. Alemseged, C. W. Marean, J. G. Wynn, and H. A. Bearat. (2010). Evidence for stone-tool-assisted consumption of animal tissues before 3.39 million years ago at Dikika, Ethiopia. *Nature* 466 (7308): 857–60.

Miller, D. J., T. Duka, C. D. Stimpson, S. J. Schapiro, W. B. Baze, M. J. McArthur, . . . and D. E. Wildman. (2012). Prolonged myelination in human neocortical evolution. *Proceedings of the National Academy of Sciences USA* 109 (41): 16480–85.

Minugh-Purvis, N., and K. J. McNamara, eds. (2002). *Human Evolution through Developmental Change*. Baltimore: Johns Hopkins University Press.

Mitteroecker, P., P. Gunz, and F. L. Bookstein. (2005). Heterochrony and geometric morphometrics: A comparison of cranial growth in *Pan paniscus* versus *Pan troglodytes*. *Evolution and Development* 7 (3): 244–58.

Montagu, M. F. A. (1955). Time, morphology, and neoteny in the evolution of man. *American Anthropologist* 57 (1): 13–27.

Naef, A. (1926). Über die Urformen der Anthropomorphen und die Stammesgeschichte des Mendschenschädels. *Naturwissenschaften* 14: 445–52.

Neubauer, S., and J.-J. Hublin. (2012). The evolution of human brain development. *Evolutionary Biology* 39 (4): 568–86.

Palagi, E., and G. Cordoni. (2012). The right time to happen: Play developmental divergence in the two *Pan* species. *PLoS One* 7 (12).

Parish, A. R. (1994). Sex and food control in the "uncommon chimpanzee": How bonobo females overcome a phylogenetic legacy of male dominance. *Ethology and Sociobiology* 15 (3): 157–79.

Parish, A. R., R. de Waal, and D. Haig. (2000). The other "closest living relative": How bonobos (*Pan paniscus*) challenge traditional assumptions about females, dominance, intra- and intersexual interactions, and hominid evolution. *Annals of the New York Academy of Sciences* 907 (1): 97–113.

Penin, X., C. Berge, and M. Baylac. (2002). Ontogenetic study of the skull in modern humans and the common chimpanzees: Neotenic hypothesis reconsidered with a tridimensional procrustes analysis. *American Journal of Physical Anthropology* 118 (1): 50–62.

Perelman, P., W. E. Johnson, C. Roos, H. N. Seuánez, J. E. Horvath, M. A. M. Moreira, . . . and J. Pecon-Slattery. (2011). A molecular phylogeny of living primates. *PLoS Genetics* 7 (3): e1001342.

Pickering, R., P. H. G. Dirks, Z. Jinnah, D. J. de Ruiter, S. E. Churchill, A. I. R. Herries, . . . and L. R. Berger. (2011). *Australopithecus sediba* at 1.977 Ma and implications for the origins of the genus *Homo*. *Science* 333 (6048): 1421–23.

Plavcan, J., and C. Van Schaik. (1994). Canine dimorphism. *Evolutionary Anthropology* 2 (6): 208–14.

Ponce de León, M. S., L. Golovanova, V. Doronichev, G. Romanova, T. Akazawa, O. Kondo, and C. P. Zollikofer. (2008). Neanderthal brain size at birth provides insights into the evolution of human life history. *Proceedings of the National Academy of Sciences USA* 105 (37): 13764–68.

Reader, S. M., and K. N. Laland. (2002). Social intelligence, innovation, and enhanced brain size in primates. *Proceedings of the National Academy of Sciences USA* 99 (7): 4436–41.

Reno, P. L., R. S. Meindl, M. A. McCollum, and C. O. Lovejoy. (2003). Sexual dimorphism in *Australopithecus afarensis* was similar to that of modern humans. *Proceedings of the National Academy of Sciences USA* 100 (16): 9404–9.

Richmond, B. G., and W. L. Jungers. (2008). *Orrorin tugenensis* femoral morphology and the evolution of hominin bipedalism. *Science* 319 (5870): 1662–65.

Rightmire, G. P. (1998). Human evolution in the Middle Pleistocene: The role of *Homo heidelbergensis*. *Evolutionary Anthropology* 6 (6): 218–27.

Rightmire, G. P. (2004). Brain size and encephalization in early to Mid-Pleistocene *Homo*. *American Journal of Physical Anthropology* 124 (2): 109–23.

Rosati, A. G., and B. Hare. (2012). Chimpanzees and bonobos exhibit divergent spatial memory development. *Developmental Science* 15 (6): 840–53.

Rosenberg, K., and W. Trevathan. (1995). Bipedalism and human birth: The obstetrical dilemma revisited. *Evolutionary Anthropology* 4 (5): 161–68.

Rosenberg, K., and W. Trevathan. (2002). Birth, obstetrics and human evolution. *BJOG: An International Journal of Obstetrics and Gynaecology* 109 (11): 1199–1206.

Rozzi, F. V. R., and J. M. B. de Castro. (2004). Surprisingly rapid growth in Neanderthals. *Nature* 428 (6986): 936–39.

Schrago, C. G., and C. A. M. Russo. (2003). Timing the origin of New World monkeys. *Molecular Biology and Evolution* 20 (10): 1620–25.

Segerdahl, P., W. Fields, and S. Savage-Rumbaugh. (2005). *Kanzi's Primal Language: The Cultural Initiation of Primates into Language.* Basingstoke, UK: Palgrave Macmillan.

Shea, B. T. (1983). Paedomorphosis and neoteny in the pygmy chimpanzee. *Science* 222 (4623): 521–22.

Shea, B. T. (1986). Scapula form and locomotion in chimpanzee evolution. *American Journal of Physical Anthropology* 70 (4): 475–88.

Shea, B. T. (1989). Heterochrony in human evolution: The case for neoteny reconsidered. *American Journal of Physical Anthropology* 80 (suppl. 10): 69–101.

Simons, E. L., E. R. Seiffert, T. M. Ryan, and Y. Attia. (2007). A remarkable female cranium of the early Oligocene anthropoid *Aegyptopithecus zeuxis* (Catarrhini, Propliopithecidae). *Proceedings of the National Academy of Sciences USA* 104 (21): 8731–36.

Smith, N. G., M. T. Webster, and H. Ellegren. (2002). Deterministic mutation rate variation in the human genome. *Genome Research* 12 (9): 1350–56.

Somel, M., H. Franz, Z. Yan, A. Lorenc, S. Guo, T. Giger, . . . and P. Khaitovich. (2009). Transcriptional neoteny in the human brain. *Proceedings of the National Academy of Sciences USA* 106 (14): 5743–48.

Stanford, C. B. (1999). *The Hunting Apes: Meat Eating and the Origins of Human Behavior.* Princeton, NJ: Princeton University Press.

Stewart, C.-B., and T. R. Disotell. (1998). Primate evolution—In and out of Africa. *Current Biology* 8 (16): R582–88.

Suwa, G., B. Asfaw, R. T. Kono, D. Kubo, C. O. Lovejoy, and T. D. White. (2009). The *Ardipithecus ramidus* skull and its implications for hominid origins. *Science* 326 (5949): 68, 68e1–7.

Suwa, G., R. T. Kono, S. W. Simpson, B. Asfaw, C. O. Lovejoy, and T. D. White. (2009). Paleobiological implications of the *Ardipithecus ramidus* dentition. *Science* 326 (5949): 69, 94–99.

Tardieu, C. (1998). Short adolescence in early hominids: Infantile and adolescent growth of the human femur. *American Journal of Physical Anthropology* 107 (2): 163–78.

Tattersall, I. (2007). "*Homo ergaster* and Its Contemporaries." In *Handbook of Paleoanthropology,* edited by W. Henke and I. Tattersall, 1633–53. New York: Springer.

Tattersall, I., and J. H. Schwartz. (2009). Evolution of the genus *Homo. Annual Review of Earth and Planetary Sciences* 37 (1): 67–92.

Tobias, P. V. (1987). The brain of *Homo habilis*: A new level of organization in cerebral evolution. *Journal of Human Evolution* 16 (7): 741–61.

Vekua, A., D. Lordkipanidze, G. P. Rightmire, J. Agusti, R. Ferring, G. Maisuradze, . . . and M. Tappen. (2002). A new skull of early *Homo* from Dmanisi, Georgia. *Science* 297 (5578): 85–89.

Walker, R., K. Hill, O. Burger, and A. M. Hurtado. (2006). Life in the slow lane revisited: Ontogenetic separation between chimpanzees and humans. *American Journal of Physical Anthropology* 129 (4): 577–83.

Wallis, J. (1997). A survey of reproductive parameters in the free-ranging chimpanzees of Gombe National Park. *Journal of Reproduction and Fertility* 109 (2): 297–307.

Wang, W., R. H. Crompton, T. S. Carey, M. M. Günther, Y. Li, R. Savage, and W. I. Sellers. (2004). Comparison of inverse-dynamics musculo-skeletal models of AL 288-1 *Australopithecus afarensis* and KNM-WT 15000 *Homo ergaster* to modern humans, with implications for the evolution of bipedalism. *Journal of Human Evolution* 47 (6): 453–78.

Ward, C. V., A. Walker, and M. Teaford. (1991). *Proconsul* did not have a tail. *Journal of Human Evolution* 21 (3): 215–20.

Weidenreich, F. (1935). The *Sinanthropus* population of Choukoutien (Locality 1) with a preliminary report on new discoveries. *Bulletin of the Geological Society of China* 14 (4): 427–68.

White, T. D., B. Asfaw, Y. Beyene, Y. Haile-Selassie, C. O. Lovejoy, G. Suwa, and G. Wolde-Gabriel. (2009). *Ardipithecus ramidus* and the paleobiology of early hominids. *Science* 326 (5949): 64, 75–86.

Williams, F., L. Godfrey, and M. Sutherland. (2003). "Diagnosing Heterochronic Perturbations in the Craniofacial Evolution of *Homo* (Neandertals and Modern Humans) and *Pan* (*P. troglodytes* and *P. paniscus*)." In *Patterns of Growth and Development in the Genus Homo*, edited by J. L. Thompson, G. E. Krovitz, and A. J. Nelson, 295. Cambridge Studies in Biological and Evolutionary Anthropology 37. Cambridge: Cambridge University Press.

Wilson, A. C., S. S. Carlson, and T. J. White. (1977). Biochemical evolution. *Annual Review of Biochemistry* 46 (1): 573–639.

Wobber, V., R. Wrangham, and B. Hare. (2010a). Application of the heterochrony framework to the study of behavior and cognition. *Communicative & Integrative Biology* 3 (4): 337–39.

Wobber, V., R. Wrangham, and B. Hare. (2010b). Bonobos exhibit delayed development of social behavior and cognition relative to chimpanzees. *Current Biology* 20 (3): 226–30.

Wood, B., and M. Collard. (1999). The human genus. *Science* 284 (5411): 65–71.

Wrangham, R., and L. Glowacki. (2012). Intergroup aggression in chimpanzees and war in nomadic hunter-gatherers. *Human Nature* 23 (1): 5–29.

Wrangham, R., and D. Peterson. (1996). *Demonic Males: Apes and the Origins of Human Violence*. Boston: Houghton Mifflin.

Wrangham, R., and D. Pilbeam. (2001). "African Apes as Time Machines." In *All Apes Great and Small*, edited by B. M. F. Galdikas, N. E. Briggs, L. K. Sheeran, G. L. Shapiro, and J. Goodall, 5–17. New York: Kluwer Academic.

Wynn, T., and W. C. McGrew. (1989). An ape's view of the Oldowan. *Man* 24 (3): 383–98.

Zollikofer, C. P. E., and M. S. Ponce de León. (2010). The evolution of hominin ontogenies. *Seminars in Cell & Developmental Biology* 21 (4): 441–52.

Chapter 14: HUMANS—PART II: SOCIALITY

NOTES

1. Cain 2013.
2. An exception is E. O. Wilson's *The Social Conquest of Earth* (Wilson 2012).
3. Actually, this is the title of his book (Tattersall 2012) and, as I have learned from hard experience, not necessarily of the author's making or even approval.
4. "It Ain't Necessarily So" is a song from *Porgy and Bess* by the Gershwin brothers in which the character Sportin' Life voices his skepticism about certain biblical stories. Richard

Lewontin (2001) used it as the title of his skeptical take on the medical "miracles" promised by proponents of the Human Genome Project. Here, in the spirit of Lewontin, it is meant to reflect skepticism about the facile applications of evolutionary biology to the human condition that are characteristic of evolutionary psychology.

5. Endler 1986; Losos, Schoener, and Spiller 2004; Losos et al. 2006; Reznick and Ghalambor 2005; Butler and King 2004.

6. Felsenstein 1985; Martins and Hansen 1997. See Leroi, Rose, and Lauder 1994 for limitations of the comparative method with regard to adaptive inferences. Brooks and McLennan 1991 is a book-length application of the comparative method to behavior.

7. Buller 2005; Hauser, Chomsky, and Fitch 2002; Fitch 2005.

8. Cosmides and Tooby 1997. Talk of cultural universals often fails to take into account the universality of certain generic cultural processes, which can lend the appearance of innateness. See Kirby, Dowman, and Griffiths (2007) for putative linguistic cultural universals. See Norenzaayan and Heine 2005 for a nuanced analysis of cultural universals and an attempt to separate the wheat from the chaff. For putative culturally universal emotional expressions, see Elfenbein and Ambady 2002. See Russell 1994 for a skeptical take.

9. For example, according to Kirby, Dowman, and Griffiths 2007, the only truly universal linguistic trait is cultural transmission of language. This analysis and that of Wimsatt and Griesemer (2007) relies on the notion of "scaffolding," which emphasizes not only the enabling but also the structuring contribution of the cultural environment to language acquisition and other cognitive behaviors. On this view, cultural universals are simply generic features of cultural scaffolding. See also Heintz et al. 2013.

10. For example, Aoki 1986.

11. Dennett (1994, 1995) is a cheerleader for the reverse-engineering approach. See also Pinker 1999. Both Dennet and Pinker, it should be noted, come from a cognitive science/philosophy of mind background in which reverse engineering is popular and, I would say, somewhat less problematic but still overused.

12. Gray 1987; Gould and Lewontin 1979.

13. Edinger 1900; Edwin and Rand 1908.

14. MacLean 1970; see also MacLean 1982, 1990.

15. Cory 2002.

16. Bruce and Neary 1995.

17. Kaas 1987.

18. See Noda, Ikeo, and Gojobori 2006 for highly conserved gene expression patterns in the nervous system. See also Mineta, Ikeo, and Gojobori 2008.

19. Linden 2008; Boero, Schierwater, and Piraino 2007.

20. Gleeson and Walsh 2000. Muscular dystrophy is one example of a disorder resulting from the perils of neural migration (Yoshida et al. 2001), as is Kallmann syndrome (Cariboni and Maggi 2006).

21. Kriegstein and Noctor 2004.

22. Marcus 2009.

23. Marcus 2009; Fitch 2012.

24. Anderson 2010.

25. Wobber, Wrangham, and Hare 2010b.

26. Denver 1999, 2009.

27. Panksepp 1982, 1988, 1998.

28. Panksepp 2003.

29. Panksepp et al. 1994.

30. Humphrey 1976; Aiello and Dunbar 1993; Dunbar 1998; Clutton-Brock and Harvey 1980; Kudo and Dunbar 2001. For a refined version of the social intelligence idea, see Dunbar and Shultz 2007. But see Healy and Rowe 2007; Holekamp 2007 for critiques. The social intelligence hypothesis should be seen as one form of the mosaic view of human cognition, which is popular among adaptationists of various sorts, including sociobiologists, evolutionary ecologists, and evolutionary psychologists. On the mosaic view, natural selection is unconstrained in optimizing specific cognitive capacities. Opposed to this is the covariance view, according to which specific cognitive capacities such as social cognition evolve as a package with selection for intelligence generally. This latter position is more in accord with evo devo, especially in its emphasis on conserved developmental mechanisms.

31. Byrne and Corp 2004.

32. Byrne and Whiten 1989; Gravilets and Vose 2006. See Gigerenzer 1997 for a partial critique.

33. Healy and Rowe 2007.

34. Francis 2004.

35. Charvet and Finlay 2012; Charvet, Darlington, and Finlay 2013.

36. Charvet and Finlay 2012.

37. Finlay, Darlington, and Nicastro 2001; Finlay and Darlington 1995 (including developmental conservation).

38. Barton and Venditti 2013.

39. Herculano-Houzel 2009.

40. Moll and Tomasello 2007, which is modified from Bratman 1992. Tomasello is an advocate of Vygotskian intelligence, named for the great Russian developmental psychologist, Lev Semyonovich Vygotsky. Central to Vygotsky's understanding of cognitive development was the role of the social/cultural environment, which is internalized in a cumulative way by the developing child. See also Tomasello 2008; Tomasello and Carpenter 2007.

41. Boesch and Boesch 1989.

42. Moll and Tomasello 2007; Hermann et al. 2010.

43. Warneken, Chen, and Tomasello 2006.

44. Warneken, Chen, and Tomasello 2006; Warneken and Tomasello 2006.

45. Tomasello 2008; Hermann et al. 2007.

46. Warneken, Grafenhain, and Tomasello 2012. On the basis of the fox domestication experiments, Hare does suggest looking for changes in the HPA axis and parts of the limbic system, but such alterations are not essential elements of the self-domestication hypothesis. The stronger form of this hypothesis would make such changes central and essential.

47. Hare et al. 2007; Tomasello et al. 2012.

48. Hare et al. 2007.

49. Tan and Hare 2013.

50. Hare 2011; Wobber, Wrangham, and Hare 2010a.
51. Wobber et al. 2012.
52. Mech 1994; Mech et al. 1998.
53. Macdonald and Carr 1995.
54. Hare 2011.
55. Wobber et al. 2010.
56. The fact that bonobos show elevated cortisol levels in competitive situations may or may not be consistent with the evidence for a reduced stress response to social stimuli in foxes, dogs, and rats.
57. Semendeferi 1998; Barger, Stefanacci, and Semendeferi 2007; Barger et al. 2012.
58. Rilling et al. 2012.
59. Bickart et al. 2012.
60. Bickart et al. 2010.

REFERENCES

Aiello, L. C., and R. I. Dunbar. (1993). Neocortex size, group size, and the evolution of language. *Current Anthropology* 34 (2): 184–93.

Anderson, M. L. (2010). Neural reuse: A fundamental organizational principle of the brain. *Behavioral and Brain Sciences* 33 (4): 245–66.

Aoki, K. (1986). A stochastic model of gene-culture coevolution suggested by the "culture historical hypothesis" for the evolution of adult lactose absorption in humans. *Proceedings of the National Academy of Sciences USA* 83 (9): 2929–33.

Barger, N., L. Stefanacci, and K. Semendeferi. (2007). A comparative volumetric analysis of amygdaloid complex and basolateral division in the human and ape brain. *American Journal of Physical Anthropology* 134 (3): 392–403.

Barger, N., L. Stefanacci, C. M. Schumann, C. C. Sherwood, J. Annese, J. M. Allman, . . . and K. Semendeferi. (2012). Neuronal populations in the basolateral nuclei of the amygdala are differentially increased in humans compared with apes: A stereological study. *Journal of Comparative Neurology* 520 (13): 3035–54.

Barkow, J., L. Cosmides, and J. Tooby, eds. (1992). *The Adapted Mind: Evolutionary Psychology and the Generation of Culture.* New York: Oxford University Press.

Barton, R. A., and C. Venditti. (2013). Human frontal lobes are not relatively large. *Proceedings of the National Academy of Sciences USA* 110 (22): 9001–6.

Bickart, K. C., M. C. Hollenbeck, L. F. Barrett, and B. C. Dickerson. (2012). Intrinsic amygdala-cortical functional connectivity predicts social network size in humans. *Journal of Neuroscience* 32 (42): 14729–41.

Bickart, K. C., C. I. Wright, R. J. Dautoff, B. C. Dickerson, and L. F. Barrett. (2010). Amygdala volume and social network size in humans. *Nature Neuroscience* 14 (2): 163–64.

Boero, F., B. Schierwater, and S. Piraino. (2007). Cnidarian milestones in metazoan evolution. *Integrative and Comparative Biology* 47 (5): 693–700.

Boesch, C., and H. Boesch. (1989). Hunting behavior of wild chimpanzees in the Tai National Park. *American Journal of Physical Anthropology* 78 (4): 547–73.

Bolhuis, J. J., G. R. Brown, R. C. Richardson, and K. N. Laland. (2011). Darwin in mind: New opportunities for evolutionary psychology. *PLoS Biology* 9 (7): e1001109.

Bratman, M. E. (1992). Shared cooperative activity. *Philosophical Review* 101 (2): 327–41.

Brooks, D., and D. McLennan. (1991). *Phylogeny, Ecology and Behavior: A Research Program in Comparative Biology*. Chicago: University of Chicago Press.

Bruce, L., and T. Neary. (1995). The limbic system of tetrapods: A comparative analysis of cortical and amygdalar populations. *Brain, Behavior and Evolution* 46 (4–5): 224–34.

Buller, D. J. (2005). *Adapting Minds: Evolutionary Psychology and the Persistent Quest for Human Nature*. Cambridge, MA: MIT Press.

Butler, M., and A. King. (2004). Phylogenetic comparative analysis: A modeling approach for adaptive evolution. *American Naturalist* 164: 683–95.

Byrne, R. W., and N. Corp. (2004). Neocortex size predicts deception rate in primates. *Proceedings of the Royal Society. B: Biological Sciences* 271: 1693–99.

Byrne, R. W., and A. Whiten, eds. (1989). *Machiavellian Intelligence: Social Expertise and the Evolution of Intellect in Monkeys, Apes, and Humans*. Oxford: Oxford University Press.

Cain, S. (2013). *Quiet: The Power of Introverts in a World That Can't Stop Talking*. New York: Random House.

Cariboni, A., and R. Maggi. (2006). Kallmann's syndrome, a neuronal migration defect. *Cellular and Molecular Life Sciences: CMLS* 63 (21): 2512–26.

Charvet, C. J., R. B. Darlington, and B. L. Finlay. (2013). Variation in human brains may facilitate evolutionary change toward a limited range of phenotypes. *Brain, Behavior and Evolution* 81 (2): 74–85.

Charvet, C. J., and B. L. Finlay. (2012) Embracing covariation in brain evolution: Large brains, extended development, and flexible primate social systems. *Progress in Brain Research* 195: 71–87.

Clutton-Brock, T., and P. H. Harvey. (1980). Primates, brains and ecology. *Journal of Zoology* 190 (3): 309–23.

Cory, G. A., Jr. (2002). "Reappraising MacLean's Triune Brain Concept." In *The Evolutionary Neuroethology of Paul MacLean. Convergences and Frontiers*, edited by Gerald A. Corey Jr. and Russel Gardner, 9–27. Westport, CT: Prager.

Cosmides, L., and J. Tooby. (1997). "Evolutionary Psychology: A Primer," Center for Evolutionary Psychology. http://www.cep.ucsb.edu/primer.html.

Dennett, D. C. (1994). "Cognitive Science as Reverse Engineering: Several Meanings of 'Top-Down' and 'Bottom-Up.'" In *Logic, Methodology and Philosophy of Science IX*, edited by D. Prawitz, B. Skyrms, and D. Westerståhl, 679–89. Studies in Logic and the Foundations of Mathematics 134. Amsterdam: North-Holland.

Dennett, D. C. (1995). *Darwin's Dangerous Idea: Evolution and the Meanings of Life*. New York: Simon & Schuster.

Denver, R. J. (1999). Evolution of the corticotropin-releasing hormone signaling system and its role in stress-induced phenotypic plasticity. *Annals of the New York Academy of Sciences* 897 (1): 46–53.

Denver, R. J. (2009). Structural and functional evolution of vertebrate neuroendocrine stress systems. *Annals of the New York Academy of Sciences* 1163 (1): 1–16.

Dunbar, R. I. M. (1998). The social brain hypothesis. *Evolutionary Anthropology* 6: 178–90.

Dunbar, R. I. M., and S. Shultz. (2007). Evolution in the social brain. *Science* 317 (5843): 1344–47.

Edinger, L. (1900). *The Anatomy of the Central Nervous System of Man and of Vertebrates in General*. Philadelphia: F. A. Davis.

Edinger, L., and H. W. Rand. (1908). The relations of comparative anatomy to comparative psychology. *Journal of Comparative Neurology and Psychology* 18 (5): 437–57.

Elfenbein, H. A., and N. Ambady. (2002). On the universality and cultural specificity of emotion recognition: A meta-analysis. *Psychological Bulletin* 128 (2): 203.

Endler, J. A. (1986). *Natural Selection in the Wild*. Princeton, NJ: Princeton University Press.

Felsenstein, J. (1985). Phylogenies and the comparative method. *American Naturalist* 125: 1–15.

Finlay, B. L., and R. B. Darlington. (1995). Linked regularities in the development and evolution of mammalian brains. *Science* 268 (5217): 1578–84.

Finlay, B. L., R. B. Darlington, and N. Nicastro. (2001). Developmental structure in brain evolution. *Behavioral and Brain Sciences* 24 (2): 263–78.

Fitch, W. T. (2005). The evolution of language: A comparative review. *Biology and Philosophy* 20 (2–3): 193–203.

Fitch, W. T. (2012). Evolutionary developmental biology and human language evolution: Constraints on adaptation. *Evolutionary Biology* 39 (4): 613–37.

Francis, R. C. (2004). *Why Men Won't Ask for Directions: The Seductions of Sociobiology*. Princeton, NJ: Princeton University Press.

Gavrilets, S., and A. Vose. (2006). The dynamics of Machiavellian intelligence. *Proceedings of the National Academy of Sciences USA* 103 (45): 16823–28.

Geary, D. C. (1995). Sexual selection and sex differences in spatial cognition. *Learning and Individual Differences* 7 (4): 289–301.

Giger, T., P. Khaitovich, M. Somel, A. Lorenc, E. Lizano, L. W. Harris, . . . and S. Pääbo. (2010). Evolution of neuronal and endothelial transcriptomes in primates. *Genome Biology and Evolution* 2: 284–92.

Gigerenzer, G. (1997). "The Modularity of Social Intelligence." In *Machiavellian Intelligence II: Extensions and Evaluations*, edited by Andrew W. Whiten and R. W. Byrne, 264–88. Cambridge: Cambridge University Press.

Gleeson, J., and C. Walsh. (2000). Neuronal migration disorders: From genetic diseases to developmental mechanisms. *Trends in Neurosciences* 23 (8): 352–59.

Gould, S. J., and R. Lewontin. (1979). The spandrels of San Marco and the Panglossian paradigm: A critique of the adaptationist programme. *Proceedings of the Royal Society. B: Biological Sciences* 205: 581–98.

Gray, R. (1987). "Faith and Foraging: A Critique of the 'Paradigm Argument from Design.'" In *Foraging Behavior*, edited by A. C. Kamil, J. R. Krebs, and H. R. Pulliam, 69–140. New York: Plenum.

Gunz, P. (2012). Evolutionary relationships among robust and gracile australopiths: An "evo-devo" perspective. *Evolutionary Biology* 39 (4): 472–87.

Hare, B. (2011). From hominoid to hominid mind: What changed and why? *Annual Review of Anthropology* 40 (1): 293–309.

Hare, B., A. P. Melis, V. Woods, S. Hastings, and R. Wrangham. (2007). Tolerance allows bonobos to outperform chimpanzees on a cooperative task. *Current Biology* 17 (7): 619–23.

Hauser, M. D., N. Chomsky, and W. T. Fitch. (2002). The faculty of language: What is it, who has it, and how did it evolve? *Science* 298 (5598): 1569–79.

Healy, S., and C. Rowe. (2007). A critique of comparative studies of brain size. *Proceedings of the Royal Society. B: Biological Sciences* 274 (1609): 453–64.

Heintz, C., L. Caporael, J. Griesemer, and W. Wimsatt. (2013). "Scaffolding on Core Cognition." In *Developing Scaffolds in Evolution, Culture, and Cognition*, edited by L. R. Caporael, J. R. Griesemer, and W. C. Wimsatt, 209–28. Cambridge, MA: MIT Press.

Herculano-Houzel, S. (2009). The human brain in numbers: A linearly scaled-up primate brain. *Frontiers in Human Neuroscience* 3: 31.

Herrmann, E., J. Call, M. V. Hernàndez-Lloreda, B. Hare, and M. Tomasello. (2007). Humans have evolved specialized skills of social cognition: The cultural intelligence hypothesis. *Science* 317 (5843): 1360–66.

Herrmann, E., B. Hare, J. Call, and M. Tomasello. (2010). Differences in the cognitive skills of bonobos and chimpanzees. *PLoS One* 5 (8): e12438.

Holekamp, K. E. (2007). Questioning the social intelligence hypothesis. *Trends in Cognitive Sciences* 11 (2): 65–69.

Humphrey, N. K. (1976). "The Social Function of Intellect." In *Growing Points in Ethology*, edited by P. P. G. Bateson and R. A. Hinde, 303–17. Cambridge: Cambridge University Press.

Kaas, J. H. (1987). The organization of neocortex in mammals: Implications for theories of brain function. *Annual Review of Psychology* 38 (1): 129–51.

Kirby, S., M. Dowman, and T. L. Griffiths. (2007). Innateness and culture in the evolution of language. *Proceedings of the National Academy of Sciences USA* 104 (12): 5241–45.

Kriegstein, A., and S. Noctor. (2004). Patterns of neuronal migration in the embryonic cortex. *Trends in Neurosciences* 27: 392–99.

Kudo, H., and R. Dunbar. (2001). Neocortex size and social network size in primates. *Animal Behaviour* 62: 711–22.

Kuhn, S. L. (2013). "Cultural Transmission, Institutional Continuity and the Persistence of the Mousterian." In *Dynamics of Learning in Neanderthals and Modern Humans*, vol. 1, *Cultural Perspectives*, 105–13. New York: Springer.

Leroi, A., M. R. Rose, and G. V. Lauder. (1994). What does the comparative method reveal about adaptation? *American Naturalist* 143 (3): 381–402.

Lewontin, R. C. (2001). *It Ain't Necessarily So: The Dream of the Human Genome and Other Illusions*, 2nd ed. New York: New York Review of Books.

Linden, D. J. (2008). Brain evolution and human cognition: The accidental mind. *Willamette Law Review* 45: 17.

Lloyd, E. A., and M. W. Feldman. (2002). Evolutionary psychology: A view from evolutionary biology [commentary]. *Psychological Inquiry* 13 (2): 150–56.

Loring, C., R. Rosenberg, and D. Hunt. (1987). Gradual change in human tooth size in the late Pleistocene and post-Pleistocene. *Evolution* 41 (4): 705–20.

Losos, J. B., T. W. Schoener, R. B. Langerhans, and D. A. Spiller. (2006). Rapid temporal reversal in predator-driven natural selection. *Science* 314 (5802): 1111.

Losos, J. B., T. W. Schoener, and D. A. Spiller. (2004). Predator-induced behaviour shifts and natural selection in field-experimental lizard populations. *Nature* 432: 505–8.

Loulergue, L., A. Schilt, R. Spahni, V. Masson-Delmotte, T. Blunier, B. Lemieux, . . . and J. Chappellaz. (2008). Orbital and millennial-scale features of atmospheric CH4 over the past 800,000 years. *Nature* 453 (7193): 383–86.

Macdonald, D., and G. Carr. (1995). "Variation in Dog Society: Between Resource Dispersion and Social Flux." In *The Domestic Dog, Its Evolution, Behaviour and Interactions with People*, edited by J. Serpell, 199–216. Cambridge: Cambridge University Press.

MacLean, P. D. (1970). "The Triune Brain, Emotion, and Scientific Bias." In *The Neurosciences: Second Study Program*, edited by F. O. Schmitt, 336–49. New York: Rockefeller University Press.

MacLean, P. D. (1982). "On the Origin and Progressive Evolution of the Triune Brain." In *Primate Brain Evolution: Methods and Concepts*, edited by E. Armstrong and D. Falk, 291–316. New York: Plenum.

MacLean, P. D. (1990). *The Triune Brain in Evolution: Role in Paleocerebral Functions*. New York: Plenum.

Marcus, G. (2009). *Kluge: The Haphazard Evolution of the Human Mind*. New York: Houghton Mifflin Harcourt.

Martins, E. P., and T. F. Hansen. (1997). Phylogenies and the comparative method: A general approach to incorporating phylogenetic information into the analysis of interspecific data. *American Naturalist* 149 (4): 646–67.

Martrat, B., J. O. Grimalt, N. J. Shackleton, L. de Abreu, M. A. Hutterli, and T. F. Stocker. (2007). Four climate cycles of recurring deep and surface water destabilizations on the Iberian margin. *Science* 317 (5837): 502–7.

Mech, L. D. (1994). Buffer zones of territories of gray wolves as regions of intraspecific strife. *Journal of Mammalogy* 75 (1): 199–202.

Mech, L. D., L. G. Adams, T. J. Meier, J. W. Burch, and B. W. Dale. (1998). *The Wolves of Denali*. Minneapolis: University of Minnesota Press.

Meisenberg, G. (2008). On the time scale of human evolution: Evidence for recent adaptive evolution. *Mankind Quarterly* 48 (4): 407–44.

Mineta, K., K. Ikeo, and T. Gojobori. (2008). "Gene Expression in the Brain and Central Nervous System in Planarians." In *Planaria: A Model for Drug Action and Abuse*, edited by R. B. Raffa and S. M. Rawls, 13–19. Austin, TX: Landes Bioscience.

Moll, H., and M. Tomasello. (2007). Cooperation and human cognition: The Vygotskian intelligence hypothesis. *Philosophical Transactions of the Royal Society. B: Biological Sciences* 362 (1480): 639–48.

Noda, A. O., K. Ikeo, and T. Gojobori. (2006). Comparative genome analyses of nervous system–specific genes. *Gene* 365: 130–36.

Norenzayan, A., and S. J. Heine. (2005). Psychological universals: What are they and how can we know? *Psychological Bulletin* 131 (5): 763.

Panksepp, J. (1982). Toward a general psychobiological theory of emotions. *Behavioral and Brain Sciences* 5 (3): 407–22.

Panksepp, J. (1988). "Brain Emotional Circuits and Psychopathologies." In *Emotions and Psychopathology*, edited by M. Clynes and J. Panksepp, 37–76. New York: Plenum.

Panksepp, J. (1998). *Affective Neuroscience: The Foundations of Human and Animal Emotions*. New York: Oxford University Press.

Panksepp, J. (2003). At the interface of the affective, behavioral, and cognitive neurosciences: Decoding the emotional feelings of the brain. *Brain and Cognition* 52 (1): 4–14.

Panksepp, J., L. Normansell, J. F. Cox, and S. M. Siviy. (1994). Effects of neonatal decortication on the social play of juvenile rats. *Physiology & Behavior* 56 (3): 429–43.

Pinker, S. (1999). How the mind works. *Annals of the New York Academy of Sciences* 882 (1): 119–27.

Reznick, D. N., and C. K. Ghalambor. (2005). Selection in nature: Experimental manipulations of natural populations. *Integrative and Comparative Biology* 45 (3): 4 56–62.

Richardson, R. C. (2007). *Evolutionary Psychology as Maladapted Psychology.* Cambridge, MA: MIT Press.

Richerson, P. J., R. Boyd, and J. Henrich. (2010). Gene-culture coevolution in the age of genomics. *Proceedings of the National Academy of Sciences USA* 107 (suppl. 2): 8985–92.

Rilling, J. K., J. Scholz, T. M. Preuss, M. F. Glasser, B. K. Errangi, and T. E. Behrens. (2012). Differences between chimpanzees and bonobos in neural systems supporting social cognition. *Social Cognitive and Affective Neuroscience* 7 (4): 369–79.

Russell, J. A. (1994). Is there universal recognition of emotion from facial expressions? A review of the cross-cultural studies. *Psychological Bulletin* 115 (1): 102.

Semendeferi, K., E. Armstrong, A. Schleicher, K. Zilles, and G. W. Van Hoesen. (1998). Limbic frontal cortex in hominoids: A comparative study of area 13. *American Journal of Physical Anthropology* 106 (2): 129–55.

Tan, J., and B. Hare. (2013). Bonobos share with strangers. *PLoS One* 8 (1): e51922.

Tattersall, I. (2012). *Masters of the Planet: Seeking the Origins of Human Singularity.* New York: Palgrave Macmillan.

Tomasello, M. (2008). "Why Don't Apes Point?" In *Variation, Selection, Development: Proving the Evolutionary Model of Language Change,* edited by R. Eckhardt, G. Jäger, and T. Veenstra, 375. Trends in Linguistics Studies and Monographs 197. New York: Mouton de Gruyter.

Tomasello, M., and M. Carpenter. (2007). Shared intentionality. *Developmental Science* 10 (1): 121–25.

Tomasello, M., A. P. Melis, C. Tennie, E. Wyman, and E. Herrmann. (2012). Two key steps in the evolution of human cooperation: The interdependence hypothesis. *Current Anthropology* 53 (6): 673–92.

Tooby, J., and L. Cosmides. (2005). "Conceptual Foundations of Evolutionary Psychology." In *The Handbook of Evolutionary Psychology,* edited by D. M. Buss, 5–67. Hoboken, NJ: Wiley.

Warneken, F., F. Chen, and M. Tomasello. (2006). Cooperative activities in young children and chimpanzees. *Child Development* 77 (3): 640–63.

Warneken, F., M. Grafenhain, and M. Tomasello. (2012). Collaborative partner or social tool? New evidence for young children's understanding of joint intentions in collaborative activities. *Developmental Science* 15 (1): 54–61.

Warneken, F., and M. Tomasello. (2006). Altruistic helping in human infants and young chimpanzees. *Science* 311 (5765): 1301–3.

Williams, G. (1966). *Adaptation and Natural Selection.* Princeton, NJ: Princeton University Press.

Wimsatt, W. C., and J. R. Griesemer. (2007). "Reproducing Entrenchments to Scaffold Culture: The Central Role of Development in Cultural Evolution." In *Integrating Evolution and Development: From Theory to Practice,* edited by R. Sansom and R. N. Brandon, 227–323. Cambridge, MA: MIT Press.

Wobber, V., B. Hare, J. Maboto, S. Lipson, R. Wrangham, and P. T. Ellison. (2010). Differential changes in steroid hormones before competition in bonobos and chimpanzees. *Proceedings of the National Academy of Sciences USA* 107 (28): 12457–62.

Wobber, V., S. Lipson, B. Hare, R. Wrangham, and P. Ellison. (2012). "Species Differences in the Ontogeny of Testosterone Production between Chimpanzees and Bonobos." Paper presented at the 81st Annual Meeting of the American Association of Physical Anthropologists.

Wobber, V., R. Wrangham, and B. Hare. (2010a). Application of the heterochrony framework to the study of behavior and cognition. *Communicative & Integrative Biology* 3 (4): 337–39.

Wobber, V., R. Wrangham, and B. Hare. (2010b). Bonobos exhibit delayed development of social behavior and cognition relative to chimpanzees. *Current Biology* 20 (3): 226–30.

Wynne-Edwards, V. C. (1963). Intergroup selection in the evolution of social systems. *Nature* 200: 623–26.

Wynne-Edwards, V. C. (1978). "Intrinsic Population Control: An Introduction." In *Population Control by Social Behavior*, edited by F. J. Ebling and D. M. Stoddart, 1–22. London: Institute of Biology.

Yoshida, A., K. Kobayashi, H. Manya, K. Taniguchi, H. Kano, M. Mizuno, . . . and T. Endo. (2001). Muscular dystrophy and neuronal migration disorder caused by mutations in a glycosyltransferase, POMGnT1. *Developmental Cell* 1 (5): 717–24.

Chapter 15: THE ANTHROPOCENE

NOTES

1. The primary advocate for the view that the Toba eruption nearly caused human extinction is Steven Ambrose (e.g., Ambrose 1998). For dissenting views, see Petraglia et al. 2007 and Endicott et al. 2009.

2. Tattersall and Schwartz 2009; Finlayson 2005.

3. Endicott et al. 2009.

4. Ibid.

5. Roberts et al. (1994) estimate the first arrival of humans in Australia as 60,000–55,000 years ago.

6. See Green et al. 2006, 2010 for the Neanderthal nuclear genome; Green et al. 2008 for the Neanderthal mitochondrial genome; Meyer et al. 2012 for the Denisovan nuclear genome.

7. For Denisovan DNA in Melanesians, see Reich et al. 2010; for Neanderthal DNA in Europeans and Asians, see Green et al. 2010; Sankararaman 2012, 2014.

8. Pickrell and Reich 2014.

9. Banks et al. 2008 is representative of this view; but see Zilhão 2006 for dissent. Zilhão has long argued hat Neanderthals were far more culturally sophisticated than most anthropologists give them credit for.

10. Shipman 2012.

11. Neanderthals and their prey were cold-adapted creatures.

12. Gupta 2004.

13. Smith 2007.

14. I prefer the term "cultural modernity" over "behavioral modernity," as the former term reflects the collective nature of whatever it was that made us modern.

15. For those interested in this topic, Sterelny 2011 is a good place to start. It's a nice review of the subject, and the conclusions are well argued.

16. For example, Tattersall 2004; Klein 2002.

17. See, for example, Marean et al. 2007 for early use of shellfish resources; Henshilwood et al. 2004 for early use of bone (Blombos Cave, South Africa); Hovers et al. 2003 for symbolic use of ocher; D'Errico et al. 2012 for shell beads. McBrearty and Brooks 2000 provide an alternative, more gradualist account of the evolution of cultural modernity.

18. Klein 1995.

19. Enard et al. 2002.

20. Ferland et al. 2003.

21. MacDermot et al. 2005.

22. Haesler et al. 2007.

23. Zhang, Webb, and Podlaha 2002.

24. Krause et al. 2007 found the human form of *FOXP2* in Neanderthals; Meyer et al. 2012 reported the same for Denisovans.

25. Maricic et al. 2013 report a novel regulatory element in *FOXP2* of humans that is not shared by Neanderthals or Denisovans. The functional significance of this difference remains to be established, though.

26. Haygood et al. 2010.

27. Pollard, Salama, King, et al. 2006.

28. Pollard, Salama, Lambert, et al. 2006.

29. Amadio and Walsh 2006.

30. Pollard, Salama, Lambert, et al. 2006.

31. Pääbo 2014.

32. See Clark 2007 for the effects of modern electronic technology on the brain/cognition of contemporary humans; Wheeler and Clark 2008 for a more general treatment of the cultural embodiment of cognitive development.

33. Perreault 2012.

34. Wilson 2012 is particularly good on this point. This book has been harshly criticized by some of Wilson's former sociobiological allies because of his criticism of kin selection and promotion of group selection. The latter induces the screaming fantods in orthodox neo-Darwinians, such as Dawkins. Dawkins is in fine vitriolic form in his review of *The Social Conquest of Earth*, entitled "The Descent of Edward Wilson" (Dawkins 2012). The pun actually doesn't work, though.

35. Weisdorf (2005) provides a useful survey of theories.

36. For representative accounts of gene-culture coevolution, see Feldman and Laland 1996; Laland, Kumm, and Feldman 1995; Richerson, Boyd, and Henrich 2010. A shortcoming of these approaches is that they are basically extensions of population genetics theory and therefore assume vertical transmission. Memetics is another quite different framework; but despite its penetration of the popular literature, few evolutionary biologists take it seriously. There is still no definition of "meme" that is not viciously circular; nor has a productive research program emerged.

37. Perry et al. 2006.

38. Diamond 1997, 2002. See also Wolfe, Dunavan, and Diamond 2007.

39. But see Pearce-Duvet (2006) for a more conservative interpretation of the available data regarding the origins of these diseases.

40. Diamond 2002.

41. Barnes 2005.
42. Greger 2007.
43. Wolfe, Dunavan, and Diamond 2007.
44. Kwiatkowski 2005.
45. See Gage and Kosoy 2005 for plague.
46. Stedman et al. 2004.
47. Organ et al. 2011.
48. Wrangham and Conklin-Brittain 2003.
49. Dor and Jablonka 2001; Jablonka, Ginsburg, and Dor 2012.
50. Pinker 1991. The modularity thesis is central to evolutionary psychology. For a critique of the modularity thesis regarding language, see Chandler 1993. For a more general critique of the modularity thesis, see Karmiloff-Smith 1995.
51. In this respect, Dor and Jablonka's thesis is in accord with a more general reaction against the modularity thesis in general and modularity as it applies to language in particular (e.g., see Bolhuis et al. 2011).
52. Lewontin 1983 is a particularly significant account of his "constructivist" thesis.
53. Traditionally, each relevant environmental variable was considered a dimension of the organism's niche, culminating in Hutchinson and MacArthur's (1959) definition of a niche as an "n-dimensional hypervolume."
54. For a comprehensive account of niche construction, and one that has greatly influenced present-day conceptions of the subject, see Odling-Smee, Laland, and Feldman 2003.
55. Bellah 2011 is a good introduction to the evolution of religion generally and the influence of social organization.
56. The role of humans in these extinctions is disputed. See, for example, Wroe and Field 2006.
57. Again there is disagreement as to the relative role of humans and climate change in the extinction of New World megafauna. Barnosky et al. (2004) provide a balanced treatment.
58. Jablonski 1986.
59. See, for example, Eldredge 2001. Kolbert 2014 is a good popular account.
60. The term "Anthropocene" has been especially championed by Paul Crutzen (e.g., see Crutzen 2006), an atmospheric chemist and Nobel Prize laureate.

REFERENCES

Aiello, L. C., and P. Wheeler. (1995). The expensive-tissue hypothesis: The brain and the digestive system in human and primate evolution. *Current Anthropology* 36 (2): 199–221.

Amadio, J. P., and C. A. Walsh. (2006). Brain evolution and uniqueness in the human genome. *Cell* 126 (6): 1033–35.

Ambrose, S. H. (1998). Late Pleistocene human population bottlenecks, volcanic winter, and differentiation of modern humans. *Journal of Human Evolution* 34 (6): 623–51.

Banks, W. E., F. d'Errico, A. T. Peterson, M. Kageyama, A. Sima, and M.-F. Sánchez-Goñi. (2008). Neanderthal extinction by competitive exclusion. *PLoS One* 3 (12): e3972.

Barnes, E. (2005). *Diseases and Human Evolution*. Albuquerque: University of New Mexico Press.

Barnosky, A., P. Koch, R. Feranec, S. Wing, and A. Shabel. (2004). Assessing the causes of late Pleistocene extinctions on the continents. *Science* 306: 70–75.

Bellah, R. N. (2011). *Religion in Human Evolution: From the Paleolithic to the Axial Age*. Cambridge, MA: Harvard University Press.

Bolhuis, J. J., G. R. Brown, R. C. Richardson, and K. N. Laland. (2011). Darwin in mind: New opportunities for evolutionary psychology. *PLoS Biology* 9 (7): e1001109.

Chandler, S. (1993). Are rules and modules really necessary for explaining language? *Journal of Psycholinguistic Research* 22 (6): 593–606.

Clark, A. (2007). Re-inventing ourselves: The plasticity of embodiment, sensing, and mind. *Journal of Medicine and Philosophy* 32 (3): 263–82.

Crutzen, P. J. (2006). *The "Anthropocene."* Berlin: Springer.

Dawkins, R. (2012). The descent of Edward Wilson. *Prospect*, June.

D'Errico, F., L. Backwell, P. Villa, I. Degano, J. J. Lucejko, M. K. Bamford, . . . and P. B. Beaumont. (2012). Early evidence of San material culture represented by organic artifacts from Border Cave, South Africa. *Proceedings of the National Academy of Sciences USA* 109 (33): 13214–19.

Diamond, J. (1997). *Guns, Germs, and Steel*. New York: Norton.

Diamond, J. (2002). Evolution, consequences and future of plant and animal domestication. *Nature* 418: 700–707.

Dor, D., and E. Jablonka. (2001). How language changed the genes: Toward an explicit account of the evolution of language. *New Essays on the Origin of Language* 133: 147–73.

Eldredge, N. (2001). "The Sixth Extinction." American Institute of Biological Sciences. http://www.actionbioscience.org/evolution/eldredge2.html.

Enard, W., M. Przeworski, S. Fisher, C. Lal, V. Wiebe, T. Kitano, . . . and S. Pääbo. (2002). Molecular evolution of *FOXP2*, a gene involved in speech and language. *Nature* 418: 369–71.

Endicott, P., S. Y. Ho, M. Metspalu, and C. Stringer. (2009). Evaluating the mitochondrial timescale of human evolution. *Trends in Ecology & Evolution* 24 (9): 515–21.

Feldman, M. W., and K. N. Laland. (1996). Gene-culture coevolutionary theory. *Trends in Ecology & Evolution* 11 (11): 453–57.

Ferland, R. J., T. J. Cherry, P. O. Preware, E. E. Morrisey, and C. A. Walsh. (2003). Characterization of Foxp2 and Foxp1 mRNA and protein in the developing and mature brain. *Journal of Comparative Neurology* 460 (2): 266–79.

Finlayson, C. (2005). Biogeography and evolution of the genus *Homo*. *Trends in Ecology & Evolution* 20 (8): 457–63.

Gage, K. L., and M. Y. Kosoy. (2005). Natural history of plague: Perspectives from more than a century of research. *Annual Review of Entomology* 50: 505–28.

Green, R. E., J. Krause, A. W. Briggs, T. Maricic, U. Stenzel, M. Kircher, . . . and S. Pääbo. (2010). A draft sequence of Neanderthal genome. *Science* 328: 710–22.

Green, R. E., J. Krause, S. E. Ptak, A. W. Briggs, M. T. Ronan, J. F. Simons, . . . and S. Pääbo. (2006). Analysis of one million base pairs of Neanderthal DNA. *Nature* 444: 330–36.

Green, R. E., A.-S. Malaspinas, J. Krause, A. W. Briggs, P. L. F. Johnson, C. Uhler, . . . and S. Pääbo. (2008). A complete Neandertal mitochondrial genome sequence determined by high-throughput sequencing. *Cell* 134: 416–26.

Greger, M. (2007). The human/animal interface: Emergence and resurgence of zoonotic infectious diseases. *Critical Reviews in Microbiology* 33 (4): 243–99.

Gupta, A. K. (2004). Origin of agriculture and domestication of plants and animals linked to early Holocene climate amelioration. *Current Science* 87 (1): 54–59.

Haesler, S., C. Rochefort, B. Georgi, P. Licznerski, P. Osten, and C. Scharff. (2007). Incomplete and inaccurate vocal imitation after knockdown of FoxP2 in songbird basal ganglia nucleus Area X. *PLoS Biology* 5 (12): e321.

Haygood, R., C. C. Babbitt, O. Fedrigo, and G. A. Wray. (2010). Contrasts between adaptive coding and noncoding changes during human evolution. *Proceedings of the National Academy of Sciences USA* 107 (17): 7853–57.

Henshilwood, C., F. d'Errico, M. Vanhaeren, K. van Niekerk, and Z. Jacobs. (2004). Middle Stone Age shell beads from South Africa. *Science* 304: 404.

Hovers, E., S. Ilani, B. Vandermeersch, L. Barham, and B. Vanermeersch. (2003). An early case of color symbolism: Ochre use by modern humans in Qafzeh cave 1. *Current Anthropology* 44 (4): 491–522.

Hutchinson, G., and R. MacArthur. (1959). A theoretical ecological model of size distributions among species of animals. *American Naturalist* 93: 117–25.

Jablonka, E., S. Ginsburg, and D. Dor. (2012). The co-evolution of language and emotions. *Philosophical Transactions of the Royal Society B: Biological Sciences* 367 (1599): 2152–59.

Jablonski, D. (1986). Background and mass extinctions: The alternation of macroevolutionary regimes. *Science* 231 (4734): 129–33.

Karmiloff-Smith, A. (1995). *Beyond Modularity: A Developmental Perspective on Cognitive Science*. Cambridge, MA: MIT Press.

Klein, R. G. (1995). Anatomy, behavior, and modern human origins. *Journal of World Prehistory* 9 (2): 167–98.

Klein, R. G. (2002). *The Dawn of Human Culture*. New York: Wiley.

Kolbert, E. (2014). *The Sixth Extinction: An Unnatural History*. New York: Holt.

Krause, J., C. Lalueza-Fox, L. Orlando, W. Enard, R. E. Green, H. A. Burbano, . . . and S. Pääbo. (2007). The derived *FOXP2* variant of modern humans was shared with Neandertals. *Current Biology* 17 (21): 1908–12.

Kwiatkowski, D. P. (2005). How malaria has affected the human genome and what human genetics can teach us about malaria. *American Journal of Human Genetics* 77 (2): 171–92.

Laland, K. N., J. Kumm, and M. W. Feldman. (1995). Gene-culture coevolutionary theory: A test case. *Current Anthropology* 36 (1): 131–56.

Lewontin, R. C. (1983). The organism as the subject and object of evolution. *Scientia* 118 (1–8): 65–95.

MacDermot, K. D., E. Bonora, N. Sykes, A.-M. Coupe, C. S. Lai, S. C. Vernes, . . . and S. E. Fisher. (2005). Identification of FOXP2 truncation as a novel cause of developmental speech and language deficits. *American Journal of Human Genetics* 76 (6): 1074–80.

Marean, C. W., M. Bar-Matthews, J. Bernatchez, E. Fisher, P. Goldberg, A. I. Herries, . . . and H. M. Williams. (2007). Early human use of marine resources and pigment in South Africa during the Middle Pleistocene. *Nature* 449 (7164): 905–8.

Maricic, T., V. Gunther, O. Georgiev, S. Gehre, M. Curlin, C. Schreiweis, . . . and S. Pääbo. (2013). A recent evolutionary change affects a regulatory element in the human *FOXP2* gene. *Molecular Biology and Evolution* 30 (4): 844–52.

McBrearty, S., and A. S. Brooks. (2000). The revolution that wasn't: A new interpretation of the origin of modern human behavior. *Journal of Human Evolution* 39 (5): 453–63.

Meyer, M., M. Kircher, M.-T. Gansauge, H. Li, F. Racimo, S. Mallick, ... and S. Pääbo. (2012). A high-coverage genome sequence from an archaic Denisovan individual. *Science* 338 (6104): 222–26.

Odling-Smee, F. J., K. N. Laland, and M. W. Feldman. (2003). *Niche Construction: The Neglected Process in Evolution.* Princeton, NJ: Princeton University Press.

Organ, C., C. L. Nunn, Z. Machanda, and R. W. Wrangham. (2011). Phylogenetic rate shifts in feeding time during the evolution of *Homo. Proceedings of the National Academy of Sciences USA* 108 (35): 14555–59.

Pääbo, S. (2014). The human condition—A molecular approach. *Cell* 157 (1): 216–26.

Pearce-Duvet, J. M. C. (2006). The origin of human pathogens: Evaluating the role of agriculture and domestic animals in the evolution of human disease. *Biological Reviews* 81 (3): 369–82.

Perreault, C. (2012). The pace of cultural evolution. *PLoS One* 7 (9): e45150.

Perry, G., J. Tchinda, S. McGrath, J. Zhang, S. Picker, A. Caceres, . . . and C. Lee. (2006). Hotspots for copy number variation in chimpanzees and humans. *Proceedings of the National Academy of Sciences USA* 39: 8006–11.

Petraglia, M., R. Korisettar, N. Boivin, C. Clarkson, P. Ditchfield, S. Jones, . . . and K. White. (2007). Middle Paleolithic assemblages from the Indian subcontinent before and after the Toba super-eruption. *Science* 317 (5834): 114–16.

Pickrell, J., and D. Reich. (2014). Towards a new history and geography of human genes informed by ancient DNA. *bioRxiv.* http://biorxiv.org/content/early/2014/03/21/003517.

Pinker, S. (1991). Rules of language. *Science* 253 (5019): 530–35.

Pollard, K. S., S. R. Salama, B. King, A. D. Kern, T. Dreszer, S. Katzman, . . . and D. Haussler. (2006). Forces shaping the fastest evolving regions in the human genome. *PLoS Genetics* 2 (10): e168.

Pollard, K. S., S. R. Salama, N. Lambert, M.-A. Lambot, S. Coppens, J. S. Pedersen, . . . and D. Haussler. (2006). An RNA gene expressed during cortical development evolved rapidly in humans. *Nature* 443 (7108): 167–72.

Reich, D., R. E. Green, M. Kircher, J. Krause, N. Patterson, E. Y. Duranc, . . . and S. Pääbo. (2010). Genetic history of an archaic hominin group from Denisova Cave in Siberia. *Nature* 468 (7327): 1053–60.

Richerson, P. J., R. Boyd, and J. Henrich. (2010). Gene-culture coevolution in the age of genomics. *Proceedings of the National Academy of Sciences USA* 107 (suppl. 2): 8985–92.

Roberts, R. G., R. Jones, N. A. Spooner, M. J. Head, A. S. Murray, and M. A. Smith. (1994). The human colonisation of Australia: Optical dates of 53,000 and 60,000 years bracket human arrival at Deaf Adder Gorge, Northern Territory. *Quaternary Science Reviews* 13 (5): 575–83.

Sankararaman, S., S. Mallick, M. Dannemann, K. Prüfer, J. Kelso, S. Pääbo, . . . and D. Reich. (2014). The genomic landscape of Neanderthal ancestry in present-day humans. *Nature* 507 (7492): 354–57.

Sankararaman, S., N. Patterson, H. Li, S. Pääbo, and D. Reich. (2012). The date of interbreeding between Neandertals and modern humans. *PLoS Genetics* 8 (10): e1002947.

Shipman, P. (2012). Do the eyes have it? Dog domestication may have helped humans thrive while Neandertals declined. *American Scientist* 100: 198–205.

Smith, B. D. (2007). Niche construction and the behavioral context of plant and animal domestication. *Evolutionary Anthropology: Issues, News, and Reviews* 16 (5): 188–99.

Stedman, H., B. Kozyak, A. Nelson, D. Thesier, L. Su, D. Low, . . . and M. Mitchell. (2004). Myosin gene mutation correlates with anatomical changes in the human lineage. *Nature* 428: 415–18.

Sterelny, K. (2011). From hominins to humans: How *sapiens* became behaviourally modern. *Philosophical Transactions of the Royal Society. B: Biological Sciences* 366 (1566): 809–22.

Tattersall I. (2004). What happened in the origin of human consciousness? *Anatomical Record. Part B, New Anatomist* 276(1): 19–26.

Tattersall, I., and J. H. Schwartz. (2009). Evolution of the genus *Homo*. *Annual Review of Earth and Planetary Sciences* 37 (1): 67–92.

Weisdorf, J. L. (2005). From foraging to farming: Explaining the Neolithic revolution. *Journal of Economic Surveys* 19 (4): 561–86.

Wheeler, M., and A. Clark. (2008). Culture, embodiment and genes: Unravelling the triple helix. *Philosophical Transactions of the Royal Society. B: Biological Sciences* 363 (1509): 3563–75.

Wilson, E. O. (2012). *The Social Conquest of Earth*. New York: Norton.

Wolfe, N. D., C. P. Dunavan, and J. Diamond. (2007). Origins of major human infectious diseases. *Nature* 447 (7142): 279–83.

Wrangham, R. (2009). *Catching Fire: How Cooking Made Us Human*. New York: Basic Books.

Wrangham, R., and N. Conklin-Brittain. (2003). "Cooking as a biological trait." *Comparative Biochemistry and Physiology. A, Molecular & Integrative Physiology* 136 (1): 35–46.

Wroe, S., and J. Field. (2006). A review of the evidence for a human role in the extinction of Australian megafauna and an alternative interpretation. *Quaternary Science Reviews* 25 (21): 2692–703.

Zhang, J., D. M. Webb, and O. Podlaha. (2002). Accelerated protein evolution and origins of human-specific features: Foxp2 as an example. *Genetics* 162 (4): 1825–35.

Zilhão, J. (2006). Neandertals and moderns mixed, and it matters. *Evolutionary Anthropology: Issues, News, and Reviews* 15 (5): 183–95.

APPENDIX A to Chapter 5

For full source citations, please refer to the References listed under the chapter which this appendix accompanies.

NOTES

1. See Hall 1992 for a good history of Waddington.
2. For example, Coyne 2005 (which is a review of Carroll 2005a); Hoekstra and Coyne 2007. Among the other ideas that Coyne objects to as violating Darwinian orthodoxy are group selection, sympatric speciation, macromutations, alternatives to the biological species concept, epigenetic inheritance, and Wright's shifting-balance theory (or any other deviation from R. A. Fisher's perspective on population genetics).
3. Craig 2009.
4. Pigliucci 2007; Pigliucci and Muller 2010; Carroll 2005b; Carroll 2008; Hall 2004; Tauber 2010.

5. See Toth and Robinson 2007 for an extension of evo-devo perspective to behavior.
6. Williams 1966; Dawkins 1976.
7. West-Eberhard 2003, 2005.
8. Lavoué et al. 2010.

APPENDIX B to Chapter 5

For full source citations, please refer to the References listed under the chapter which this appendix accompanies.

NOTES

1. It is important to note a distinction between synonymous and nonsynonymous point mutations. Because of the redundancy in the genetic code, some mutations do not result in a change of amino acid. These are called synonymous mutations. These are overwhelmingly likely to be selectively neutral. Nonsynonymous mutations, by contrast, do result in amino acid substitution. These amino acid substitutions may or may not be selectively neutral.
2. Noncoding sequences that are transcribed into functional RNAs are often referred to as "RNA genes," but this term greatly expands the original conception of genehood.
3. Dermitzakis, Reymond, and Antonarakis 2005.
4. Keller 1984; McClintock 1987.
5. These are short tandem repeats of two to six base pairs.

APPENDIX to Chapter 7

For full source citations, please refer to the References listed under the chapter which this appendix accompanies.

NOTES

1. Ginja et al. 2010; Magee et al. 2002.
2. See also Cymbron et al. 1999 for zebu-Portuguese hybrids; see Negrini et al. 2007 for African influence on Italian cattle prior to their transport to the New World.
3. Miretti et al. 2004.
4. Mirol et al. 2003.
5. Ginja et al. 2010.
6. Kidd and Cavalli-Sforza 1974.
7. Ajmone-Marsan, Garcia, and Lenstra 2010; Del Bo et al. 2001; Maudet, Luikart, and Taberlet 2002.
8. Mukesh et al. 2004.
9. N. R. Joshi and Phillips 1953; B. Joshi, Singh, and Gandhi 2001. More purely dairying breeds are being constructed in India through crossbreeding with European dairy breeds (McDowell 1985).
10. Harris 1992.
11. Wurzinger et al. 2006.
12. Gautier, Laloë, and Moazami-Goudarzi. 2010.

13. Hybrids are also common in the Near East, parts of Europe, South America, and even North America.
14. Manwell and Baker 1980.
15. Felius et al. 2011. These Podolian breeds are sometimes referred to as gray steppe cattle.
16. Pellecchia et al. 2007.
17. Zebu-taurine hybrids include the Yakutian, which is adapted to the extreme cold conditions of the Yakut region of Siberia (Kantanen et al. 2009).
18. Pellecchia et al. 2007. Herodotus was an early proponent of the Etruscan connection.
19. Maretto et al. 2012; Pariset et al. 2010; see also Negrini et al. 2007 and D'Andrea et al. 2011 for autochthonous Tuscan breeds; Achilli et al. 2009, for evidence that many Italian breeds reflect local maternal auroch contribution.
20. Blott, Williams, and Haley 1998.

APPENDIX to Chapter 10

For full source citations, please refer to the References listed under the chapter which this appendix accompanies.

NOTES

1. Dagg 1974.
2. Webb 1972. Janis, Theodor, and Boisvert (2002) argue that the pacing gait allows for greater stride length without interference from other legs.
3. Janis, Theodor, and Boisvert 2002; Pfau et al. 2011.
4. Promerová et al. 2014.
5. Becker, Stock, and Distl 2011; Zwart 2012.
6. Andersson et al. 2012.
7. Webb 1972.
8. The relatively narrow chest of camels is also thought to be an adaptation to the pacing gait.

APPENDIX A to Chapter 11

For full source citations, please refer to the References listed under the chapter which this appendix accompanies.

NOTES

1. Froelich 1999, 2002. In this brief description of horse evolution I benefited greatly from an online essay by Kathleen Hunt, which I recommend to the interested reader: "Horse Evolution," TalkOrigins Archive, last modified January 5, 1995, http://www.talkorigins.org/faqs/horses/horse_evol.html.
2. MacFadden 2005.
3. MacFadden and Hulbert 1988.
4. MacFadden and Hulbert 1988; MacFadden 1976.
5. Prothero and Shubin 1989.
6. *Parahippus* and *Merychippus* are representative of this period of horse evolution.

7. These traits can be found in *Merychippus* and a number of related species that soon radiated on this branch (Simpson 1961).
8. MacFadden and Hulbert 1988.
9. *Dinohippus*, one of the most common equids in North America at that time, is representative (Hulbert 1989).
10. Janis, Damuth, and Theodor 2002.
11. George and Ryder 1986; Lindsay, Opdyke, and Johnson 1980; McFadden 2005; Steiner and Ryder 2011; Steiner et al. 2012.

APPENDIX B to Chapter 11

For full source citations, please refer to the References listed under the chapter which this appendix accompanies.

NOTES

1. Leroy et al. 2009.
2. Petersen et al. 2013.

APPENDIX to Chapter 12

For full source citations, please refer to the References listed under the chapter which this appendix accompanies.

NOTES

1. Francis 2011.
2. Wong, Gottesman, and Petronis 2005; Feinberg and Irizarry 2010.
3. Jablonka and Lamb 1998, 2005.
4. Coyne 2009. See also Coyne's blog for ad hominem criticisms of Jablonka: Why Evolution Is True, http://whyevolutionistrue.wordpress.com.
5. See Danchin et al. 2011; Richards, Bossdorf, and Pigliucci 2010; Bonduriansky and Day 2008.
6. Trut, Oskina, and Kharlamova 2009.
7. Rakyan and Whitelaw 2003.
8. Jablonka and Raz 2009.

APPENDIX to Chapter 14

For full source citations, please refer to the References listed under the chapter which this appendix accompanies.

NOTES

1. Here I am referring to the Santa Barbara school of evolutionary psychology promulgated by Tooby and Cosmides—that is, Evolutionary Psychology (with caps). A future evolutionary psychology need not suffer from these defects.

2. Buller 2005; Richardson 2007; Francis 2004; Lloyd and Feldman 2002; Bolhuis et al. 2011.

3. See, for example, Gunz 2012; Giger et al. 2010; Kuhn 2013.

4. Barkow, Cosmides, and Tooby 1992.

5. Loulergue et al. 2008; Martrat et al. 2007.

6. Tooby and Cosmides 2005.

7. Richerson, Boyd, and Henrich 2010; Meisenberg 2008. See Loring, Rosenberg, and Hunt 1987 for changes in tooth size.

8. Cosmides and Tooby (1997) view Williams 1966 (*Adaptation and Natural Selection*) as the most important foundational text.

9. Evolutionary psychologists misread Williams as arguing that only group-level adaptations are onerous. Williams did have a particular ax to grind with regard to group selection. But Williams's notion of the onerousness of adaptive claims was not confined to group selection (see Williams 1966, 4). Williams was on my dissertation committee, and I asked him about this in 1979, because at the time many sociobiologists—like present-day evolutionary psychologists—had adopted Williams as a patron saint. He replied that he meant the onerousness of adaptive claims to apply across the board, not just to group selection. Some of his actions, though, belied this assertion. At the time, he was editor of the *American Naturalist*, which became, under his stewardship, the in-house journal for some of the most extreme adaptationists of a sociobiological bent.

10. A particular target of Williams was Wynne-Edwards (e.g., 1963, 1978).

11. Gould and Lewontin 1979.

12. See, for example, Geary 1995 for an example of unbridled evolutionary psychology-based speculation on human sex differences.

APPENDIX to Chapter 15

For full source citations, please refer to the References listed under the chapter which this appendix accompanies.

NOTES

1. Wrangham 2009.

2. See, for example, Aiello and Wheeler 1995.

3. Wrangham and Conklin-Brittain 2003.

INDEX

Page numbers in *italics* refer to illustrations and figures.
Page numbers beginning with 355 refer to endnotes.